William Hughes

The geography of the British Empire

Physical, political, commercial

William Hughes

The geography of the British Empire
Physical, political, commercial

ISBN/EAN: 9783337175597

Printed in Europe, USA, Canada, Australia, Japan

Cover: Foto ©Andreas Hilbeck / pixelio.de

More available books at **www.hansebooks.com**

𝔓𝔥𝔦𝔩𝔦𝔭𝔰' 𝔊𝔢𝔬𝔤𝔯𝔞𝔭𝔥𝔦𝔠𝔞𝔩 𝔐𝔞𝔫𝔲𝔞𝔩𝔰.

THE BRITISH EMPIRE.

THE WORLD
ON MERCATOR'S PROJECTION
British Possessions colored Red

DON & LIVERPOOL.

Philips' Geographical Manuals

THE GEOGRAPHY

OF

THE BRITISH EMPIRE,

PHYSICAL—POLITICAL—COMMERCIAL

BY

WILLIAM HUGHES, F.R.G.S.,

AND

J. FRANCON WILLIAMS, F.R.G.S.

LIVERPOOL :

GEORGE PHILIP & SON, 32 FLEET STREET. E.C

LIVERPOOL: 15 TO 51 SOUTH CASTLE STREET.

PART I.

THE BRITISH ISLES.

CONTENTS.

THE BRITISH ISLES.

The British Isles consist of Great Britain, Ireland, and numerous smaller adjacent islands, situated in the Atlantic Ocean, off the western side of the European continent.

Great Britain consists of England, Wales, and Scotland, and is the largest island in Europe, being 600 miles in length, and having an area of nearly 89,000 square miles.

Ireland lies to the west of Great Britain, and is divided from it by the Irish Sea. Great Britain is nearly *three times* the size of Ireland, the area of which is 32,500 square miles.

Of the numerous islands and islets adjoining Great Britain and Ireland, the principal are the *Isle of Wight*, on the south ; the *Orkney* and *Shetland Islands*, on the north ; the *Hebrides*, off the west coast of Scotland ; *Anglesey* and the *Isle of Man*, in the Irish Sea ; and *Achil* and *Aran Islands*, off the west coast of Ireland.

England, Wales, Scotland, Ireland, and the adjacent islands, constitute politically the **United Kingdom of Great Britain and Ireland.**

Wales, Scotland, and Ireland were formerly distinct countries from England. The conquest of Ireland commenced in 1170, and virtually ended when Limerick was surrendered in 1691. Wales was conquered in 1282, and formally annexed in 1536. The crowns of England and Scotland were united in 1603 ; in 1707 England and Scotland were united under the same Parliament ; and in 1801 the Parliaments of Great Britain and Ireland were united ; hence the name, " The United Kingdom of Great Britain and Ireland."

ENGLAND AND WALES.

ENGLAND AND WALES,[1] which together form the **southern division** of Great Britain, constitute by far the most important portion of the British Islands.

Although in times long past **Wales** was a distinct country from England, yet the two are now so inseparably connected, and have been so long under the same government, that it is most convenient to describe them under one head and to speak of them as a single country.

---

[1]. England, *i.e., Angle-land,* the land of the 'Angles," the most numerous of the Saxon Invaders of Britain. The name " Wales" is derived from the Anglo-Saxon *Wealhas,* foreigners.

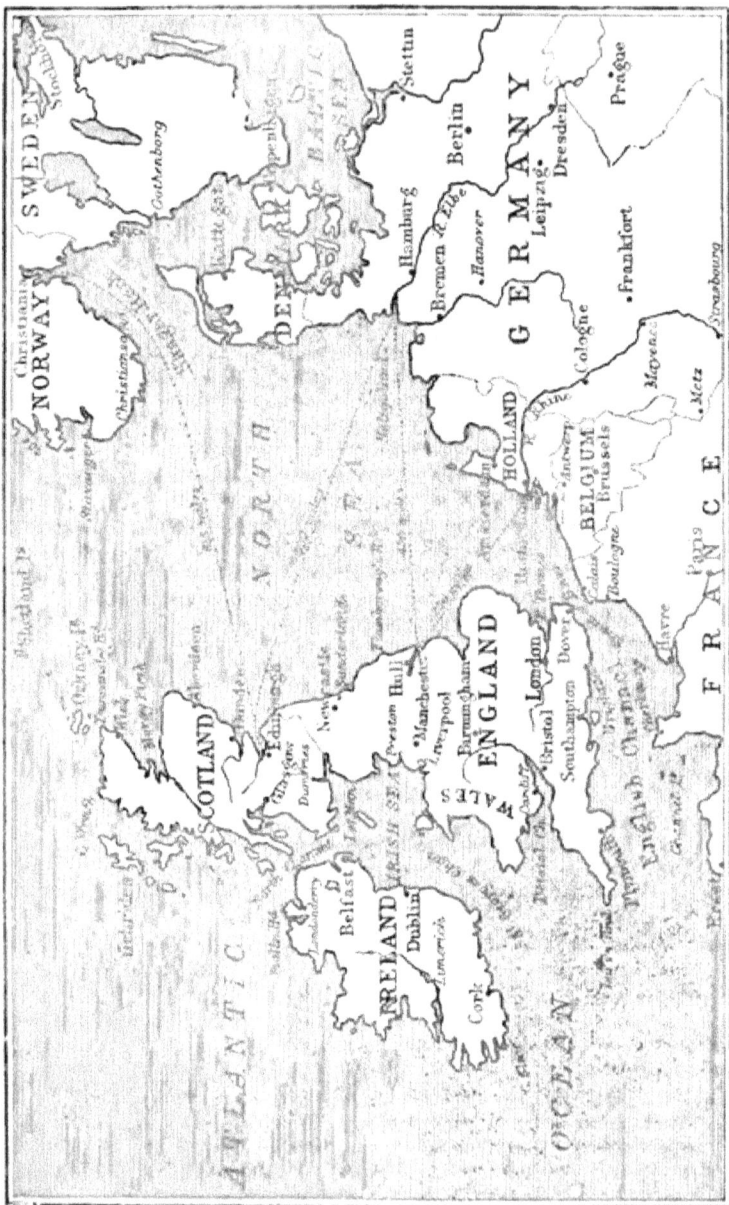

BRITISH ISLES IN RELATION TO THE CONTINENT

England makes nearer approach to the mainland of Europe than any other portion of the British Islands. The Strait of Dover, which divides the shores of England from those of France, is only twenty-one miles across.

BOUNDARIES.—England is bounded on all sides, except the north, by the sea; Wales also has the sea on all sides except the east, where it adjoins England.

ENGLAND is bounded on the *north* by Scotland; on the *east* by the North Sea; on the *south* by the English Channel; on the *west* by the Irish Sea, the principality of Wales, and the Atlantic Ocean.

WALES is enclosed on three sides—the *north, west*, and *south*—by the Irish Sea, St. George's Channel, and the Bristol Channel; on the *east* it adjoins England.

England is divided from Scotland by the *Tweed*, the *Cheviot Hills*, and the *Solway Firth;* from Ireland, by the *Irish Sea* and *St. George's Channel;* from France, by the *English Channel* and the *Strait of Dover;* and from Belgium, Holland, and Germany by the *North Sea*.

EXTENT.—Taken together, England and Wales are but a small country, little more than one-sixtieth part of Europe, or a nine-hundredth part of the total land-area of the globe.

The area of England is a little more than 50,000 square miles, and Wales a little over 7,000 square miles.[1] The greatest length of England, from Berwick-on-Tweed to the Lizard, in Cornwall, is 420 miles, and the greatest breadth, from Lowestoft Ness, in Suffolk, to Land's End, in Cornwall, is 360 miles. The least breadth, from the Tyne to the Solway Firth, is 60 miles; the average breadth south of the Wash is 200 miles; north of that inlet, 120 miles. The mean length, along the meridian of 2° west, is 360 miles.

Shape.—In shape England is triangular, the south coast being the base, and Berwick-on-Tweed the apex.

This fact may be made use of by the student in drawing the map of England; for if, in addition to the lines of latitude and longitude, a triangle be described by straight lines joining Land's End and the South Foreland, and from each of these points to Berwick, the sinuosities of the coast may be much more easily and accurately drawn.

NATURAL FEATURES: The coasts of England are indented by numerous inlets and estuaries; its mountains are but moderately elevated; while many of its rivers, though short, and of small volume compared with continental rivers, are of the highest commercial importance.

It may indeed be said that no other country of like extent presents such a variety of natural features as England; its surface is diversified by mountain, hill, and valley; its uplands are drained and its lowlands watered by numerous rivers and streams; while, in a few localities, the mountain scenery is enriched by most beautiful and picturesque lakes.

---

1. The above are round numbers. The actual | 58,000 square miles—will be taken as a *standard of* area of England is 50,823 square miles; of Wales, | *comparison* for all the larger countries, lakes, &c.; 7,363 square miles; and of both together, 58,186 | the size of smaller divisions of land or water will be square miles. The area of England and Wales— | compared to that of Wales.

COASTS : Few countries have a longer coast-line in proportion to their area than England, and the numerous inlets and estuaries, many of which form splendid harbours, penetrate so far inland that no part of the country is more than 60 miles from the sea.

Excluding minor indentations, the coast-line measures about 1,500 miles, but including tidal inlets and estuaries, it extends to upwards of 3,000 miles, or an average of 1 mile of coast to every 20 square miles of area, a proportion exceeded by very few countries in the world.

The numerous inlets and estuaries, some of them forming magnificent natural harbours, and others rendered accessible to the largest sea-going vessels by *dredging*, bring all parts of the country within easy distance of the sea, and thus facilitate the *inflow* of raw materials and food-stuffs, and the *outflow* of manufactured goods and minerals. And as water carriage is the cheapest possible form of transport, the extent and accessibility of the coasts of England have powerfully influenced the development of its trade and commerce with all parts of the world.

∴ England being, as we have already said, triangular in shape, and bounded by the sea on all sides except the north, its coasts are naturally divided into three sections, the *east coast*, the *south coast*, and the *west coast*, each of which presents peculiarities of form and character.

The East Coast of England is the least indented : there are no great bays or estuaries in the bold and rocky shores which terminate at Flamborough Head ; thence to the North Foreland, the coast is bordered by low cliffs, flat marshy lands, sand-hills or sandy levels, and is broken only by three large inlets, one of which, the estuary of the Thames, is the most important commercial waterway in the world.

In some parts of the east coast the sea is wearing away the land ; between Flamborough Head and Spurn Point whole parishes have disappeared, and farms and villages have been engulfed. Further south, however, this loss is to some extent balanced ; on the Lincolnshire coast the land is gaining on the sea, while the Wash is gradually silting up.

The inlets on the East Coast are few in number, the largest of them being the *Mouth of the Tees*, the *Humber*, the *Wash*, and the *Estuary of the Thames*. The estuary of the Tees is important, commercially, as the outlet of the great iron district of Cleveland. The Humber is the estuary of the Ouse and Trent, and is navigable for sea-going vessels to the ports of Hull and Grimsby. The Wash, which receives the drainage of the Fen district, is shallow, and the navigation of its tortuous channels is rendered difficult and dangerous by shifting sandbanks. The Estuary of the Thames is the most important inlet in Britain, and, indeed, in the world. This great waterway is traversed day and night, all the year round, by ocean steamers and merchant vessels to and from London—the metropolis of the world's commerce.

Of the few capes on the East Coast the following are the most prominent :— Flamborough Head, the loftiest promontory on the north-eastern coast, so named from the *beacon-fires* formerly lit on it ; Spurn Point, a low and sandy "spit" at the mouth of the Humber ; Lowestoft Ness, on the coast of Suffolk, the most easterly point of England ; and the Naze and Foulness, low capes on the flat and marshy shores of Essex. At the North Foreland, south of the estuary of the Thames, the "White Cliffs" of England commence, and again rise prominently in the South Foreland, on the south-east coast of Kent.

There are no large islands, and only a few small ones, on the East Coast. These are :—Holy Island or Lindisfarne, the Farne Islands, and Coquet Island, off the Northumbrian coast ; Thanet and Sheppey, on the Kentish side of the estuary of the Thames—both "islands" only in name, the formerly wide passages having silted up, and these islands are now separated from the mainland only by narrow streams, kept open for drainage.

The South Coast of England is remarkable for its *magnificent natural harbours* and lofty *chalk cliffs*.

Lofty chalk cliffs fringe the eastern, and tolerably regular portion of the south coast ; the western half is sinuous and deeply indented by numerous inlets, many of which form safe and commodious harbours.

Of the greater inlets on the South Coast, the most important are Portsmouth Harbour, a splendid natural harbour, and strongly fortified naval station, capacious enough to hold the entire British navy at one time ; Southampton Water, an important commercial waterway, traversed by steamers to and from all parts of the world ; Weymouth Bay, a harbour of refuge protected by Portland breakwater ; Plymouth Sound, protected by a *breakwater* a mile in length, and large enough to hold the British fleet ; and Falmouth Bay and Mount's Bay, on the southern coast of Cornwall.

The chief capes on the South Coast are the South Foreland and Dungeness, in Kent ; Beachy Head and Selsea Bill, in Sussex ; St. Catherine's Point and the Needles, in the Isle of Wight ; St. Alban's Head and Portland Bill, on the Dorsetshire coast ; Start Point in Devon ; and the Lizard, the most southerly point of England, in Cornwall.

The only noteworthy island on the South Coast is the Isle of Wight, which is extremely beautiful and fertile. The lovely scenery and genial climate of the island, well named the "Garden of England," attract thousands to cross the narrow channel which divides it from the mainland. The eastern part of this channel is called Spithead ; the western, the Solent.

The West Coast of England is far more irregular and deeply indented than the east, or even the south coast. The Cornish and Welsh coasts are bold and rocky, those of Lancashire and Cumberland are flat and sandy.

The larger inlets on the West Coast are :—Barnstaple Bay, on the north coast of Devon ; the Bristol Channel, an important waterway leading into the estuary of the Severn, and forming, as it were, the outer harbour of the three great ports of Bristol, Cardiff, and Swansea. The largest indentation on the coast of Wales is Cardigan Bay : other notable inlets are St. Bride's Bay ; Milford Haven, the finest natural harbour in England, capable of accommodating the entire British navy at one time ; Carmarthen Bay and Swansea Bay, on the south coast of South Wales. In North Wales, Carnarvon Bay leads into the Menai Strait, which separates Anglesey from the mainland.

Further north the coast is indented by three river estuaries—those of the Dee, the Mersey, and the Ribble—and two great inlets, Morecambe Bay and the Solway Firth. The approaches to, and navigation within all these inlets are impeded by sandbanks, and they can only be entered by large vessels at high water. The *bar* at the mouth of the Mersey is now being dredged, and ultimately a permanent deep water channel may be formed, so that the "Atlantic liners" and other large vessels will no longer have to wait outside the bar, but will be able to enter the river at all states of the tide.

The chief headlands on the West Coast are the Land's End, in Cornwall, the most westerly point of England; Hartland Point, in Devonshire; Worms Head, in Glamorganshire; St. David's Head, in Pembrokeshire; Great Orme's Head, the loftiest headland in England and Wales, in Carnarvonshire; Point of Aire, in Flintshire; Formby Point, in Lancashire; and St. Bees Head, on the coast of Cumberland.

The chief islands on the West Coast are the *Scilly Isles*, a numerous group of islands and islets off the Cornish coast; *Lundy*, off the coast of Devon; *Anglesey, Holyhead,* and *Bardsey,* off the Carnarvonshire coast; *Walney,* off the Furness coast; and, 70 miles out from the coast, the *Isle of Man.*

The Scilly Isles lie 30 miles south-west of Land's End. Of the 145 islands and islets 6 only are inhabited. St. Mary, 10 miles in circumference, is the largest island. Anglesey is a large island separated from the mainland of North Wales by the narrow Menai Strait, across which two famous bridges— the Suspension Bridge and the Tubular Railway Bridge—have been thrown. Holyhead Island is connected with Anglesey by two huge embankments. The Isle of Man is in the middle of the Irish Sea, and *Douglas,* the capital, is at about the same distance from England as *Ramsey* is from Scotland, and *Peel* from Ireland. The interior of the island is hilly, rising in *Snaefell* to 2,000 feet above the sea. *Lead* and *slate* are the chief mineral products. The climate is healthy and invigorating, and, during the summer, thousands of people from the manufacturing towns of Lancashire and Yorkshire visit the island, to which large steamers run from Liverpool, Fleetwood, and Barrow.

STRAITS AND ROADSTEADS : The straits and roadsteads on the coasts of England and Wales are few in number, but are nearly all of the highest commercial importance.

The principal straits are the Strait of Dover, 21 miles in width, between the south-eastern coast of Kent and the northern coast of France, the chief water-way for most of the foreign and colonial trade of London; Spithead and Solent, the eastern and western parts of the channel which separates the Isle of Wight from the mainland, both affording safe anchorage for men-of-war and merchant vessels; and the Menai Strait, a beautiful channel, 14 miles long, between Anglesey and Carnarvonshire, crossed by Telford's Suspension Bridge and Stephenson's Tubular Railway Bridge.

The principal roadsteads are the Downs, between the Kentish coast and the Goodwin Sands, the largest natural harbour of refuge in the world, and, during severe storms especially, crowded with shipping; Yarmouth Roads, between the coast of Norfolk and a line of sandbanks, the only safe anchorage between the Thames and the Humber. Spithead, the Solent, Portland Roads, Plymouth Sound, and the Menai Strait are all resorted to as roadsteads, and in stormy weather, vessels of all kinds may be seen riding safely at anchor.

∴ The coasts of England are, of course, minutely mapped out; the *Admir-alty Charts* show not only the configuration of the coast, but also every rock, shoal, or sandbank on or near the coast, with the landmarks, lighthouses, light-ships, and buoys, which enable the mariner to pursue his course from port to port in safety. But in spite of all these "aids" to navigation, the number of *shipwrecks* which occur every year on the coasts of England is appalling, and the loss of life and property is very great.

SURFACE : England exhibits generally a gently-sloping or undulating surface, which, however, rises in some places into lofty
hills.　Wales is chiefly mountainous.

The high grounds of England and Wales lie principally upon the western
side of the island, and form a succession of elevated regions which stretch
nearly from the borders of Scotland to the Land's End, and are seldom far
removed from the western coasts.　The eastern slope of England is thus longer
than the western slope, which is so short and rapid that, with the exception of
the Severn, not a single river flowing west attains a length of 100 miles, while
several streams on the eastern side of England have a course of between 150
and 200 miles.

∴ " It is important to realise how low and level a great part of the country really is.　If
the island were sunk 500 feet below its present level, England would be reduced to a
scattered group of islands, the largest of which would extend from near Derby to
Hexham.　Wales would form a second island of about the same size.　The uplands of
eastern Yorkshire would make a third, and a scattered archipelago would run from Cornwall eastwards to Kent, northwards to Shropshire, and north-eastwards to Lincolnshire.
If the depression were only to the extent of 230 feet, the sea would spread over all the
low grounds from the Tees to the Thames, and from Westmoreland to Shropshire." [1]

MOUNTAIN-SYSTEMS : The mountains of England may
be divided into three sections or systems, namely, the Northern,
Cambrian, and Devonian.

The Northern System includes the *Cheviot Hills, Pennine Range*, and
the *Cumbrian Group*.

The Cheviot Hills are on the borders of England and Scotland ; Cheviot
Peak, in Northumberland, is 2,676 feet above the sea.　The Pennine Range [2]
extends from the Cheviot Hills to the Peak, nearly along the dividing line between the six northern counties, and constitutes the most continuous elevated
tract in England.　This range, the longest in the island south of the Tweed,
and locally known as the "*back-bone of England*," has no well-defined continuous ridge-line, but consists of a series of huge moorlands, from 10 to 20
miles broad, cut up with valleys, and interspersed with mountainous masses.
The highest points are :—Cross Fell, [3] in Cumberland, 2,892 feet ; Whernside,
2,414 feet ; Ingleborough, 2,373 feet, and Pen-y-gant, 2,273 feet, in Yorkshire.

∴ The Pennines are, for the most part, bleak and treeless, but the wild and
romantic district of the Peak, in Derbyshire, is one of the most picturesque in
Britain.

The Cumbrian [4] Mountains, a group in the counties of Cumberland, Westmoreland, and northern Lancashire, near the coast of the Irish Sea, are more
rugged and somewhat loftier than the adjoining Pennine Range, and contain
the highest elevation in England—Scaw Fell, situated in the centre of the
group, and reaching 3,208 feet above the level of the sea.　Skiddaw, 3,054 feet,
and Helvellyn, 3,118 feet, are in the same group.

∴ The Cumbrian mountain-region includes the beautiful and much-frequented " Lake District."　The Cumbrian lakes—Windermere, Ulleswater,
Derwent Water, &c.—picturesquely embosomed in long and narrow valleys,
are celebrated for their beauty.

1. Geikie: Geography of the British Isles (Mac-
millan).
2. Pennine, from Celtic, *pen*, a hill.

3. Cross Fell, Danish, *fell*, a hill.
4. Cumbrian, from *Cumbria*, the old name of
Cumberland.

The **Cambrian**[1] **System** includes all those mountains situated between the basin of the Severn and the Irish Sea, and spread over the greater part of the surface of Wales, reaching in their highest point a greater elevation than any of the English mountains.

'The mountains of Wales consist neither of a single range nor a succession of mountain-chains. They form rather a high mountain-region, in some places spreading into broad masses of tableland, intersected by deep valleys, and in others forming huge mountain-summits, which rise conspicuously above the surrounding ground.'

The culminating point of the system is Snowdon,[2] in the county of Carnarvon, which rises to a height of 3,570 feet above the sea, and is thus the highest mountain in England and Wales. The spurs from the main range rise in Cader Idris[3] to 2,929 feet, and in Plinlimmon to 2,469 feet. In South Wales, the heather-covered Black Mountains culminate in the Beacons of Brecknock, two majestic summits over 2,900 feet in height.

The magnificent scenery of the Snowdon Range surpasses, in some respects, the finest mountain scenery in other parts of the island. Words can scarcely convey an adequate idea of the beauty and sublimity of the scene revealed from the summit of Snowdon. In fine weather the view is extensive; in stormy weather, mists and fogs gather round the Wyddfa. Pennant witnessed it under both aspects. "I saw," he observes, "the county of Chester, the high hills of Yorkshire, part of the north of England, Scotland, and Ireland; a plain view of the Isle of Man, and that of Anglesey, lay extended like a map beneath us, with every rill visible." Of his second ascent he remarks: "On this day the sky was obscured very soon after I got up. A vast mist enveloped the whole circuit of the mountain. The prospect down was horrible. It gave the idea of a number of abysses, concealed by a thick smoke, furiously circulating round us. Very often a gust of wind formed an opening in the clouds, which gave a fine and distinct vista of lake and valley. Sometimes they opened only in one place; at others in many places at once, exhibiting a most strange and perplexing sight of water, fields, rocks, or chasms, in fifty different places. They then closed at once, and left us involved in darkness."

The **Devonian System** includes the hills and highlands of Devonshire and Cornwall. They are less elevated than the Welsh hills, but still impart a varied and often rugged surface to the south-west corner of the island.

Brown Willy, in Cornwall, 1,368 feet; Yes Tor, 2,040 feet; Cawsand Beacon, 1,802 feet, on Dartmoor, in Devonshire, and Dunkerry Beacon, 1,707 feet, on Exmoor, in Somerset, are the highest points in this system.

∵ To the eastward of the above tracts the elevations are much less conspicuous; few points reach more than a thousand feet above the sea-level, and most of them are considerably below that altitude.

HILLS: These lower heights may be arranged in two groups, according to their formation, namely, the *chalk hills*, and the *limestone* or *oolitic ranges*.

The chief *chalk ranges* are the Yorkshire and Lincolnshire Wolds,[1] the East Anglian Heights, which extend from the Chiltern Hills to the coast of Norfolk, and terminate in Hunstanton Cliffs, near to the Wash; Gog Magog Hills, in Cambridgeshire; the North and South Downs, with the Salisbury Plain, the Marlborough Downs, and other lower heights, to the south of the Thames.

Of the *oolitic hills*, the principal are the Cotswold Hills, in Gloucestershire, between the head-waters of the Thames and the lower course of the Severn; the Malvern Hills, between the Severn and the Wye; the Mendip, Quantock, and Blackdown Hills, in Somersetshire; the Clee Hills and the Wrekin, in Shropshire; and the Clent Hills, in Worcestershire.

MOORS[2] : Bleak and treeless wastes of sterile land still cover considerable areas, although large tracts of *moorland* have been reclaimed and rendered available for cultivation.

The most extensive moors and moorlands are the bleak and barren North York Moors, in the north-east of Yorkshire; the boggy, peat-covered Lancashire Moorlands, between the Irwell and the Wyre; the wild and elevated waste of Exmoor, in Somerset and Devon; and the granitic, "Tor-" crowned tableland of Dartmoor, in Devon.

PLAINS : The most extensive plains in England are the York Plain, the Cumbrian and Cheshire Plains, the Central Plain, the district of the Fens, and the Eastern Plain.

The York Plain, between the Pennine Range and the Yorkshire Wolds, is the most extensive in England. The Cumbrian and Cheshire Plains lie to the north and south of the Cumbrian Group, and on the west side of the Pennine Range. Both are fertile, the latter, especially, being admirably adapted for grazing and dairy-farming. The Central Plain is from 200 to 400 feet above the sea, and extends from the Thames on the south to the Ouse on the north, and from the Severn on the west to the Trent on the east. The district of the Fens,[3] which lies round the shores of the Wash, includes parts of the counties of Lincoln, Northampton, Huntingdon, Cambridge, Norfolk, and Suffolk, and forms the lowest and most perfectly level portion of the island. The whole tract has been converted by drainage into a highly productive district. The Eastern Plain includes the sea-board of Essex, Suffolk, and Norfolk, and is separated from the Fens by the East Anglian Heights. The so-called Salisbury Plain,[4] in Wiltshire, is a treeless expanse of moderate elevation (400 feet).

VALLEYS : Most English river valleys are beautiful and fertile, and many of them are the seats of great industries and a world-wide commerce.

The dales of the north of England, and the vales of the southern counties, are famous for their quiet beauty and extreme fertility. The valley of the upper Thames displays all the charms of rural beauty; the lower valley is the scene of an industrial and commercial activity unsurpassed in any part of the world. In England, as in all countries, manufacturing towns and commercial centres naturally gravitate to the river valleys and estuaries.

1. Wold, A.-S. *weald*, a forest. *Cf.* German *wald*.
2. Moor, A.-S. *mor*, waste land.
3. Also called the "*Bedford Level*," from the Duke of Bedford, who reclaimed large portions of it in the reign of Charles II. The coast is in some parts protected from inundation by *dykes*, as in Holland.
4. On this plain, about eight miles from Salisbury, is *Stonehenge*, a Druidical or Danish Circle.

RIVERS: As all the higher elevations of land lie nearer the western than the eastern shores, the longest rivers are, with one exception, on the eastern side of England, and flow into the North Sea. The main slope of the country is therefore towards the east; the shorter slopes are towards the south and west. The numerous streams and rivers of England may be classified in four groups, according to the *inclination of their basins.*

I.   Rivers flowing into the North Sea, from the
1. *Pennine Range,* the Tyne, Wear, Tees, Ouse, and Trent.
2. *Watershed of the Central Plain,* the Witham, Welland, Nen, and Great Ouse, all of which enter the Wash.
3. *East Anglian Heights,* the Yare, Orwell, Stour (Essex), Colne, and Blackwater.
4. *Cotswold and other hills,* the Thames and its tributaries.
5. *Wealden Heights,* the Stour (Kent).

II.   Rivers flowing into the English Channel, from the
1. *Wealden Heights,* the Rother, Ouse (Sussex), and Arun.
2. *The Downs* of Hants and Wilts, the Itchen, Test, and Avon.
3. *Devonian Range,* the Stour (Dorset), Frome, Axe, Otter, Exe, Teign, Dart, Tamar, and Fal.

III.   Rivers flowing into the Bristol Channel, from the
1. *Devonian Range,* the Torridge, Taw, and Parret.
2. *Cotswold Hills,* the Bristol Avon.
3. *Watershed of the Central Plain,* the Avon (tributary of the Severn).
4. *Welsh Mountains,* the Severn, Wye, Usk, Taff, Neath, Tawe, and Towy.

IV.   Rivers flowing into the Irish Sea, from the
1. *Welsh Mountains,* the Teify, Dyfi, Conway, Clwyd, and Dee.
2. *Pennine Range,* the Mersey, Ribble, Wyre, Lune, and Eden.
3. *Cumbrian Group,* the Kent and the Derwent.

The rivers of England may be also arranged in three groups according to the *watershed* which they drain.

The Eastern Watershed is drained by the rivers which enter the North Sea; of these the most important are the *Tyne, Tees, Ouse,* and *Trent,* the *Great Ouse,* and the *Thames.*

The Tyne is formed by the confluence of the North Tyne, which rises in the Cheviots, and the South Tyne, rising on Cross Fell, and has a course of 73 miles. Area of basin, 1,100 square miles. The Tyne, as the outlet of the great Northumbrian coalfield, is commercially one of the most important; the tonnage of vessels entering and leaving the *Tyne ports* exceeding 4¾ million tons a year.

The Tees is similarly important as the outlet to the *Cleveland iron district.*

The Ouse, sometimes distinguished as the Yorkshire Ouse, is formed by the junction of the *Swale* and the *Ure,* and becomes navigable at York, and enters the Humber after a course of 150 miles. Area of basin, 5,500 square miles. Tributaries—on right bank, *Nidd, Wharfe, Aire, Don;* on the left, the *Derwent.*

The Trent rises in Mow Cop Hill on the borders of Staffordshire, and falls into the Humber after a course of 180 miles. Area of basin, 4,000 square miles.

Tributaries—on right bank, *Tame* and *Soar;* on left, the *Dove* and *Derwent.*
The Trent is navigable for vessels of 200 tons to Gainsborough, and for boats
and barges to Burton, 105 miles from its confluence with the Humber.

The so-called **River Humber** is a great estuary, formed by the junction of the
Trent and the Ouse, and is a most important commercial highway, Hull, one
of the chief ports in the kingdom, being on its northern shore.

The **Great Ouse** is the longest river entering the Wash, having a length of
156 miles, and draining an area of 2,960 square miles. Tributaries—*Cam,
Lark,* and *Little Ouse,* all on right bank. Navigable to Bedford, 90 miles
from the sea.

The **Thames**[1] is, with two exceptions—the Shannon and the Severn—the
longest river in the British Islands, and it ranks first in order of importance,
since it has London, the metropolis of the British Empire, upon its banks. It
is formed by the junction of the Thame and the Isis, both of which rise in the
Cotswold Hills, in Gloucestershire, and flow east into the North Sea after a
course of 215 miles. Area of basin, 6,160 square miles. Tributaries—on right
bank, *Kennet, Loddon, Wey, Mole, Darent,* and *Medway;* and on the left,
Windrush, Evenlode, Cherwell, Thame, Colne, Brent, Lea, and *Roding.* The
Thames is navigable for the largest vessels nearly to London, and for smaller
craft to Lechlade, 160 miles from the sea. The tide is felt as far as Teddington
(hence the name), about 80 miles from the sea. From the Nore Lightship at
the mouth of the Thames, the river is navigable at all times for merchant
vessels and steamers up to London Docks, a distance of 47 miles.

The Southern Watershed is so narrow that the rivers draining
it are all short, and most of them shallow and rapid. The inlets
which some of them enter are, however, fine natural harbours, of
immense value and much frequented by men-of-war and merchant
vessels.

The largest of these small rivers are the **Tamar** (45 miles), which rises in
Dartmoor and falls into Plymouth Sound ; the **Exe,** which enters the sea at
Exmouth ; the **Fal,** which falls into Falmouth Harbour ; the **Dart,** the " English
Rhine;" the **Itchin,** which discharges into Southampton Water, and other
smaller streams which may be traced upon the map.

The Western Watershed is drained by the numerous rivers that
fall into the Bristol Channel and the Irish Sea ; of these the *Severn*
and *Mersey* are by far the most important. Of the smaller rivers on
the western slope, the Bristol *Avon* is an important commercial
waterway, while the winding *Wye,* which falls into the same
estuary, is one of the most beautiful of British rivers. The far-
famed river *Dee* flows from Bala Lake and winds past the ancient
city of Chester, entering the sea by a broad but shallow estuary.
North of the Mersey, the *Ribble* enters the sea at Preston. The
Lune flows into Morecambe Bay, while the *Eden* falls into the
Solway Firth.

The **Severn**[2] rises on the east side of Plinlimmon, and enters the Bristol
Channel after a course of 240 miles. It drains an area of 4,500 square miles.

---

1. Thames, from *Tamesis,* the broad Isis: Gaelic | 2. Severn or Havren. Being joined by the *Se* at
uisge, water. | Llanidloes, it becomes the *Seavren* or Severn.

Tributaries—on the right bank, the *Teme ;* on the left, the *Vyrnwy, Tern, Stour,* and *Upper Avon.*[1]   The Severn is navigable to Welshpool, a distance of 170 miles from the sea.

The **Mersey** is a small river (68 miles in length), but its estuary forms the "Liverpool Channel," one of the most important harbours in the world.   It drains an area of 1,706 square miles, and is navigable to its junction with the *Irwell,* on which Manchester is situated.

LAKES : There are few lakes in England and Wales, and they are nearly all situated within the region of the Cumbrian Mountains—the far-famed " English Lake District"—and in North Wales.

In the English Lake District, the most picturesque part of the country, are several lakes, of which are **Windermere,**[2] **Ulleswater, Coniston, Derwent Water, Bassenthwaite, Crummock,** and **Wastwater.**   The chief Welsh lakes are **Bala Lake,** the **Lakes of Llanberis, Llyn Conway,** in North Wales; and **Brecknock-mere,** in South Wales.   There are also a few *meres* in Cheshire and the Fen District.

∴ The largest English lake is **Windermere,** 14 miles long and 1 mile broad ; in Wales, **Bala Lake,** 4 miles long and ⅔ mile broad, is the largest.

CLIMATE : The climate of England is temperate and healthy. The average temperature of the year is rather higher than that of the adjacent shores of the continent, while the summers are not so hot nor the winters so severe as those experienced on the mainland in similar latitudes.

The general moisture of the atmosphere, and the frequent occurrence of rain, as well as the above-mentioned characteristics, are explained by the insular position of Britain.   The western side of the island has a rather higher temperature than the neighbourhood of the eastern coasts, and has also a greater fall of rain.   The coasts of Cornwall and Devon, and the shores of South Wales, are especially distinguished by the mildness of their winter.

The mean temperature of summer in London is 63.8°; of winter, 37.3°. Mean annual temperature, 50.55°.   The lowest average winter temperature is about 35° or 36°, and the highest average summer about 64°.   The number of days in the year upon which west and east winds blow has been observed to be in the ratio of 225 to 140 ;   north and south winds as 192 to 173.   The average annual rainfall at Dover is 30 inches ; London, 24 ; Coniston, 85 ; Liverpool, 35 ; and at Plymouth, 40.   The summer of 1888 was the rainiest and coldest for 125 years, and the winter of 1890-1 has been exceptionally severe.

But in spite of variable winds, and sudden changes of weather, *the climate of England is, on the whole,* the most favourable for industry in the world. Charles II. had evidently formed a correct estimate of the climate of his kingdom, when, " in reply to some who were reviling it and extolling those of Italy, Spain and France, he said he thought that the best climate where he could be abroad in the air with pleasure, or at least without trouble or inconvenience, *the*

---

1. Three rivers named **Avon** have been mentioned above, and there are several other streams in various parts of Britain that bear the same name. *Avon* (or *afon*) is the Celtic term for a stream or running water. Ouse, which is also a name attached to several English rivers, is from the French *eaux* (waters), and is a record of the Norman conquest of our island.

2. **Windermere** (British), *gwyn,* bright ; *dwr,* water ; and A.-S. *mere,* a lake.

most days in the year and the most hours in the day; and this he thought he could be in England more than any other country in Europe."

MINERALS.—As regards mineral products, England is distinguished by the extraordinary abundance of those which are most necessary to civilized man—*coal* and *iron*—together with *copper*, *lead, zinc, tin,* and other ores.

The coalfields of England occur chiefly in the northern and midland counties, and yield an inexhaustible supply of fuel—necessary alike for the purposes of manufacturing industry and for household use. South Wales includes a rich coalfield of large extent, and North Wales contains smaller carboniferous areas.

Iron-ore occurs abundantly within the limits of nearly all the coal districts, and is most extensively worked in Yorkshire and South Wales, together with the counties of Stafford, Shropshire, Derbyshire, Durham, and Northumberland. Lead is principally worked in Derbyshire, Northumberland, Cumberland, North and South Wales, and Devonshire. Copper and tin are found chiefly (the latter entirely) in the counties of Cornwall and Devon.

Of salt, Cheshire furnishes an abundant supply from the brine-springs and mines of rock-salt in the valley of the River Weaver, a tributary of the Mersey. Good marble and building-stone are largely quarried in the northern and north-midland districts of the country, as well as in the south-western peninsula. The eastern and south-eastern counties are deficient in this material; but the valuable clay with which they abound supplies the material for making *bricks*, of which the metropolis and other cities in these counties are chiefly built. Slate is extensively quarried in Wales and in the mountain-region of Cumberland and Westmoreland.

PLANTS AND ANIMALS: The large extent to which the land has been brought under culture has greatly diminished the size of the *forests* with which considerable portions of England were formerly covered, while many of the *wild animals* which its woods once sheltered have become altogether banished from within its limits.

The wild grasses, flowers, and shrubs, with the numerous smaller members of the animal kingdom, though interesting to the naturalist, are of less real importance than the grains, fruits, and vegetables, the domestic cattle and various farm-yard stock which engage the attention of the agricultural portion of the English population. Few, even of those which thrive most upon its soil, were originally native to England, and several have been introduced within a comparatively modern date.

Among trees, the *oak, elm, birch, poplar, alder, aspen, yew, mountain-ash,* and *Scotch-fir,* are probably indigenous to the soil; as also are the *apple,* the *hazel-nut, willow,* black and white *thorn, black-berry,* and common *dog-rose.* The trees, shrubs, and roots that are most common in England are, for the most part, the same that belong to similar latitudes of the European continent.

The vegetables which compose our common salads (as lettuces, radishes, &c.) were not grown in England until the reign of Henry VIII. The potato

—a native of the New World—was first introduced into England in the reign of Elizabeth. The peach, and other fruits of like kind, have been derived (by way of Southern Europe) from the countries of Western Asia. We owe some of our most common garden-flowers, as the ranunculus and the damask-rose, to the Crusaders and their companion pilgrims. The various roses, the narcissus, iris, jonquil, mignonette, and many other well-known ornaments of our garden, have been derived from Western Asia or the coasts of the Mediterranean.

INHABITANTS: England and Wales, with an area of 58,000 square miles, contain a population of over 29 millions,[1] so that, in the proportion of inhabitants to extent of surface, our country is more populous than any other European country, with the exception of Belgium.

In 1891, England contained 27,482,104, and Wales 1,518,914 inhabitants, equal to an average of 540 persons to the square mile in England, and 206 in Wales.

The density of population varies exceedingly in different parts of the country ; thus, while the County of London has over 350,000 inhabitants to the square mile, Westmoreland has only about 85. In Wales, the most thickly-peopled county is Glamorgan, with about 850 persons to the square mile ; and the least populous is Radnor, with 50. The great centres of population are London, South Lancashire, the West Riding of Yorkshire, Birmingham, and Newcastle, in England ; and Merthyr-Tydvil in Wales.

RACE: The people of England belong to the *Teutonic* race, those of Wales and Cornwall are of *Celtic* origin. Of the Celtic race, the *Gaels* were evidently the first settlers in the south of England, and were subsequently driven north to Scotland and west to Ireland by the *Kymri* (Cymru), another section of the Celtic family, and the ancestors of the present Welsh. Of the Teutonic Race, the *Angles, Jutes*, and *Saxons* first invaded Britain, and in time dislodged the Kymri from every part of the country except Cornwall and Wales. Many, belonging to two other sections—the *Danes* and the *Normans*—of the Teutonic family, also settled in England, and by the gradual admixture of all these elements the *English* nation was formed.

LANGUAGE : The languages spoken also differ ; that of the *Welsh*, and to a comparatively recent date that of the Cornish people, being purely Celtic ; that of the *English* is mainly derived from the Anglo-Saxon, but with a large admixture of Latin, Greek, Norman-French, and other words of foreign origin. In the Isle of Man a peculiar Celtic dialect, called *Manx*, is still spoken. In the Channel Islands, *French* is the common language.

INDUSTRIES : *Manufactures* and *trade* constitute the chief national industries of England. Until the close of the last century, England was essentially an *agricultural country.*

The rapid extension in the use of machinery, and the amazing growth of the great manufacturing industries, have effected a striking change in the general character of the national industry. England ranks first among the nations of

1. In 1881, England and Wales contained a population of 25,974,439—12,639,9 2 males, and 13,334,537 females : an increase since 1871 of 3,272,173, equal to an average daily addition of 930 persons through- | out the decade. In 1881 the density in England and Wales was 446 per square mile ; in 1851, 308 ; and in 1801, 153. In 1831, according to the Census Returns, the density was 498 per square mile

the world, in regard both to the quantity as well as quality of its manufactured products and the amount of its foreign trade. Its agricultural produce, though considerable, is unequal to the consumption of its teeming population, and the deficiency is supplied by the importation of food-stuffs from other lands.

AGRICULTURE : The more strictly agricultural districts of England are found chiefly in the eastern and southern portions of the island.

The manufacturing districts of England belong to the northern, north-midland, and western counties, and in these the farming pursuit is devoted in great measure to the rearing of stock, which is also (from the hilly nature of its surface, unsuited to the plough) the case in Wales. Of the total area of England more than two-thirds are under cultivation or in permanent pasture. In Wales, only about half the land is in pasture or under cultivation.

OBJECTS OF CULTURE : Wheat, oats, barley, and rye are the grains most largely grown ; wheat most extensively in the south-eastern counties, barley in the eastern and midland counties, oats within the district of the Fens and in the north. Hops are cultivated chiefly in Kent and Surrey (within the tract of country known as the Weald), and in the counties of Worcester and Hereford. The potato is very largely grown in Lancashire, Cumberland, and Cheshire, the turnip chiefly in Norfolk. Rape is much cultivated in Lincolnshire and Cambridgeshire ; hemp and flax in the counties of Lincoln and Suffolk. Garden vegetables are grown, on the most extensive scale, in the vicinity of the metropolis, and near the large towns in general. The counties of Hereford and Devon are distinguished for the extensive culture of the apple, from which cider is largely made.

MANUFACTURES : Cotton, wool, and iron are the three great staples of the manufacturing industry of England.

Cotton, which is a vegetable material—the pod of the cotton plant—is derived by import, chiefly from the United States of America.

Wool, an animal product, is furnished by the sheep reared upon our plains and downs, together with enormous quantities imported from distant lands—principally from our colonies in Australia and South Africa.

Iron, a mineral ore, is supplied in exhaustless abundance by the English soil, and is also largely imported from other countries.

The southern division of Lancashire and the adjoining part of Cheshire are the great seat of the cotton manufacture, which, though of comparatively recent origin, employs a much larger number of artisans than any other single branch of British industry. *Manchester* is the centre and capital of the cotton manufacture and *Liverpool* is its port.

The West Riding of Yorkshire is the chief seat of the woollen manufacture, and *Leeds* and *Bradford* are the principal centres of this industry.

The south part of Staffordshire, and the adjacent portions of Warwick, Worcester, and Shropshire, are the chief seats of the manufacture of iron and hardware goods, and *Birmingham* is the manufacturing capital of this district. *Sheffield*, in Yorkshire, is the chief seat of one branch of the hardware trade—the making of cutlery.

Other manufactures are carried on extensively in England, but none upon a scale of such magnitude as the three above-named. The silk and linen manufactures, the making of *hosiery* and *lace*, the *leather* manufacture, those of *earthenware* and *glass*, of *watches* and *clocks*, of *paper*, and a vast variety of others, are all of importance. The making of earthenware, on an extensive scale, is almost peculiar to a district in the north of Staffordshire, which is hence called "the Potteries."

TRADE : The internal trade of England is very extensive, and its development is facilitated by good *roads* and *railways, navigable rivers* and *canals ;* while the numerous inlets and estuaries, which form such splendid *harbours,* enable thousands of *coasting steamers* and sailing vessels to trade regularly from port to port on the coasts—east, south, and west.

Good roads and well-kept canals traverse every part of England, and lines of railway supply the means of rapid communication between all the principal towns.

Roads.—There are about 25,000 miles of turnpike roads, and more than 100,000 miles of cross roads. The former especially are well made, and are always kept in good condition.

Canals and **navigable rivers** formed a most important means of communication before railways were introduced, and are still largely used for the conveyance of heavy goods and coals. There are above 3,000 miles of canals, and nearly 1,800 miles of navigable rivers, so that the available water-communication in England is nearly 5,000 miles.

Railways.—All the great railway lines of England radiate from London, which is thus directly connected by rail with every part of England. The railways of England carry over 600 millions of passengers annually, besides enormous quantities of goods, and have done more than anything else to extend the trade and commerce of the country.

Posts and Telegraphs.—The postal and telegraphic services of England are the most complete and efficient in the world.

COMMERCE : The foreign commerce of England extends to every part of the globe ; her ships traverse every sea, and her flag is seen in the harbours of every land. The *import* of raw materials, and the *export* of manufactured goods, are distinguishing features of English commerce.

Imports.—*Sugar, coffee, spices,* and other productions of tropical regions, foreign to the English soil, are imported from the East and West Indies ; *tea* from British India and China ; *tobacco* from the United States and elsewhere ; *timber* from Canada and the countries lying around the Baltic Sea ; *wines* from France, Spain, Portugal, Germany, Hungary, and Australia ; *hides, skins,* and *tallow* from South America, South Africa, and Russia ; *raw cotton* from the United States, India, Egypt, and other countries ; *wool* from Australia, South Africa, &c. ; *corn* and *flour* from the United States, India, Australia, Russia, Austria-Hungary, &c.

Exports.—The most important articles exported are *cotton, woollen, linen,* and *silk* goods, *iron* and *steel, earthenware, tin, machinery, stationery* and *books,* and *coal.*[1]

---

1. A fuller account of the productions, manufac- | those of Scotland and Ireland) is given in the sec-
tures, and commerce of England (together with | tion devoted to the " United Kingdom " as a whole.

The countries to which the largest quantities of British manufactures and other produce are exported are the British Possessions abroad, the United States of America, the East Indies, France, Germany, Holland, Russia, Belgium, Italy, Africa, Brazil, Turkey, China, Spain, Denmark, Japan, and the various States of South America.

PORTS : Of the great ports of England, **London** has by far the largest *general trade*, but its *foreign trade* is not much larger than that of **Liverpool.** More than **two-thirds of the entire foreign** trade of England passes through these two great ports, which also yield considerably more than **two-thirds of the total customs** receipts.

The other chief ports (in order of tonnage entered and cleared) are the **Tyne Ports** (Newcastle, Gateshead, North and South Shields), **Cardiff, Hull, Sunderland, Newport, Portsmouth, Southampton, Swansea, Bristol, Middlesborough, Plymouth, and Hartlepool.**

TOWNS : Three-fifths of the people of England and Wales live in towns. **London** alone contains one-sixth of the total population of the country, and two other cities, **Liverpool and Manchester,** each contain upwards of seven hundred thousand inhabitants.

According to the Census Returns for 1891, there are, besides *London, Liverpool,* and *Manchester,* 21 English towns with over 100,000 inhabitants, seven of which—Birmingham, Leeds, Sheffield, Bristol, Bradford, Nottingham, and Hull, contain a population of over 200,000. There are altogether no less than 62 towns with a population of over 50,000.

The ten largest towns in England and Wales are the following :—

London, with a population of 5,500,000 ; Liverpool, 730,000 ; Manchester and Salford, 703,000 ; Birmingham, 429,000 ; Leeds, 368,000 ; Sheffield, 324,000 ; Bristol, 222,000 ; Bradford, 216,000 ; Nottingham, 212,000 ; Hull, 200,000. Each of these ten cities thus contain over 200,000 inhabitants, and together contain over 8½ millions of people, or considerably more than one-fourth of the total population of England and Wales.

Twelve other towns contain over 100,000 inhabitants, namely :—

Newcastle, 186,000 ; Portsmouth, 159,000 ; Leicester, 142,000 ; Oldham, 131,000 ; Sunderland, 131,000 ; Cardiff, 129,000 ; Blackburn, 120,000 ; Bolton, 115,000 ; Brighton, 115,000 ; Preston, 108,000 ; Norwich, 101,000 ; Birkenhead, 100,000.

Fifteen other towns contain over 70,000 inhabitants, namely :—

Huddersfield, 95,000 ; Derby, 94,000 ; Swansea, 90,000 ; Ystradyfodwg, 88,000 ; Burnley, 87,000 ; Gateshead, 86,000 ; Plymouth, 84,000 ; Halifax, 83,000 ; Wolverhampton, 83,000 ; South Shields, 78,000 ; Middlesborough, 76,000 ; Walsall, 72,000 ; Rochdale, 71,000 ; St. Helens, 71,000 ; Stockport, 70,000.

COUNTIES : England and Wales are divided into **52 counties** or Shires, of which there are **40** in England, and **12** in Wales.

All these divisions are very irregular in shape, and, as the map shows, very unequal in size. Rutlandshire, *the smallest of the English counties,* is hardly more than one-fortieth part of the size of Yorkshire, which is the *largest.* Lincoln, Devon, and Norfolk come next to Yorkshire in order of magnitude. Middlesex, Huntingdon, and Bedford are next to Rutland, the smallest in extent.

The English counties are divided, with reference to relative situation, into six Northern, six Western, five Eastern, nine Southern, and fourteen Midland (six North-Midland and eight South-Midland) Counties.

(1.) The six Northern Counties are: Northumberland, Durham, Yorkshire, Cumberland, Westmoreland, and Lancashire. The first three border on the North Sea, the other three are on the western side of England.

(2.) The six Western Counties are: Cheshire, Shropshire, Herefordshire, Monmouthshire, Gloucestershire, and Somerset. The first four border on Wales; the last two are on the English side of the Severn and its estuary.

(3.) The five Eastern Counties are: Lincoln, Cambridge, Norfolk, Suffolk, and Essex. All, except Cambridge, are washed by the waters of the North Sea.

(4.) The nine Southern Counties are: Kent, Surrey, Sussex, Berkshire, Hampshire, Wiltshire, Dorsetshire, Devon, and Cornwall. They stretch (with the exception of Surrey, Berkshire, and Wiltshire, which are inland) along the south coast of England, from the South Foreland on the east, to the Land's End on the west.

(5.) The six North-Midland Counties are: Stafford, Derby, Nottingham, Leicester, Warwick, and Worcester.

(6.) The eight South-Midland Counties are: Oxford, Buckingham. Middlesex, Hertford, Bedford, Huntingdon, Northampton, and Rutland. *Middlesex* is termed the "Metropolitan County," since it contains the greater part of *London*, the capital of England and of the British Empire.

(7.) The six Counties in North Wales are: Anglesey, Carnarvon, Denbigh, Flint, Merioneth, and Montgomery.

(8.) The six Counties in South Wales are: Cardigan, Radnor, Brecknock, Carmarthen, Pembroke, and Glamorgan.

ADMINISTRATIVE COUNTIES: Besides the 52 counties, each of which is now governed by a **County Council**, *London* and 61 *provincial Boroughs* have, for all purposes of *Local Government*, been formed into **Administrative Counties**, absolutely independent of the counties of which they geographically form a part.

These County Boroughs are Barrow, Bath, Birkenhead, Birmingham, Blackburn, Bolton, Bootle-cum-Linacre, Bradford (Yorks.), Brighton, Bristol, Burnley, Bury, Canterbury, Cardiff, Chester, Coventry, Croydon, Derby, Davenport, Dudley, Exeter, Gateshead, Gloucester, Great Yarmouth, Halifax, Hanley, Hastings, Huddersfield, Ipswich, Kingston-upon-Hull, Leeds, Leicester, Lincoln, Liverpool, Manchester, Middlesborough, Newcastle-upon-Tyne, Northampton, Norwich, Nottingham, Oldham, Plymouth, Portsmouth, Preston, Reading, Rochdale, St. Helens, Salford, Sheffield, Southampton, South Shields, Stockport, Sunderland, Swansea, Walsall, West Bromwich, West Ham, Wigan, Wolverhampton, Worcester, York.

The County of London has an area of nearly 120 square miles, and a population of over 4¼ millions.

For Administrative purposes, the County of London is controlled by a County Council. For Parliamentary purposes, London is divided into 59 Boroughs, each of which return one member.

I.—SIX NORTHERN COUNTIES.[1]

NORTHUMBERLAND, the most northerly of the English counties, contains the towns of *Newcastle, Tynemouth* and *North Shields, Berwick, Alnwick, Morpeth,* and *Hexham.* The south-eastern portion of the county contains a rich coalfield, and has coal-mines, ironworks, and various manufactures. Its western and northern divisions are hilly and pastoral.

NEWCASTLE (186),[2] on the River Tyne, is the seat of an enormous *coal trade,* and is also the county town. Hexham, a few miles west of Newcastle, was the scene of a victory gained by the Yorkists over the Lancastrian forces in 1464. Berwick, at the mouth of the River Tweed, adjoins the Scottish frontier, and is celebrated in the history of early Border warfare. *Halidon Hill,* the scene of a victory gained by the English over the Scots in 1333, is immediately to the north of Berwick. Northumberland includes the site of the battle of *Flodden,* so disastrous to the Scots, fought (in 1513) near the village of Flodden, a few miles distant from the small town of Wooler, at the eastern foot of the Cheviot Hills. *Homildon,* the scene of Harry Percy's victory over Douglas (in 1402), lies only a mile distant from Wooler.

DURHAM adjoins Northumberland, and resembles that county in its eastern part, which is a rich coalfield, with numerous iron and other works, and busy seaport towns.

The cathedral city of DURHAM (15), on the Wear, is the capital of the county ; but Sunderland (131), at the mouth of the same river, a great *coal* port, with large *ship-building* yards, is a much more important place. Stockton on the Tees, Hartlepool, and South Shields are also in this county. There are large *ship-building* yards and *chemical works* at Jarrow. Gateshead, on the Durham side of the Tyne, is practically a suburb of Newcastle, to which it is joined by several bridges.[3]

YORKSHIRE, the largest English county, is divided into three *Ridings*—the North, East, and West Ridings. The two former are agricultural and pastoral ; the West Riding embraces part of an extensive coalfield, and is a populous district, the chief seat of the English woollen manufacture.

YORK (67), the capital of the county, and an archbishop's See, stands on the River Ouse, in the centre of a fertile plain, at the junction of the three Ridings. A few miles west of York is the village of *Long Marston,* near which the army of Charles I. was defeated by Cromwell in 1644. Further to the south-west, near the banks of the Wharfe, is *Towton,* the scene of the bloodiest engagement fought during the Wars of the Roses (1461). *Stamford Bridge,* on the River Derwent, a few miles east of York, is noteworthy for the victory gained there by Harold over his brother Tostig and a Norwegian army, a few days prior to the battle of Hastings.

---

1. The areas and population of the Six Northern Counties are as follows :—
(1.) Northumberland, 2,016 sq. m., pop. 506,096.
(2.) Durham, 1,011 sq. m., pop. 1,016,449.
(3.) Yorkshire, 6,067 sq. m., pop. 3,208,813.
(4.) Cumberland, 1,515 sq. m., pop. 266,550.
(5.) Westmoreland, 760 sq. m., pop. 66,098.
(6.) Lancashire, 1,889 sq. m., pop. 3,926,768.
2. As an indication of their absolute and relative importance, the population of the largest towns are given in thousands between brackets, thus—Newcastle (186), i.e., 186,000. All populations of the counties and towns of the United Kingdom are taken from the Census Returns for 1891, except in the case of the smaller towns in Ireland.
3. The battle of *Neville's Cross,* gained over the Scottish army in 1346, was fought in this county, a few miles distant from the city of Durham.

The chief towns in the WEST RIDING of Yorkshire are Leeds, Sheffield,
Bradford, Huddersfield, Halifax, Wakefield, Barnsley, Dewsbury, Doncaster,
and Ripon. Leeds (367), on the River Aire ; Bradford (216) ; Huddersfield (95) ;
and Halifax (83), are the great seats of the *woollen* and *cloth manufactures.*
Sheffield (324), on the River Don, is the seat of the *cutlery trade.* Ripon, on
the Ure, is a cathedral city. Wakefield (33), also a cathedral city, on the River
Calder, was the scene of a victory gained by the Lancastrians over the Yorkists
in 1460.

The EAST RIDING contains the towns of Hull, Beverley, and Bridlington.
Hull (200) stands on the north bank of the Humber, at the entrance of the little
River Hull, and is an important seaport.

The NORTH RIDING contains Scarborough and Whitby, famous sea-side
resorts, with Malton, Richmond, and Northallerton in the interior. The *Battle
of the Standard,* between the English and Scottish armies in 1138, was fought
near Northallerton. The port of Middlesborough (although its foundation only
dates from 1829) contains over 75,000 inhabitants, and is the centre of the important *iron* and *salt* district of Cleveland.

CUMBERLAND is for the most part **mountainous**, but its
northern division includes a plain of some extent, along the Solway
Firth and the lower course of the Eden. In the west, along the
coast, is a small but highly-productive **coalfield**. The chief towns
are *Carlisle, Whitehaven, Workington, Maryport, Cockermouth,
Penrith,* and *Keswick.*

CARLISLE (39), the capital, is a cathedral city, on the River Eden. Whitehaven, on the coast, is the chief seat of the *coal* trade. Workington and Maryport are other coal-exporting ports. Keswick is in the heart of the Lake District.

WESTMORELAND is for the most part **mountainous and pastoral**, and contains the towns of *Kendal* and *Appleby.*

APPLEBY is the county town ; but Kendal (14), on the River Kent, which
flows into Morecambe Bay, is of larger size, and has *woollen* and other manufactures. Ambleside is much resorted to on account of the beauty of the scenery.

LANCASHIRE includes the great seats of the **cotton manufacture**. This branch of industry is pursued through all the southern
portion of the county, which includes a valuable **coalfield**. A small
detached portion of the county, known as Furness, lies to the north
of Morecambe Bay, and belongs physically to the Lake District.

Manchester (505, or including the adjoining town of Salford, 703), on the
Irwell, and Liverpool (730), at the mouth of the Mersey, are the largest towns
in England, next to the metropolis ; the former is the great centre of the *cotton
trade,* and the latter the chief *cotton port.* Manchester and Liverpool are
cathedral cities. Liverpool is connected with Birkenhead, on the opposite side
of the Mersey, by a railway tunnel under the river. Preston, on the Ribble,
Bolton, Oldham, Blackburn, Wigan, Rochdale, Ashton-under-Lyne, Warrington,
Colne, Chorley, and Lancaster, are all busy *cotton* towns. Barrow-in-Furness
(52), is the port of the rich *iron-ore* district of North Lancashire, and has the
largest *steel-works* in the kingdom. LANCASTER (31), on the River Lune,
ranks as the capital of the county. Part of Stalybridge is in this county.

II.—SIX WESTERN COUNTIES.[1]

CHESHIRE is chiefly an **agricultural** county, but includes part of the **cotton-manufacturing** district.

The cathedral city of **CHESTER** (37), on the River Dee, is the capital, and in many respects one of the most interesting cities in England. The Cathedral, Roman walls, the Castle, and the picturesque " Rows " are altogether unique. On *Rowton Heath*, to the west of this city, the troops of Charles I. were defeated by the Parliamentary forces in 1645—a few months after the battle of Naseby. **Stockport** (70), a cotton town ; **Macclesfield** (36), with silk manufactures, and **Birkenhead** (100), with great *docks* and *ship-building* yards, opposite Liverpool (with which it is connected by a railway tunnel under the Mersey), are important towns. **Nantwich, Middlewich,** and **Northwich,** in the valley of the River Weaver, are famous for their *saltworks*. The manufacturing town of **Stalybridge** (26) is principally in this county.

SHROPSHIRE is chiefly **agricultural,** but includes a small **coalfield,** and has some **iron** and other **manufactures** at *Coalbrookdale* and elsewhere.

SHREWSBURY (27), on the Severn, is the county town. The battle between the army of Henry IV. and the insurgent forces under Hotspur was fought in its immediate vicinity in 1403. There are *coal* and *iron* mines at **Wenlock** and **Madeley.** The other towns are **Bridgnorth, Wellington, Ludlow,** and **Oswestry.**

HEREFORDSHIRE is entirely **agricultural,** and is famous for its *hop gardens* and its *orchards.*

The city of **HEREFORD** (20), its capital, stands on the River Wye, and has an ancient Cathedral. Important cattle and cheese fairs are held in October each year. **Leominster, Ledbury,** and **Ross** are small towns in this county. A few miles to the north-west of Leominster is *Mortimer's Cross,* one of the battle-fields of the Wars of the Roses, A.D. 1461.

MONMOUTHSHIRE is a **mining** and **manufacturing** county. It includes a portion of the South Wales **coalfield,** and has numerous **ironworks.**

The chief town, **MONMOUTH** (6), is on the Wye, at the junction of the small River Monnow, hence its name. At **Tredegar** (17) are important *coal* and *iron* mines. The other towns are **Newport** (55), a considerable seaport at the mouth of the Usk, **Abergavenny, Pontypool,** and **Chepstow.**

GLOUCESTERSHIRE includes the long chain of the Cotswold Hills, and is in great part **agricultural.** But it possesses two small **coalfields**—one of them in the Forest of Dean, to the west of the Severn, the other near the Avon, on the south border of the county. The manufacture of **woollen cloth** is pursued extensively at Stroud and other places near the Cotswold Hills.

GLOUCESTER (40), the capital, on the Severn, is a cathedral city. Cheltenham, **Stroud, Cirencester,** and **Tewkesbury** are in this county. Cheltenham

1. The area and population of the Six Western Counties are as follows:—
(1.) Cheshire, 1,026 sq. m., pop. 730,652.
(2.) Shropshire, 1,319 sq. m., pop. 259,324.
(3.) Herefordshire, 833 sq. m., pop. 115,086.
(4.) Monmouthshire, 578 sq. m., pop. 252,660.
(5.) Gloucestershire, 1,224 sq. m., pop. 601,604.
(6.) Somersetshire, 1,640 sq. m., pop. 484,376.

is famed for its mineral waters. Tewkesbury, on the Severn, is historically noteworthy on account of the victory gained there by Edward IV. over the army of Queen Margaret, in 1471, three weeks after the battle of Barnet.

Bristol (222), an important port on the Avon, is partly in Gloucestershire and partly in Somersetshire, but has long had the privileges of a county in itself.

SOMERSETSHIRE is almost wholly agricultural.

TAUNTON (18), on the River Tone, is the county town.   Bath (52), on the Avon, was in the time of the Romans, and still is, famous for its hot *mineral springs*.   Wells is an ancient city, lying at the foot of the Mendip Hills. Frome (10), has some *cloth* factories ; Bridgwater (12) is a port on the Parret, and Yeovil has *glove* manufactures.   Between Bridgwater and Taunton is *Sedgemoor*, the scene of Monmouth's defeat by the troops of James II. in 1685. *Athelney*, the temporary retreat of Alfred, was in former ages a marshy tract of ground, lying near the junction of the rivers Tone and Parret, by which it was insulated.

III.—FIVE EASTERN COUNTIES.[1]

LINCOLNSHIRE is an agricultural county.

Its capital, LINCOLN (41), is an ancient cathedral city on the River Witham. Boston (15), also on the Witham, a few miles above its mouth, and Grimsby (52), on the south bank of the Humber, are flourishing ports.   Gainsborough (on the Trent), Stamford (on the Welland), Grantham, with an important corn trade, and Louth, are inland towns.

CAMBRIDGESHIRE is an agricultural county.   Its northern half is called the Isle of Ely, from its having in former times been insulated by marshes, and is within the level region of the Fens.

The county town, Cambridge (37), is on the River Cam, which joins the Ouse, and is the seat of one of the two ancient universities of England.   Wisbeach, Ely, and Newmarket are in this county.   Ely, on the Great Ouse, is a cathedral city.

NORFOLK is the only one of the eastern counties that possesses any considerable manufactures ; but by far the larger portion of the county is agricultural.   Wild fowl and sea birds flock to the "Broads" of Norfolk in enormous numbers.

The ancient city of NORWICH (101), its capital, was an early seat of woollen manufacture, which it still retains.   Norwich stands on the River Wensum, immediately above its junction with the Yare.   Yarmouth (49), at the mouth of the Yare, is the chief English *herring port*.   Off the coast are the famous *Yarmouth Roads*, a much frequented roadstead.   Lynn, or King's Lynn (18), which also possesses considerable trade, is at the mouth of the Great Ouse.

SUFFOLK is an entirely agricultural county.

Its chief town, IPSWICH (57), stands on the River Orwell, a few miles above its mouth.   Bury St. Edmunds, Lowestoft, and Sudbury are the other principal

---

1. The areas and population of the Five Eastern Counties are as follows :—
   (1.) Lincolnshire, 2,762 sq. m., pop. 472,778.
   (2.) Cambridgeshire, 820 sq. m., pop. 188,862.

(3.) Norfolk, 2,118 sq. m., pop. 456,474.
(4.) Suffolk, 1,475 sq. m., pop. 369,301.
(5.) Essex, 1,542 sq. m., pop. 705,399.

towns. Lowestoft is a seaport, situated at the most easterly extremity of Great Britain, and is an important station of the *herring fishery*.

ESSEX is also chiefly agricultural.

Its county town is **CHELMSFORD** (11), on the Chelmer. **Colchester** (35), on the Colne, has *oyster fisheries*. Harwich is an important packet station for passengers and goods to and from the continental ports on the opposite side of the North Sea. **Stratford** (43) is a part of East London. **Saffron-Walden, Braintree,** and **Maldon** are all small inland towns.

IV.—NINE SOUTHERN COUNTIES.[1]

KENT is chiefly an agricultural county, but it contains numerous seaports, some of which, from their proximity to the coast of France, were early of historical importance, and several are still important **packet stations** or favourite **watering-places.**

The county town is **MAIDSTONE** (32), on the River Medway; but Chatham, Woolwich, and Dover are of larger size. **Chatham** (32), which lies near the mouth of the Medway, is an important naval arsenal—the second in the kingdom. It adjoins Rochester, an ancient cathedral city, and Strood—the three forming really but one town. **Woolwich**, the great *military arsenal* of England, is on the south bank of the Thames, below London. **Dover** is at the south-east corner of the island, immediately opposite to the coast of France; a few miles west of it is Folkestone, also a seaport of ancient date. Dover and Folkestone are now the chief packet stations for the continent, steamers running regularly from Dover to Calais and Ostend, and from Folkestone to Boulogne. **Canterbury** (23), on the River Stour, is an ancient cathedral city—the ecclesiastical metropolis of England[2]—and surpasses any other place in Kent in historic dignity. It was here that the Saxon king, Ethelbert, in A.D. 597, embraced Christianity, on its re-introduction into Britain by the agency of Augustine.

Along the shores of Kent (beginning at the eastern suburbs of London) there occur in succession the following places :— Greenwich, Woolwich, Gravesend, Rochester, and Chatham ; Sheerness and Queenborough (on the Isle of Sheppey), Whitstable, Margate, Ramsgate, Sandwich, Deal, Dover, Folkestone, and Hythe.

Margate and **Ramsgate** are popular *watering-places* on the eastern coast of Kent. Opposite Deal are the Goodwin Sands, between which and the shore is the much-frequented roadstead called the *Downs*.[3]

---

1. The areas and population of the Nine Southern Counties are as follows :—
(1.) Kent, 1,555 sq. m., pop. 1,142,287.
(2.) Surrey, 758 sq. m., 1,730,871.
(3.) Sussex, 1,458 sq. m., pop. 550,442.
(4.) Berkshire, 722 sq. m., pop. 238,416.
(5.) Hampshire, 1,621 sq. m., pop. 659,686.
(6.) Wiltshire, 1,354 sq. m., pop. 264,969.
(7.) Dorsetshire, 980 sq. m., pop. 194,487.
(8.) Devonshire, 2,586 sq. m., pop. 631,767.
(9.) Cornwall, 1,349 sq. m., pop. 322,589.
2. Canterbury and York are each the seat of an archbishopric. But Canterbury ranks first in point of dignity. The Archbishop of York is a primate of England; the Archbishop of Canterbury is primate of *all* England.

3. Some of the ports on the Kentish and Sussex coasts had peculiar privileges granted them in the later Saxon and early Norman times, from the importance of their position with reference to the opposite shores of the continent. The CINQUE PORTS, as they were called—originally five in number—were *Sandwich, Deal, Dover, Hythe,* and *Romney,* all in Kent. Three others, *Rye, Winchelsea,* and *Hastings,* in the adjoining county of Sussex, were afterwards added. DOVER, DEAL, and HASTINGS are now the most considerable of the Cinque Ports, the harbours of the others having become choked up by sand. These ports were bound by charter to provide a certain number of ships for the defence of the coast.

SURREY includes that portion of the metropolis situated to the south of the Thames, but by far the greater part of the county is agricultural.

GUILDFORD (14), the county town, is on the Wey, an affluent of the Thames; Croydon is really a suburb of London. **Richmond** and **Kingston**, riverside resorts on the Thames, are in this county. Kingston was important in Saxon times, and seven of our Saxon kings were crowned there. *Runnymead*, where the Great Charter was signed by King John, at the instance of his armed barons (A.D. 1215), is within the north-western border of Surrey, immediately adjoining the south bank of the Thames, and near the small town of Egham.

SUSSEX, an agricultural county, includes an extensive line of coast lying along the English Channel. From Beachy Head westward this coast is backed by the range of chalk hills called the *South Downs*.

LEWES (11), on the Ouse, is the county town, and gave its name to a battle fought in the vicinity between Prince Edward (afterwards Edward I.) and the rebellious barons, during the reign of Henry III. (A.D. 1264). **Brighton** (115), on the coast, is a large and flourishing watering-place and pleasure resort, attracting vast numbers of visitors from the metropolis. **Hastings**, also on the coast, is further to the eastward. A few miles west of Hastings (near Bulverhithe, on the shore of Pevensey Bay) is the place where William the Conqueror landed in 1066; the small town of *Battle*, to the north-west of Hastings, marks the scene of the engagement which, a few days afterwards, transferred the dominion of England from Saxon to Norman hands. **Shoreham** and **Worthing** are on the coast, to the west of Brighton. **Chichester**, still further west, is a cathedral city.

BERKSHIRE is an agricultural county.

READING (60), its capital, stands at the junction of the Kennet with the Thames. It is famous for its *biscuits*. **Windsor Castle**, the chief *royal residence* of the sovereign of England, is on the south bank of the Thames. **Abingdon**, **Maidenhead**, **Newbury**, and **Wantage** are in this county. Two engagements between the armies of Charles I. and the Parliament occurred near *Newbury*, in 1643 and the following year. Wantage is distinguished as the birthplace of Alfred the Great.

HAMPSHIRE is an agricultural county.

It possesses two important seaports—Portsmouth (159), the *chief naval arsenal* of England, and Southampton (65), an important packet station; but **WINCHESTER** (19), an ancient cathedral city, in the fertile valley of the Itchen, is the capital. Winchester was the chief city of the West Saxon kings, and continued to be regarded, in early Norman times, as the capital of the kingdom. **Andover**, **Lymington**, and **Basingstoke** are small towns. **Bournemouth** is a charming seaside resort, built on the shores of a beautiful bay, and sheltered by fine woods.

The Isle of Wight lies to the south of Hampshire, and forms a portion of that county. **Ryde**, **Ventnor**, and **Cowes** are favourite seaside resorts and yachting stations. **Newport**,[1] on the Medina, 5 miles south of Cowes, is the capital of the island and a great tourist resort. **Osborne** is a favourite residence of the Queen.

---

1. Near Newport is *Carisbrooke Castle*, at one time the prison of King Charles I.

WILTSHIRE is chiefly agricultural, but the manufacture of woollen cloth is carried on in its westerly division.

SALISBURY (16), the capital of the county, and a cathedral city, is on the Avon. The manor-house of *Clarendon*, where the well-known statutes called "The Constitutions of Clarendon" were drawn up in the reign of Henry II., is a few miles east of Salisbury. *Stonehenge*, a supposed Druidical remain— among the most ancient monuments of our island—is upon the high chalk tract of Salisbury Plain. At Bradford, Trowbridge, and Westbury, the famous "West of England" cloth is made. Devizes, Warminster, and Marlborough are among the other towns in this county. *Roundway Down*, near Devizes, was the scene of a skirmish between the Royalist and Parliamentary forces in 1643.

DORSETSHIRE is an agricultural county, and has several small seaports on the shores of the Channel.

DORCHESTER (8), the county town, is on the Frome. Weymouth (14) is a packet station for the Channel Islands and a fashionable watering-place. Poole and Bridport are among the other towns. The peninsular tracts known as the *Isle of Purbeck* and the *Isle of Portland* are both within this county.

DEVONSHIRE, which is chiefly an agricultural county, includes the high tract of *Dartmoor*, and (in the north) part of an elevated region called *Exmoor*, on the borders of Somerset. It has numerous seaports, some on the shore of the English Channel, and others on the side of the Bristol Channel.

EXETER (38), the capital, is a cathedral city, on the River Exe. Plymouth (84) and Devonport (55) are adjacent towns, situated on the fine estuary of Plymouth Sound, which is one of the chief stations of the British navy. The Sound is protected from heavy seas by a huge breakwater, and 14 miles out at sea, to the south-west, stands the famous *Eddystone Lighthouse*. The following, and numerous other smaller towns, are in this county : Barnstaple, on the northern coast ; Bideford, the home scene of Kingsley's "Westward Ho !"; Tiverton and Tavistock, two small inland towns ; Dartmouth, at the mouth of the "English Rhine"; Torquay, a beautiful seaside resort ; Teignmouth, and Exmouth.

CORNWALL is chiefly a mining county, but its mackerel and pilchard fisheries are also of considerable importance. Its tin mines have been worked from a very early age—some centuries before the Christian era. Its numerous copper mines are now virtually abandoned.

BODMIN (5) is the county town, but Truro (11), a cathedral city, ranks as the capital of the *mining* district. Penzance, Falmouth, St. Austell, and Launceston are among the other towns. *Stratton*, near the northern extremity of the county, was the scene of a victory gained by the Royalists over the Parliamentary forces in 1643.

The Scilly Islands lie off the coast of Cornwall, at the entrance of the English Channel. Hugh Town, the capital, is on St. Mary's, the largest of the six inhabited islands.

V.—SIX NORTH-MIDLAND COUNTIES.[1]

STAFFORDSHIRE, a mining and manufacturing county, includes two **coalfields**—one (that of South Staffordshire) the seat of the iron and hardware manufacture ; the other (in the northern part of the county) embracing the district of the **Potteries**.

The county town is **STAFFORD** (20), on the River Sow, an affluent of the Trent. But Wolverhampton (83), **West Bromwich** (60), **Walsall** (72), Bilston (23), and **Wednesbury** (25), within the coal and iron district of the south (and in the vicinity of Birmingham), are all, except the last named, of larger size. Stoke-upon-Trent (24), Hanley (55), and **Etruria** are in the Pottery district. Lichfield (8), towards the eastern border of the county, is a cathedral city. Burton-on-Trent (46) is famous for its *ale* and *beer*, which are exported in enormous quantities to all parts of the world. The site of the battle of *Blore Heath* (A.D. 1459) is within this county, about eleven miles to the north-west of Stafford, and close to the Shropshire border.

DERBYSHIRE is partly a manufacturing county, but embraces the rugged and elevated district of **The Peak**, which forms its northerly division. It has numerous **lead and iron** mines, and includes part of an extensive **coalfield**, the larger portion of which is within the adjacent county of York.

The chief town, **DERBY** (94), is on the River Derwent, which joins the Trent, and is a great seat of the silk, cotton, and lace manufactures, and is, besides, an important railway centre. **Chesterfield, Belper, Wirksworth**, and **Ashborne** are among the other towns. **Buxton and Matlock**, in the Peak district, are famous for their mineral waters, and for the romantic beauties of the scenery in the neighbourhood.

NOTTINGHAMSHIRE is partly **manufacturing, but is more** generally an **agricultural** district.

The chief town, **NOTTINGHAM** (212), on the Trent, has extensive manufactures of lace, cotton stockings, &c. **Newark, Mansfield**, and **Worksop** are smaller towns. A few miles from Newark, and near the south bank of the Trent, is the village of *Stoke*, the scene of a battle fought (A.D. 1487) between the army of Henry VII. and the followers of the impostor Lambert Simnel.

LEICESTERSHIRE has extensive **manufactures**, though a great portion of the county is **agricultural**.

The county town, **LEICESTER** (142), on the River Soar, which joins the Trent, is noted for its extensive manufacture of woollen hosiery, and boots and shoes. *Bosworth*, near which the battle that terminated the Wars of the Roses was fought in 1483, is a few miles west of Leicester. **Ashby-de-la-Zouch**, in the north-west part of the county, has a small coalfield in its neighbourhood. The other towns are Loughborough, Hinckley, Melton Mowbray, **Market Harborough**, and **Lutterworth**.

1. The areas and population of the Six North-Midland Counties are as follows :—
(1.) Staffordshire, 1,169 sq. m., pop. 1,083,273.
(2.) Derbyshire, 1,029 sq. m., pop. 527,806.
(3.) Nottinghamshire, 824 sq. m., pop. 445,502.
(4.) Leicestershire, 799 sq. m., pop. 373,603.
(5.) Warwickshire, 884 sq. m., pop. 805,072.
(6.) Worcestershire, 738 sq. m., pop. 413,755.

WARWICKSHIRE is a manufacturing county.

The county town, WARWICK (12), is on the banks of the Upper Avon. Adjoining it is Leamington, famous for its mineral waters. Birmingham (429), in the north-west part of the county, close to the Staffordshire border, is the fourth town in England in order of population, and the chief centre of the English *iron* trade, and the greatest *hardware* manufacturing town in the world. Coventry (53), further to the east, has a small coalfield in its vicinity, and was formerly noted for its manufacture of *ribbons*, but is now the principal seat of the *cycle manufacture* in the kingdom. Rugby (famous for its great public school), Nuneaton, and Stratford-on-Avon (the birthplace of Shakespeare), are in this county. The battle of *Edgehill* (A.D. 1642) was fought on the rising ground of that name, within the southern extremity of Warwickshire.

WORCESTERSHIRE has extensive iron and other manufactures in its northerly division, but is chiefly an agricultural county.

Its capital, WORCESTER (42), on the Severn, is a cathedral city, and is noted for its *porcelain* and *glass-works*, as well as for many events of historical fame—the chief among them being the victory of Cromwell over the adherents of Charles II. in 1651. Kidderminster (25), on the Stour, has extensive carpet factories. Dudley (46), further to the north, though belonging to this county, is locally within the iron and coal district of South Staffordshire, and is a populous seat of the *hardware* trade. Bromsgrove, Stourbridge, Stourport, and Evesham are within this county. *Evesham* (within the fertile vale of that name, watered by the Avon) was the scene of a battle between Prince Edward and the barons under Simon de Montfort in A.D. 1265.

VI.—EIGHT SOUTH-MIDLAND COUNTIES.[1]

OXFORDSHIRE is also an agricultural county.

This county, often called Oxon, has for its capital the diocesan city of Oxford (46), seated at the junction of the Cherwell and the Thames. Oxford is a celebrated seat of learning, its famous university is the most ancient of the two great universities of England. Among the other towns are Banbury, Witney, Henley-on-Thames, and Woodstock. *Chalgrove Field*, the scene of a skirmish between the forces of Charles I. and the Parliament, in which Hampden was mortally wounded (1643), is in this county.

BUCKINGHAMSHIRE is entirely an agricultural county.

The town of Buckingham (3), lies on the Great Ouse, in the northern part of the county. AYLESBURY (9), the county town, is noted for the manufacture of *condensed milk;* Eton has a famous college.

MIDDLESEX, which contains by far the largest portion of London, is, therefore, the most populous of the counties of England. It stretches along the north bank of the River Thames, by which it is divided from the neighbouring county of Surrey.

LONDON, with its suburbs, stretches over a vast space, and contains altogether fully 5½ millions of inhabitants, by far the largest population

---

1. The areas and population of the Eight South-Midland Counties are as follows:—
(1.) Oxford, 745 sq. m., pop. 185,938.
(2.) Buckinghamshire, 743 sq. m., pop. 185,194.
(3.) Middlesex, 283 sq. m., pop. 3,251,702.
(4.) Hertfordshire, 633 sq. m., pop. 220,325.
(5.) Bedfordshire, 460 sq. m., pop. 160,729.
(6.) Huntingdonshire, 358 sq. m., pop. 257,772.
(7.) Northamptonshire, 984 sq. m., pop. 302,184.
(8.) Rutlandshire, 148 sq. m., pop. 20,659.

any other city on the globe.   Besides its rank as the metropolis of the kingdom, London is a great manufacturing and commercial city, a centre of art, litera-ture, general refinement, and wealth.   Westminster, which is now included within the western limits of the metropolis, was formerly separated from London by intervening fields.   Southwark, the southern division of London, is to the south of the Thames, and within the county of Surrey.   London con-tains the cathedral church of St. Paul's, and constitutes a bishop's See.

The towns of Brentford (14), Uxbridge, and Staines are in this county; Brentford is the county town.   At Enfield is a *small-arms* factory; at Harrow-ou-the-Hill is one of the great public schools.

HERTFORDSHIRE is an agricultural county, and has no towns of large size.

HERTFORD, the county town, on the River Lea, is not so large as St. Albans.   Hitchin, Watford, Bishop Stortford, and Barnet are small places in this county.   *St. Albans* was the scene of two of the battles fought (1455 and 1461) during the Wars of the Roses.   *Barnet*, which lies on the borders of Hertford and Middlesex, witnessed a more important event belonging to the same disastrous period—the battle in which the famous Earl of Warwick was slain, A.D. 1471.

BEDFORDSHIRE is a small agricultural county, with some manufacture of *straw-plait*.

It has for its capital the town of BEDFORD (28), on the Great Ouse, the birthplace of John Bunyan.   Among its other towns are Luton (30) and Dun-stable (4), both noted for *straw-plait* and *straw-hat* manufacture.

HUNTINGDONSHIRE is an agricultural county, partly within the district of the Fens.

Its county town, HUNTINGDON (4), on the River Ouse, was the birthplace of Oliver Cromwell.   St. Ives and St. Neots are small towns in this county. Stilton is famous for its cheese.

NORTHAMPTONSHIRE is chiefly an agricultural county.

Its county town, NORTHAMPTON (61), on the River Nen, has, however, an extensive manufacture of *boots* and *shoes*.   Peterborough (25), also on the Nen, is a cathedral city, and, being an important railway centre, has considerable trade.   Wellingborough and Kettering are both engaged in the staple urban industry of the county—the boot and shoe manufacture.   *Naseby*, the scene of the decisive victory gained by Cromwell over the army of Charles I. (A.D. 1645), is in this county, twelve miles distant from Northampton, to the north-west.   A battle was fought near the town of *Northampton* in 1460, during the Wars of the Roses.

RUTLANDSHIRE, the smallest of the counties of England, is entirely agricultural.

It contains the small towns of OAKHAM (2), the county town, and Upping-ham, famous for its public school.

VII.—SIX COUNTIES IN NORTH WALES.[1]

ANGLESEY is chiefly pastoral and agricultural, but has a small coalfield. Its formerly valuable deposits of copper ore are now worked out.

BEAUMARIS, a much-frequented pleasure resort, finely situated at the entrance to the Menai Strait, is the county town ; but Holyhead (9), the western terminus of the London and North-Western Railway and the principal packet station for Ireland, is much larger and more important.

CARNARVONSHIRE, famous for its magnificent mountain and coast scenery, is also the centre of the slate industry.

At CARNARVON (10), the county town, and Conway (3) are fine castles, both built by Edward I. Bangor (10) has an ancient cathedral, and is the chief *slate port* of North Wales. Llandudno is a beautifully situated watering-place at the foot of Great Orme's Head, in the north-west of the county. Large passenger steamers run daily during the season between Llandudno and Liverpool. The drive round Great Orme's Head is much admired.

DENBIGHSHIRE includes the far-famed Vale of Clwyd, and the even more beautiful Vale of Llangollen, and the charming River Dee. The coalfield in the north-east is largely worked, and there are also mines of lead, iron, and slate.

Wrexham (13) and Ruabon (15) are the *mining*, and DENBIGH (6), the county town, in the Vale of Clwyd, and Llanrwst, on the banks of the Conway, are the *agricultural* centres of the county. Beddgelert and Ruthin are favourite tourist resorts.

FLINTSHIRE, the smallest of the Welsh counties, has valuable mines of coal and lead.

MOLD (4), Flint (5), and Holywell are the chief towns on the rich *coalfield* of Flint, and of which Mostyn, on the Dee, is the port. Rhyl, a favourite pleasure resort, stands on the coast near the mouth of the Clwyd. A few miles higher up the river is the cathedral town of St. Asaph.

MERIONETHSHIRE is wild and hilly, but well wooded and with some fine scenery. Its woollen manufactures and slate quarries are of considerable importance, and some gold has been obtained from the mines in the valley of the Mawddach.

The *slate quarries* are at Festiniog, and the gold mines are near DOLGELLY (3), the county town, and also the centre of the *Welsh flannel* manufacture in the county. At Bala, on Bala Lake, is the chief theological college of the Welsh Calvinistic Methodists.

MONTGOMERYSHIRE is a wild and hilly county, but rich in lead, copper, and slate, and has important woollen manufactures.

Montgomeryshire is noted for the manufacture of *Welsh flannel* at Welshpool (7), Newtown, Llanidloes, and other towns. MONTGOMERY is the county town.

---

1. The areas and population of the Six Counties of North Wales are as follows :—
(1.) Anglesey, 3 2 sq. m., pop. 50,079.
(2.) Carnarvonshire, 577 sq. m., pop. 118,203.

(3.) Denbighshire, 664 sq. m., pop. 117,950.
(4.) Flintshire, 257 sq. m., pop. 77,189.
(5.) Merionethshire, 601 sq. m., pop. 49,704.
(6.) Montgomeryshire, 773 sq. m., pop. 58,003.

VIII.—SIX COUNTIES IN SOUTH WALES.[1]

CARDIGANSHIRE is mainly agricultural, but contains rich deposits of lead, zinc, and copper.

Lead, silver, zinc, and copper ores are largely exported from Aberystwith (7), the chief port. **CARDIGAN** (3) is the county town.

RADNORSHIRE is the smallest, least populous and interesting of the six southern counties.

Its few towns are small and unimportant. **PRESTEIGN** (1) is the county town. The famous *Offa's Dyke* passes through the beautiful town of Knighton.

BRECKNOCKSHIRE, also an inland county, is wild and mountainous, with some fine scenery along the Wye and in the Brecon uplands.

The only considerable town is **BRECON** (6), on the Usk, which is also the county town. Brecon is noted for its *horse fairs*.

CARMARTHENSHIRE, the largest of the Welsh counties, is low towards the sea, rising inland into lofty hills and barren uplands. The eastern division of the county includes part of the great *South Wales coalfield*, and **coal, tin,** and **copper** mines are numerous.

Llanelly (24) is an important coal, iron, and copper port. **CARMARTHEN** (10), the county town, is on the **Towy.** Higher up the river are the little market towns of Llandeilo and Llandovery, "whose names," says Geikie, "are now known all over the world, because they have been given to certain fossiliferous rocks originally found there."

PEMBROKESHIRE, the most westerly of the Welsh counties, has a small coalfield, and possesses in Milford Haven a magnificent natural harbour.

PEMBROKE, Milford, Tenby, and **Haverfordwest** are the chief towns. Pembroke (15), the county town, and Milford, are situated on Milford Haven. There is a Government dockyard at Pembroke. **St. David's** and its cathedral are historically interesting : St. David being the patron saint of Wales.

GLAMORGANSHIRE is the most populous and important county in Wales, with **coal mines** and **iron-works** on a scale of great magnitude.

Merthyr Tydvil (58), Aberdare (39), Ystradyfodwg (88), Dowlais, and Neath, are the great mining and metal centres, while **CARDIFF** (129) and Swansea (90), are the chief ports of this rich region. Cardiff, the county town, is the principal outlet of the rich coalfield of South Wales, and annually exports millions of tons of *coal* to all parts of the world. Enormous quantities of *tin-plate* are manufactured here for export, chiefly to the United States. Barry, near Cardiff, has the largest single dock in the world. Large quantities of *steel* are produced at Merthyr Tydvil, but most of the iron ore smelted in this district is imported from Spain. Swansea is the head-quarters of the *copper-smelting* industry, which is also carried on at and near the port of **Neath.**

1. The area and population of the Six Counties of South Wales are as follows:—
(1.) Cardiganshire, 692 sq. m., pop. 62,596.
(2.) Radnorshire, 432 sq. m., pop. 21,791.
(3.) Brecknockshire, 719 sq. m., pop. 57,031.
(4.) Carmarthenshire, 928 sq. m., pop. 130,574.
(5.) Pembrokeshire, 611 sq. m., pop. 80,125.
(6.) Glamorganshire, 807 sq. m., pop. 687,147

SCOTLAND.

SCOTLAND is the northern portion of the island of Great Britain.

The name Scotland means the "land of the Scots." The *Scots* were a Celtic tribe, from the north of Ireland, who passed over and settled in Cantire about the year 503, and gradually extended their conquests until, in 843, the king of the Scots ruled over nearly the whole of Scotland north of the Forth and the Clyde. The name "Scotland" came into general use about the year 950. The native Picts and their descendants, the modern Highlanders, call the country *Albyn*. To the Romans it was known as *Caledonia*.

BOUNDARIES.—Scotland is bounded on all sides by the sea, except on the south-east, where it adjoins England.

Scotland is bounded on the *north* and *west* by the Atlantic Ocean; on the *south* by England and part of the Irish Sea; and on the *east* by the North Sea.

Scotland is divided from England by the Solway Firth, the Cheviot Hills, and the Tweed; from Ireland by the North Channel; and from Denmark and Norway by the North Sea.

EXTENT.—The area of Scotland (inclusive of its numerous islands) is over 30,000 square miles, or a little more than half that of England and Wales. The mainland alone embraces an area of about 27,000 square miles.

The greatest length, from *Dunnet Head* to the *Mull of Galloway*, is 288 miles; the greatest breadth, from *Buchan Ness* to *Ardnamurchan Point*, is 175 miles; the least breadth, between the *Firths of Forth* and *Clyde*, is only 32 miles.

COASTS.—The coasts of Scotland are more indented than those of England, especially on the *west* and *north*, and have a total length of not less than 2,500 miles, equivalent to 1 mile of coast to every 12 square miles of area. Owing to the numerous indentations of the coast-line, no part of the mainland is more than 40 miles from the sea.

The broader indentations by which the sea penetrates the land are generally called Firths—the narrower inlets bear the name of Lochs.

The term *loch* is uniformly given to lakes in Scotland, as it also is to the narrow inlets of the sea upon the western and northern coasts, such as Loch Fyne and others. There is, however, an important difference between the two. The inland *lochs*, such as Loch Lomond, have *fresh water*, like the *lakes* of England and other countries. The lochs that lie along the coast, such as Loch Fyne, are arms of the sea, and consist, consequently, of salt water.

The West Coast of Scotland presents a bewildering succession of deep inlets, bold and rocky headlands, and long peninsulas. South of Ardnamurchan Point, the coast, though very irregular, is generally low; north of that "headland of the great sea," the deep lochs and channels are fringed by lofty cliffs and numerous islands. The wild and desolate scenery of the northern coast culminates in huge sea-cliffs, which rise steeply from the sea like a mountain-wall, attaining near Cape Wrath a height of over 600 feet.

The East Coast, from Tarbet Ness to the Firth of Forth, is on the whole flat and generally sandy, but from St. Abb's Head to the Tweed it is extremely bold and rocky.

The western coast of Scotland is thus much more irregular than the eastern coast, and a glance at a map of England will show the same contrast between the broken form of the western coast, and the rounded, flowing curves of the eastern coast. The reason is that, on the west, the huge billows of the Atlantic dash upon the coast with irresistible force: all the softer parts have been thus removed, leaving the harder rocks to defy the baffled waves, although they also are slowly, but surely, wasting away.

The force of the Atlantic breakers on the western coasts of the British Isles is enormous. Immense blocks have been displaced, sea-walls and breakwaters broken down, and even lighthouses swept away. The spray is frequently driven right over the lantern of the Eddystone Lighthouse, on the south, while the lantern at Dunnet Head on the north, although two hundred and seventy feet above the level of the sea, has been cracked by pebbles hurled from the beach by the waves; and, during very violent storms, the Atlantic waves dash up the sides of the cliff at Hoy Head, in Orkney, to a height of nearly six hundred feet!

On the eastern coast, however, the waves are not, on the whole, nearly so large or so powerful. The North Sea is so shallow that, were the water drained away, we should hardly notice any downward slope at all; and, in fact, if its bed were raised only a hundred and fifty feet, we should be able to walk dryshod from Britain to Belgium or Holland. Indeed, geologists tell us that Great Britain was once joined to the Continent, and that there was then no German Ocean. They suppose that the land sank very slowly, and that the Atlantic burst in from the south-west and the north, and thus formed a great sea.

Further, the eastern coasts of both England and Scotland are largely composed of much softer materials than the broken and rugged western coasts; and, besides, on the eastern side of the island, the land generally slopes gently seawards, while on the western side, it rises abruptly from the water. It is, of course, evident that large waves dashing against a steep coast will not break it up so regularly as smaller waves beating upon a gently sloping shore.

CAPES.—The principal capes on the coasts of Scotland are the following :—

On the *North Coast*, Dunnet Head and Cape Wrath; on the *East Coast*, Duncansbay Head, Tarbet Ness, Kinnaird's Head, Buchan Ness, Buddon Ness, Fife Ness, and St. Abb's Head; on the *West Coast*, Ardnamurchan Point, the Mull of Cantire, Corsewall Point; on the *South Coast*, the Mull of Galloway, and Burrow Head.

The most northerly point is *Dunnet Head;* the most southerly, the *Mull of Galloway;* the most easterly, *Buchan Ness;* the most westerly, *Ardnamurchan Point.*

INLETS.—The most important inlets are the following :—

On the *north coast*, Dunnet Bay, Kyle[1] of Tongue, Loch Eriboll; on the *south coast*, Glenluce Bay, Wigtown Bay, and the Solway Firth; on the *east coast*, the Firth of Forth, the Firth of Tay, the Moray Firth, Cromarty Firth, and Dornoch Firth; and on the *west coast*, Loch Broom, Loch Carron, Loch Linnhe, Loch Fyne, the Firth of Clyde, Loch Long, and Loch Ryan.

---

1. Kyle, Gaelic a ferry.

Of these inlets the most important, commercially, are the **Firth of Clyde**, through which flows the great tide of commerce to and from Glasgow, the commercial metropolis of Scotland ; the **Firth of Forth**, connected with the Firth of Clyde by a canal, and having on its southern shores the great seaport of Leith, the port of Edinburgh ; and the **Firth of Tay**, on which stands the busy manufacturing and commercial city of Dundee.

CHANNELS AND SOUNDS.—The principal are the following :—

The **Pentland Firth**, between the Orkneys and the mainland ; the **Sound of Sleat**, between Skye and the mainland ; the **Sound of Mull**, between the Isle of Mull and the mainland ; the **Sound of Jura**, between Jura and the mainland ; the **Minch**, between Lewis and the mainland ; the **Little Minch**, between the Outer Hebrides and Skye ; the **Sound of Islay**, between Jura and Islay ; the **North Channel**, between Scotland and Ireland ; **Kilbrennan Sound**, between the peninsula of Cantire and Arran ; and **Bute Sound**, between the islands of Arran and Bute.

ISLANDS.—The coasts of Scotland are fringed by a far greater number of islands than England. They form four distinct groups, and have a total area of about **3,700** square miles.[1]

The **Orkneys** lie immediately to the north of the mainland of Great Britain, and are divided from it by the Pentland Firth. There are altogether 67 islands, of which the principal are **Pomona** (or Mainland), **Hoy**, **North and South Ronaldsha**, and **Westra**. *Kirkwall* on the eastern, and *Stromness* on the western, coast of Pomona, are the largest towns.

The **Shetlands**[2] lie north-east of the Orkneys, and consist of about 100 islands, of which 24 are inhabited. The largest islands are Mainland, Yell, and Unst. The chief town is *Lerwick*, on the east coast of Mainland Island.

The **Hebrides** include a great number of islands lying off the west side of Scotland, and surrounded by the waters of the Atlantic Ocean. Some of them, as the islands of **Skye, Mull**, and **Jura**, are near the mainland, and only divided from it by narrow channels. Others, as **Lewis, North Uist, Benbecula, South Uist**, and **Barra**, are farther off to seaward. The channel between Lewis and the mainland is called the Minch. *Lewis, Skye, Mull, Jura*, and *Islay* are the largest of the Hebrides. The small islets of **Iona** and **Staffa**, lying off the west side of Mull, are famous—Iona for its remains of ancient churches, and Staffa for the famous basaltic cavern known as Fingal's Cave.

The **Islands in the Firth of Clyde** include two large islands, **Arran** and **Bute**, and the islets of **Great and Little Cumbray**.

SURFACE.—Scotland is naturally divided into the **Highlands** and the **Lowlands**. The Highlands embrace the northern and western portions of the country ; the Lowlands, its southern and eastern districts.

The Lowlands of Scotland, however, are by no means level. They embrace numerous hilly tracts, but the hills are less elevated, and of more rounded form,

---

1 A few detached islets off the coasts of Scotland deserve notice. The *Bass Rock*, and *Inchkeith* are in the Firth of Forth ; the *Bell Rock* is about fourteen miles east of the mouth of the Firth of Tay ; the *Ailsa Craig*, in the Firth of Clyde, rises over 1,000 feet above the sea ; the *Pentland Skerries*, in the firth of that name.   2. About midway between the Shetlands and the Orkneys is *Fair Island*, on which the admiral of the Spanish Armada was wrecked in 1588.

with broader valleys between, than is the case in the Northern Highlands. These uplands are often distinguished as the *Southern Highlands*.

The division between the Highlands and the Lowlands is marked by a broad plain called **Strathmore** (that is, the "great strath" or valley), which stretches across the country in the direction of north-east and south-west, from near Stonehaven on the North Sea, to Dumbarton on the Clyde. A narrower valley, called Glenmore, the "great glen," extends through the Highland region, and forms a complete natural division across the country. It is through this valley that the Caledonian Canal has been formed, by joining the waters of the lakes which occupy a large portion of its bed.

MOUNTAINS.—The mountains of Scotland are naturally divided into three groups or systems—the *Northern Highlands*, the *Grampians* or *Central Highlands*, and the *Southern Highlands*.

In Scotland, as in England, the higher groups lie chiefly on the western side of the country. But mountains cover a much larger proportional extent of Scotland than is the case with the English hills, and they reach a greater height. England is chiefly a level country, and mountains are exceptional to its general character; Scotland is principally mountainous, and its plains are of limited extent. Geikie says:—"If Scotland were submerged 500 feet, a much less extent of it would be under water than in the case of England. A broad strait would then cover the country between the Firth of Clyde and the North Sea, a narrower sound would run from Loch Linnhe to the Moray Firth, and the low-lands along the east side of the country would disappear. But the southern and northern tracts of the kingdom would still rise hundreds of feet above the sea."

THE NORTHERN HIGHLANDS include the mountain-ranges and groups north of Glenmore. The highest points are Ben' Wyvis, near Cromarty Firth, 3,400 feet above the sea; Ben Attow, 4,000 feet.

THE CENTRAL HIGHLANDS, as the **Grampians** may be termed, are the highest mountains in Scotland, and stretch across the country in the direction of east and west. Ben Nevis, which is the loftiest summit in the Grampians, reaches 4,406 feet above the sea, and is the highest mountain, not only in Scotland, but in the British Islands. The other principal heights are *Ben Macdhui*, 4,300 feet; *Cairntoul*, 4,200 feet; *Ben Avon*, 4,000 feet; *Ben More*, 3,900 feet; *Ben Lomond*, 3,200 feet. To the south of the Grampians are the minor hill-ranges—the Sidlaw Hills, the Ochil Hills, and the **Campsie Fells**—which form the southern boundary of Strathmore.

THE SOUTHERN HIGHLANDS include the Cheviot Hills, the Moffat Hills, and the Lowthers, and also the Pentland, Moorfoot, and Lammermoor Hills. The highest points are:—*Broadlaw*, 2,700 feet, and *Hart Fell*, 2,260 feet, at the head of Tweeddale, in the Lowther Hills; *Carnethy*, in the Pentlands, 1,800 feet; and *Says Law*, in the Lammermoor Hills, 1,750 feet. *Cheviot Peak*, 2,676 feet, the highest point in the Cheviots, lies within the English border.

PLAINS.—Owing to the broken nature of the country, there are no plains of any great extent.

The principal are the Plain of Caithness, in the extreme north; the Plain of Cromarty, along both sides of the Cromarty Firth, and on the north side as

1. The following etymologies of the names of Highland peaks may be useful:—*Ben*, mountain; *more* (Welsh, *mawr*), great; *Ben More*, great mountain; *Ben Dearg*, red mountain; *Ben Wyvis*, mountain of terror; *Ben Attow*, rash mountain; *Ben Macdhui*, black swine mountain (*f.* Welsh, *mich*, swine, and *du*, black); *Ben Nevis*, mountain of death; *Cairngorm*, blue mountain.

far as Dornoch Firth ; Strathmore, between the Grampians and the Sidlaw and Ochil Hills ; the Carse of Gowrie, between the Sidlaw Hills and the Tay.

Besides the above, we may notice also the valleys or dales of southern Scotland, and the glens of the Highlands. Of the former, the most noted are *Clydesdale, Tweeddale, Teviotdale, Eskdale*, through which flow the rivers so named. Of the latter, the most important is *Glenmore*, which extends right across the country from Loch Linnhe to the head of Moray Firth.

RIVERS.—Most of the larger rivers of Scotland belong to the east side of the country, and discharge their waters into the North Sea, but the most important river is on the western side.

The chief rivers are the **Clyde** on the western slope, and the **Forth, Tay,** and **Dee** on the eastern slope.

The **Clyde,** 98 miles in length, rises in the Lowthers, and drains about 1,580 square miles. Near Lanark are the celebrated " Falls of Clyde"—the finest in Britain. The Clyde is navigable to Glasgow, and is, commercially, the most important of the Scottish rivers, its lower part having been deepened sufficiently to admit ships of the largest size. The numerous manufacturing and mining towns and villages in the basin of the Clyde, sustain an extensive trade with all parts of the world.

The **Tay** is the largest river in Scotland, and carries more water to the sea than any other river of Great Britain. It has a length of 105 miles, only 8 miles of which are navigable (that is, from the head of its estuary to Perth), and, with its tributaries, drains an area of about 2,400 square miles.

The **Forth** rises on Ben Lomond, and at Alloa falls into the firth to which it gives its name. Its windings form the " Links of Forth." The Forth is 60 miles in length, and drains an area of 645 square miles. The Firth of Forth is about 50 miles long, and is navigable for the largest vessels. At Queensferry, it is crossed by the *Forth Bridge*, one of the greatest engineering works in the world.

The **Dee,** 87 miles in length, has its source on the Cairngorm, 4,060 feet above the sea, considerably higher than any other British river. The **Spey,** 96 miles long, is the most rapid of British rivers, and is unnavigable throughout. The **Tweed** rises in the Lowthers, and has a length of 96 miles and a drainage area of 1,870 square miles. It is noted for its salmon fisheries, and is unnavigable above Berwick.

LAKES.—Scotland, especially its Highland division, abounds in lakes, celebrated for their beauty, many of them embosomed in long and narrow valleys, and surrounded by the wildest and most picturesque mountain scenery.

Loch Lomond is the largest lake in Scotland, and also in Great Britain. It is 24 miles long, 7 miles broad, and has an area of 45 square miles. It contains about 30 islands, and is unquestionably " the pride of Scottish lakes," exceeding all others in extent and beauty. **Loch Katrine,** 9 miles long and ½ mile broad, is the chief attraction of the beautiful tract known as the *Trossachs* (that is, *narrows*), and is the scene of Scott's " Lady of the Lake." **Loch Awe,** 23 miles long and nearly 1½ mile broad, is surrounded by lofty mountains, and is one of the most beautiful of Scottish lakes. **Loch Ness,** in the eastern part of Glenmore, is 24 miles long, and Loch Shin, in the centre of the Northern Highlands, is 20 miles in length, neither being more than about a mile in breadth. Loch Maree has a curious cluster of 24 islands near its centre. On one of the islands in Loch Leven (in Kinross) stood the castle in which Mary Queen of Scots was imprisoned.

CLIMATE.—The climate of Scotland is slightly colder than that of England, owing to its more northerly situation. Within the Highlands, especially, the winter is more severe ; rain is also more abundant there than in the Lowland region.

The *mean annual temperature* of Edinburgh is 47.1° ; Aberdeen, 49.1° ; and Wick, 46.9°. The mean winter temperature of the Shetlands is about the same as that of the Isle of Wight. The annual *rainfall* is from 22 to 23 inches on the east coast, and from 30 to 44 inches on the west coast.

PRODUCTIONS.—The natural productions of Scotland are, with a few exceptions, the same as those of England.

ANIMALS.—The domestic and wild animals of Scotland are the same as those of England. *Deer*, however, are much more abundant, chiefly in the game forests of the Highlands. Clydesdale is noted for a fine breed of draught *horses*, the Shetland Islands for the hardy "Shetland" *ponies*, and Ayrshire for its cattle. The seas around Scotland swarm with *fish*, and enormous quantities of *herring*, *cod*, and *haddock* are caught, principally off the east coast. There are also important fisheries in the western Highlands, and, throughout the country, the rivers and lakes yield large supplies of *salmon* and *trout*.

PLANTS.—The vegetation consists, for the most part, of plants of hardier growth than those of England. The *Scotch fir*, and other members of the *coniferæ*, or cone-bearing trees, are abundant upon the mountain-sides, and the heather imparts its purple colour to their lower slopes. Several of the richer fruits and plants that thrive on the southern coasts of England, as the peach and apricot, will not come to perfection in Scotland ; and the hardier grains— *oats* and *barley*—are those most generally grown. Fine crops of *wheat*, however, are grown in many districts of southern Scotland.

MINERALS.—The mineral resources of Scotland are very great. As in England, coal and iron are the staples of its wealth in this regard, and they form the basis of its manufacturing prosperity. Both coal and iron occur in vast abundance within an extensive district of the Lowlands—that which stretches across the country from Fifeshire on the east side to Ayrshire on the west, embracing the extensive plain between the Firths of Forth and Clyde. *Lead* is worked in some districts of southern Scotland. Good *building-stone* also occurs there. *Granite* is obtained from the Grampians, and also from the Isle of Arran, and some other localities.

INHABITANTS.—Scotland contains over 4,000,000 inhabitants, an average of 134 to the square mile.

Scotland is thus much less populous than England—both *absolutely* (that is, according to the actual number of its inhabitants) and *relatively* (or in the ratio of population to extent of surface). The Lowlands are much more densely-peopled than the Highlands.

The number in 1881 was 3,735,573, an average of only 125 per square mile, a little more than a fourth of that of England, and considerably lower than that of Wales or Ireland. The population, according to the Census Returns, in 1891, was 4,033,103, an average of 134 per square mile.

Race and Language.—The people of the Highlands and of the Lowlands are two distinct races, speaking different languages. The language of the Lowlands resembles the English tongue ; that of the Highlanders is a distinct dialect, called the *Gaelic*. But the Highland population are gradually growing accustomed to the use of the English language, and their native tongue is becoming, with each succeeding generation, less prevalent.

It is commonly supposed that the Lowland Scotch is merely a dialect of, or a corruption of, the *English* language. But while the latter is based chiefly upon the Anglo-Saxon, the former is essentially derived from the Norse or Scandinavian. Now, both the Anglo-Saxon and the Norse were branches of the same old Teutonic ; hence the similarity between the Lowland Scotch and the English. The *Gaelic* is a Celtic dialect, and is allied to the *Welsh*.

INDUSTRIES :—Scotland is principally a manufacturing and commercial country ; but agriculture, though the productive area is limited, is in a flourishing condition.

AGRICULTURE is perhaps nowhere more skilfully practised than in the Lowlands. Only one-fourth of the country is arable, and about one-half of this is in permanent pasture or in grass (clover, &c.) The chief objects of culture are *oats, barley, rye, wheat, potatoes,* and *turnips*. But the staple crop throughout Scotland is *oats*. *Turnips* are largely grown in Haddingtonshire, and *potatoes* throughout the eastern counties.

In the Highlands the rearing of cattle, with the extensive fisheries pursued off the coasts, are the principal branches of industry. In the southern Lowlands, also, great numbers of cattle are reared, chiefly for the supply of the English markets.

MANUFACTURES are largely pursued in Lowland Scotland, chiefly within the coal and iron district between the Forth and the Clyde, and in the counties of Fife and Forfar, on the east coast.

The **Cotton** manufacture is carried on chiefly in Glasgow and Paisley. The latter is also famous for its thread factories, and the manufacture of silks and shawls. The **Linen and Jute** manufacture is pursued chiefly in the east coast towns—principally at Dundee and Dunfermline. The **Woollen** manufacture is pursued in many parts of the Lowlands, especially in the counties of Aberdeen, Stirling, Ayr, Fife, Renfrew, Forfar, and Lanark, and the manufacture of tweeds and tartans in Galashiels, Selkirk, Hawick, &c. Ship-building, especially of iron and steel steamships, is more extensively carried on at Glasgow, Greenock, and Port Glasgow than in any other part of the world.

COMMERCE : The commerce of Scotland resembles that of England. The imports are chiefly *raw materials* for manufacture and *colonial produce ;* the exports are *manufactured goods* and *agricultural produce, coal, iron,* and *fish*.

The commercial metropolis of Scotland is Glasgow. Other important ports are Leith, Greenock, Aberdeen, Dundee, Irvine, and Montrose.

The internal trade is facilitated by excellent roads (even in the Highlands), several canals, and over 3,000 miles of railways.

TOWNS: One-third of the inhabitants of Scotland live in towns, only two of which contain over a quarter of a million people.

Scotland has only one city—Glasgow—with over half-a-million inhabitants. Edinburgh, the next in size, has a population of over a quarter of a million. Besides Edinburgh and Glasgow, two towns—Dundee and Aberdeen—contain over 100,000 inhabitants, while three towns—Greenock, Leith, and Paisley—each contain over 50,000 people. Perth, the next in size, has a population of 30,000.

COUNTIES: Scotland is divided into 33 counties, which are very unequal in size, more so than the English counties. Clackmannanshire, the smallest, is only one-third part the size of Rutland. Inverness-shire, the largest among them, is nearly four-fifths the size of Yorkshire.

Thirteen of the counties are within that portion of Scotland which lies to the south of the Firths of Forth and Clyde, and are entirely within the Lowlands. These are:—Edinburgh, Linlithgow, Haddington, Berwick, Roxburgh, Selkirk, Peebles, Dumfries, Kirkcudbright, Wigtown, Ayr, Lanark, and Renfrew. *Three* of the counties that are to the northward of the Firth of Forth are also wholly within the Lowland region, namely, Fife, Kinross, and Clackmannan. In all, therefore, *sixteen* of the counties are comprised entirely within the Lowlands. Of the others, several are partly within the Lowlands and partly within the Highlands. All the counties that extend along the east coast of the country, Perth, Forfar, Kincardine, &c., belong in part to the Lowland region; but they stretch westward into the rugged mountain region of the interior, and their larger portion falls within the Highland limits. Bute, Argyle, Inverness, Ross and Cromarty, and Sutherland are almost exclusively *Highland counties.* Stirling, Dumbarton, Perth, Forfar, Kincardine, Aberdeen, Banff, Elgin, Nairn, and Caithness are partly Highland. Buteshire consists of the two large islands of Arran and Bute, in the Firth of Clyde, with a few smaller islets. The Orkney and Shetland Islands form a distinct county. Of the Hebrides, some belong to Argyle, some to Inverness, and others to Ross.[1]

I.—THIRTEEN SOUTH-LOWLAND COUNTIES.[2]

EDINBURGHSHIRE, or Midlothian, is distinguished by its agricultural industry, and also as the metropolitan county of Scotland. It is level and well cultivated in the north, but hilly in the south.

The city of **EDINBURGH** (261) occupies a striking situation near the southern shore of the Firth of Forth (about two miles distant), and its famous Castle

---

1. In several parts of Scotland the ancient territorial names of particular districts are still familiarly used. Thus the three counties that extend along the southern shore of the Firth of Forth are known as the *Lothians*—Edinburgh corresponding to Mid-Lothian, Haddington to East-Lothian, and Linlithgow to West-Lothian. The counties of Kirkcudbright and Wigtown, in the south-west, are popularly known as *Galloway*. Forfarshire is still often referred to by its former name of *Angus*, and the county of Elgin by that of *Moray* (or Murray).

2. The areas and population (in 1891) of the thirteen South-Lowland Counties are as follows:—

(1.) Edinburgh, 362 sq. m., pop. 444,055.
(2.) Linlithgow, 120 sq. m., pop. 52,789.
(3.) Haddington, 270 sq. m., pop. 37,491.
(4.) Berwick, 460 sq. m., pop. 32,398.
(5.) Roxburgh, 665 sq. m., pop. 53,726.
(6.) Selkirk, 257 sq. m., pop. 27,349.
(7.) Peebles, 354 sq. m., pop. 14,761.
(8.) Dumfries, 1,162 sq. m., pop. 74,308.
(9.) Kirkcudbright, 897 sq. m., pop. 39,979.
(10.) Wigtown, 455 sq. m., pop. 36,048.
(11.) Ayr, 1,128 sq. m., pop. 224,2.2.
(12.) Lanark, 881 sq. m., pop. 1,045,787.
(13.) Renfrew, 244 sq. m., pop. 290,790.

—a strong fortress of ancient date, and the scene of numerous events of importance in Scottish annals—crowns a commanding rock, which rises high above the **Old Town**. A broad valley forms a well-marked natural division between the **New Town** of Edinburgh and the older portion of the city. Holyrood, the ancient palace of the Scottish sovereigns, is within the Old Town, at the opposite extremity to the Castle Hill. Edinburgh is the seat of one of the Scottish Universities, and ranks as the literary metropolis of the north. Arthur's Seat, a lofty crag rising to 822 feet above the sea, overlooks the city from the south-east.

Leith (63), on the coast of the neighbouring Firth of Forth, forms the port of Edinburgh, and is joined to that city by continuous lines of building. On either side of Leith are numerous thriving fishing and trading ports—**Newhaven** and **Granton** to the westward, **Portobello** and **Musselburgh** on its eastern side. The battle of *Pinkie* (1547) was fought in the neighbourhood of Musselburgh, near the right bank of the little River Esk, which enters the Firth of Forth at that point. **Dalkeith**, in the interior of the county, is a small town at the junction of the two rivers—the North Esk and the South Esk. There are large paper mills at **Penicuik** and other villages in the valley of the Esk.

LINLITHGOWSHIRE, or West Lothian, is a small agricultural county, but its surface is generally irregular. The western part is rich in coal, and the eastern district in oil-shale.

Its chief town, **LINLITHGOW** (4), possesses the remains of an ancient palace, in which Mary Queen of Scots was born in 1542. Bathgate is a great mining and manufacturing town in the interior. **Bo ness** trades in coal and iron.

HADDINGTONSHIRE, or East Lothian, is agriculturally the foremost county in Scotland, being for the most part level and fertile. In the west are coal mines.

The town of **HADDINGTON** (4), its capital, stands on the River Tyne— a less important stream than the English river of that name. Dunbar, a thriving port on the coast of this county, is of great note in Scottish annals, frequently besieged, and alternately in Scottish and English hands. Two battles fought in the immediate vicinity add to the chequered interest of its fortunes—one a victory gained by Edward I. over the army of Baliol in 1296, the other a more important victory which Cromwell obtained over the Scottish army in 1650. **North Berwick**, on the coast of Haddington, lies at the entrance of the Firth of Forth. **Prestonpans**, also on the shore of the same firth, and a few miles to the east of Edinburgh, is noteworthy for the defeat of the English forces by the troops of Prince Charles Edward in 1745.

BERWICKSHIRE adjoins the English border, reaching from the Lammermoor Hills to the banks of the Tweed. Its industry is chiefly agricultural.

The county town is **GREENLAW**. Duns is the centre of the agricultural district. Coldstream was formerly important "as commanding the only ford by which armies could cross the Tweed." Eyemouth is an important fishing station situated a short distance to the south of St. Abb's Head.

ROXBURGHSHIRE stretches from the banks of the Tweed to the summits of the Cheviot Hills, including the fine pastoral district of Teviotdale—watered by the River Teviot, an affluent of the Tweed. It is undulating in the north, and hilly in the south. Its industries comprise the woollen manufacture, agriculture, and the rearing of cattle and sheep.

JEDBURGH (3), its county town, is on the little stream of the Jed, which joins the Teviot. A few miles north-west of Jedburgh is *Ancrum Moor*, where the Earl of Angus defeated an English army in 1545. Hawick, in upper Teviotdale, the centre of the manufacture of "tweeds" and other woollen goods, and Kelso, on the Tweed, are both of larger size. Melrose, famous for the ruins of its well-known abbey, and Abbotsford, formerly the residence of Sir Walter Scott, are in this county, both on the south bank of the Tweed.

SELKIRKSHIRE, a pastoral and hilly region, includes the tract of country called Ettrick Forest, watered by the Ettrick, which joins the Tweed, receiving on its way the tributary stream of the Yarrow.

The county town, SELKIRK (6), on the right bank of the Ettrick, has large *woollen* mills. On the opposite bank of the river is *Philiphaugh*, the scene of Montrose's surprise and defeat in 1645. Galashiels (on the Gala Water) also manufactures "*tweeds,*" *tartans, shawls,* &c.

PEEBLESSHIRE, a pastoral and hilly region, embraces the upper portion of Tweeddale, and is entirely agricultural.

The county town, PEEBLES (5), is on the north bank of the Tweed. Innerleithen has a mineral spring—St. Ronan's Well, and *tweed* manufactures.

DUMFRIESSHIRE, which is agricultural in its lower grounds, and pastoral towards the interior, includes the greater part of Nithsdale (or the valley of the River Nith, which enters the Solway Firth), and also the valleys of the Annan and the Esk—Annandale and Eskdale.

The town of DUMFRIES (17), its capital, stands on the left bank of the Nith; it is the largest town in the south-west portion of Scotland, and a great market for agricultural produce. Burns, the Scottish poet, died at Dumfries in 1796. To the north of Annan is Ecclefechan, the birthplace of Carlyle. Moffat is a favourite summer resort.

KIRKCUDBRIGHTSHIRE is hilly and pastoral in the north, and agricultural along the shores of the Solway Firth on the south.

The county town, KIRKCUDBRIGHT (3), lies near the mouth of the River Dee, which enters the Solway Firth. There are large granite quarries at Creetown and Dalbeattie; the former supplied the granite for the Liverpool Docks, and the latter that for the Thames Embankment.

WIGTOWNSHIRE, at the south-west extremity of Scotland, is also a pastoral region.

The small town of WIGTOWN (2), its capital, is on the shore of Wigtown Bay. Stranraer, at the head of Loch Ryan, and Port-Patrick, on the shore of the North Channel, are small ports. Port-Patrick is only 22 miles distant from Donaghadee, on the coast of Ireland.

AYRSHIRE, sometimes called the "Dairy of Scotland," embraces a pastoral tract of country in the south and east, but includes a manufacturing and coal-mining district in the north and along the sea-coast.

Its county town, AYR (25), at the mouth of the River Ayr, has a considerable trade. Robert Burns was born in its vicinity—a short way to the southward, in a cottage beside the stream of the Doon. Kilmarnock, on the River Irvine, has extensive *woollen manufactures*, large *ironworks*, and *machinery* factories. Ardrossan, Irvine, Saltcoats, Troon, Girvan, and Ballantrae are thriving ports on the Ayrshire coast.

LANARKSHIRE includes Clydesdale, the upper part of which is a pastoral region. But its lower portion is a populous seat of manufacturing and commercial industry.

GLASGOW (565 including suburbs 771), on the banks of the Clyde, within the north-western border of Lanarkshire, is the centre at once of the *cotton* manufactures, the *iron* trade, and the *foreign commerce* of Scotland. It is, indeed, the commercial metropolis of North Britain, and surpasses all other cities in Scotland in number of inhabitants. Glasgow is, besides, the seat of an ancient university. A short distance to the southwards is the battlefield of *Langside*, the scene of the last contest on behalf of the ill-fated Mary Queen of Scots (A.D. 1568).

Airdrie, to the east of Glasgow, is in the midst of extensive *coal mines* and large *ironworks*, and has also *cotton* and other *manufactures*. Hamilton has large textile factories, and Coatbridge, Rutherglen, and Motherwell, have numerous *coal* and *iron* mines. Below Hamilton, on the Clyde, is *Bothwell Bridge*, the scene of a well-known skirmish between the Covenanters and the Royal forces in 1679. LANARK (5), in Upper Clydesdale, is the county town. Below it are the picturesque Falls of Clyde, formed by the descent of the river over successive ledges of rock.

RENFREWSHIRE is manufacturing and commercial. It includes part of the coal and iron district, and possesses several flourishing ports on the lower Clyde.

RENFREW (6), the county town, has *silk* and *muslin* factories. Paisley (66), seven miles west of Glasgow, shares in the manufacturing industry of that city, and is famous for its *thread* and *shawls*. Port-Glasgow (15), and Greenock (63), are both great *ship-building* ports on the Clyde; the latter is the seat of an extensive foreign trade. Johnstone, three miles south-west of Paisley, has *flax* and *cotton mills*, and *engineering works*.

II.—SEVEN NORTH-LOWLAND COUNTIES.[1]

DUMBARTONSHIRE is Highland and pastoral in its northerly division, including the chief part of the shores of Loch Lomond. In the south it reaches to the banks of the Clyde. A detached portion of the county is in the plain further to the eastward.

The chief town, DUMBARTON (13), an important port with large *ship-building* yards, stands on the Clyde, at the mouth of the little River Leven,

---

1. The areas and population of the Seven North Lowland Counties are as follows:—
(1.) Dumbarton, 241 sq. m., pop. 64,511.
(2.) Stirling, 447 sq. m., pop. 1,5,604.
(3.) Clackmannan, 47 sq. m., pop. 21,433.
(4.) Kinross, 72 sq. m., pop. 6,788.
(5.) Fife, 492 sq. m., pop. 187,300.
(6.) Forfar, 870 sq. m., pop. 277,788.
(7.) Kincardine, 383 sq. m., pop. 35,647.

which forms the outlet of Loch Lomond. The vale of Leven is a busy scene of manufacturing industry, with cotton-print and calico works, &c. Kirkintilloch is a manufacturing town of considerable importance, to the north-east of Glasgow.

STIRLINGSHIRE is mining and manufacturing in its eastern division, which is within the Lowland region ; but its westerly portion, which is pastoral and agricultural, stretches into the Highlands. The fertile "Carse of Stirling" is highly cultivated.

The county town, STIRLING (17), stands on the south bank of the Forth, its fine castle, on a lofty rock which overlooks the town, has been the scene of many events important in Scottish annals. Stirling has manufactures of *woollen* and *cotton* fabrics. St. Ninians and Bannockburn, both populous villages in its immediate vicinity (forming, in fact, suburbs of the town), also possess extensive *woollen* manufactures. Bannockburn recalls the memory of Bruce's great victory over the English in 1314. Falkirk, in this county, is a great *coal* and *iron* centre, and an important cattle market, and is historically noteworthy on account of two engagements which have taken place in its vicinity—one a victory gained by Edward I. over the Scottish army (1298); the other a defeat sustained by the royal forces at the hands of Prince Charles Edward's followers in 1746. Grangemouth, an important seaport, with large shipbuilding yards, is on the Firth of Forth, at the eastern terminus of the Forth and Clyde Canal. Larbert, between Falkirk and Stirling, is an important railway junction. Kilsyth, on the southern border of the county, was the scene of Montrose's brilliant (though fruitless) victory in 1645. Slamannan, near the southern border of the county, has extensive *coal* and *ironstone* mines.

CLACKMANNANSHIRE is enclosed by the counties of Perth and Stirling, and is the smallest county in Scotland. It is partly within the great Lowland coalfield.

Alloa (10), its largest town (on the left bank of the Forth), has some trade in *coal* and *iron*. CLACKMANNAN, a mining village, is the county town. Alva (5) has manufactures of shawls and tweeds.

KINROSS-SHIRE is enclosed between the counties of Fife and Perth, and is, on the whole, well cultivated. It includes Loch Leven—the largest lake within the Lowland region.

Its county town, KINROSS (2), is on the west shore of the lake. The castle of Lochleven, on an island in the lake, was the temporary prison of Mary Queen of Scots, and the scene of her romantic escape in 1568, immediately prior to the battle of *Langside*. Milnathort has manufactures of *tartan shawls* and *plaids*, and has also important live-stock sales.

FIFESHIRE, a Lowland county, forms a peninsula, lying between the Firths of Forth and Tay. Its interior is hilly and pastoral, but a broad and fertile belt of country stretches along the coast,[1] on which are numerous seaport and fishing towns.

CUPAR (5), on the Eden, is the county town. Kirkcaldy is a busy seaport, and has numerous *coal* mines and manufactures of *linen, oilcloth*, &c. Dunfermline, in the centre of the coalfield, is a great seat of the *linen* manufacture, and has an ancient abbey church, within which are the remains of Robert Bruce.

---

[1] James VI. compared Fife to a "beggar's mantle fringed with gold," alluding to the contrast between its fertile sea-board and comparatively sterile interior.

St. Andrews is the seat of the oldest of the Scottish universities. Along the coast are a number of seaports and seaside resorts—Inverkeithing, **Aberdour, Burntisland, Kinghorn, Largo, Elie, Pittenweem, Anstruther**, and Crail.

FORFARSHIRE, or Angus, is principally within the Lowland region, and is (with Fifeshire) the chief seat of the linen and jute manufacture.

FORFAR (13), the county town, lies in the heart of the great plain of Strathmore. Dundee is not only a great seaport, with an extensive coasting and foreign trade, but an important manufacturing city—the chief centre of the *linen* and *jute* manufacture. Montrose, also a thriving seat of trade, is on the coast of the North Sea. Midway between Montrose and Dundee is Arbroath (or Aberbrothock), the nearest port to the celebrated Bell Rock, or Inchcape. Brechin and **Coupar-Angus** (so called to distinguish it from the county town of Fifeshire) are in this county.

KINCARDINE, or the Mearns, extends from Forfar to the banks of the Dee, and includes the most eastward portion of the Grampians. But the coast division of the county is Lowland.

STONEHAVEN (4), the chief town, is on the shore of the North Sea, and is an important herring-fishing station, as also are Bervie, Findon, and other fishing ports. From Findon, or Finnan, the name "Finnan haddocks" is derived.

III.—THREE SOUTH-HIGHLAND COUNTIES.[1]

BUTESHIRE, the most southward of the Highland counties, consists of the islands of *Bute* and *Arran*, in the Firth of Clyde. Arran is hilly, but Bute is level and fertile. The latter has the most salubrious climate in Scotland.

The county town, **ROTHESAY** (9), a large watering-place on Bute, has an exceptionally mild climate, and is much frequented. The beautiful channel which divides Bute from the mainland is called the *Kyles of Bute*. Arran has a more rugged surface, and furnishes some granite; its chief town is Brodick a small place on the east coast. Lamlash is a small fishing village in Arran.

ARGYLESHIRE embraces a rugged Highland tract on the Scottish mainland, and includes many of the adjacent islands— amongst them Mull, Jura, and Islay, with Coll, Tiree, Colonsay, and many of smaller size. *Staffa* and *Iona*, off the west coast of Mull, are of the number.

The county town is **INVERARAY**, near the head of Loch Fyne. Campbeltown, on the peninsula of Cantire, is of larger size. Oban, the "Capital of the West Highlands," and the north-western terminus of the Highland Railway, is a rising place on the coast, near the entrance of Loch Linnhe. *Glencoe*, the scene of the infamous massacre of the Macdonalds in 1692, is a wild pastoral valley, which adjoins the south shores of Loch Leven,[2] one of the estuaries of the western coast.

---

1. The areas and population of the Three South-High and Counties are as follows:—
(1.) Bute, 271 sq. m., pop. 18,408.
(2.) Argyle, 3,213 sq. m., pop. 75,945.
(3.) Perth, 2,527 sq. m., pop. 126,128.

2. There are two lochs bearing this name—one in Kinross (the scene of Queen Mary's confinement and escape, etc; the other, that referred to above, on the border-line between the counties of Argyle and Inverness. This latter is an arm of the sea.

PERTHSHIRE is Lowland in the east and centre ; but its western division includes an **extensive and rugged portion of the Highlands.** It comprehends, in the south-west, the romantic district of the Trossachs, within which are embraced the wooded heights of Ben A'an and Ben Venue, with the winding shores of Loch Katrine, Loch Achray, and Loch Vennachar.

The city of **PERTH** (30) lies on the right bank of the Tay, immediately above the estuary which the river forms in its lower course. Near Perth, (upon the opposite bank of the river) is Scone, the ancient coronation-place of the kings of Scotland. *Tippermuir*, the scene of one of the victories gained by Montrose (in 1644), during his brilliant but evanescent career of success, is a short distance to the south-west of Perth. The other towns of Perthshire are all of small size. **Dunkeld** and **Dunblane** are noted for their ancient cathedrals. Two miles to the eastward of Dunblane is *Sheriffmuir*, the scene of an indecisive engagement between the royalist and rebel forces in **1715**. **Crieff** and **Callander** are favourite summer resorts. Callander is at the entrance to the far-famed Trossachs. The *Pass of Killiecrankie*, on the romantic banks of the Garry, which joins the Tummel[1] below the defile, is in the northern part of Perthshire. It was here that the leader of the Highland clans, Viscount Dundee, fell in the moment of victory over the forces of King William in 1689.

IV. NINE NORTH-HIGHLAND COUNTIES.[2]

ABERDEENSHIRE is Lowland towards the coast ; but its interior belongs to the Highlands. Large numbers of cattle are reared in this county for the English markets, and there are extensive herring and salmon fisheries, and large granite works near Aberdeen and Peterhead.

The " granite city " of **ABERDEEN** (113), its capital, situated between the mouths of the rivers Dee and Don, is one of the most important commercial towns in Scotland, and is distinguished for its university. At the bridge of Dee, two miles above Aberdeen, Montrose defeated the Covenanters in 1644. Peterhead, a flourishing port, is on the coast to the northward, and is the chief station of the Greenland seal and whale-fishery. Fraserburgh, another seaport, still further north, is a great centre of the Scottish herring-fishery. Inverury, at the junction of the little River Ury with the Don, is a small inland town. Balmoral Castle, the Highland home of Queen Victoria, is in this county, within the beautiful valley of the Upper Dee, near its southern bank.

BANFFSHIRE is chiefly Lowland, but penetrates the Highland region in its southerly division. The chief industries are **agriculture and fishing.**

The county town, **BANFF** (7), is near the mouth of the River Deveron. Portsoy and Cullen are small fishing towns on the coast. The noted distilleries of Glenlivet are in this county.

---

1. The Tummel is an affluent of the Tay.
2. The areas and population of the Nine North-Highland Counties are as follows :—
(1.) Aberdeen, 1,972 sq. m., pop. 281,331.
(2.) Banff, 686 sq. m., pop. 64,167.
(3.) Elgin, 475 sq. m., pop. 43,448.
(4.) Nairn, 178 sq. m., pop. 10,019.

(5.) Inverness, 4,088 sq. m., pop. 88,362.
(6.) Ross & Cromarty, 3,129 sq. m., pop. 77,751.
(7.) Sutherland, 2,027 sq. m., pop. 21,940.
(8.) Caithness, 685 sq. m., pop. 37,161.
(9.) { Orkney, 375 sq. m., pop. 30,438.
      { Shetland, 551 sq. m., pop. 28,711.

ELGINSHIRE, or Moray, is Lowland in the north, but Highland in its southwardly portion. Its climate is remarkably mild.

The town of ELGIN (8), its capital, is a few miles distant from the coast, on the little River Lossie. Forres is farther to the west, near the River Findhorn.

NAIRNSHIRE, a small county, is partly Lowland, but becomes hilly in the south. The level districts are along the coast, and are generally fertile.

The town of NAIRN (4) is on a small river of that name, at its entrance into the Moray Firth. *Auldearn*, a village lying a few miles south of Nairn, was the scene of one of Montrose's victories in 1645.

INVERNESS-SHIRE, the largest of the Scottish counties, is entirely a Highland county, and most of its surface is occupied as sheep-runs and deer forests.

Ben Nevis, the highest mountain in Britain, is in this county, which includes a succession of bleak moorlands, high mountains, and narrow glens. The line of the *Caledonian Canal* crosses the county from north-east to south-west, passing through the narrow valley of *Glenmore*. The large island of *Skye* belongs to this county, as also do *Harris, North* and *South Uist, Benbecula*, and *Barra*, among those of the Hebrides lying farther to the west.

The town of INVERNESS (19), regarded as the "Capital of the Highlands," stands at the entrance of the River Ness into Beauly Firth (as the upper extremity of the Moray Firth is called). A few miles east is *Culloden Moor*, the scene of Prince Charles Edward's final defeat in 1746. Near Fort William, at the south-western extremity of the Caledonian Canal, and at the outlet of the River Lochy into Loch Eil, is *Inverlochy*, where Montrose gained, in 1645, the most brilliant of his victories. Portree is a small fishing town on the east coast of Skye.

ROSS and CROMARTY (now united into one shire) comprehends a rugged Highland tract, which stretches across the country from the Moray Firth to the Atlantic coast. It includes **Lewis**, the largest of the Hebrides.

The county town is TAIN (2), on Dornoch Firth. Dingwall, further south, is a royal burgh and railway junction of some importance at the head of Cromarty Firth. Stornoway is a fishing station on the island of Lewis.

The town of CROMARTY (1) is situated at the entrance of the magnificent estuary called Cromarty Firth—one of the finest of natural harbours. Hugh Miller, the geologist, was born in this town.

٭٠ The formerly separate shire of Cromarty consisted of several small and detached portions of country, enclosed by Ross and the adjacent county of Sutherland.

SUTHERLAND is entirely Highland, and is the most thinly-populated county in Scotland. Immense numbers of sheep are reared in this county, and there are large deer forests.

DORNOCH, the county town, is on the east coast, on the northern side of the firth to which its name is given. Golspie, near which is *Dunrobin Castle*— the seat of the Duke of Sutherland, and Helmsdale, further north, at the mouth of the river of the same name, are small seaports. Lairg, at the south end of Loch Shin, is a great centre for sportsmen and tourists.

CAITHNESS includes the north-eastern extremity of the Scottish mainland. The county is level and generally sterile.

Its chief town, **WICK** (5), is the capital of the Scottish herring fishery. Thurso is on the north coast of the island. Near Duncansbay Head is the site of the famous *John o' Groat's House*, the most northerly dwelling in Scotland. Hence the popular saying, "From Land's End to John o' Groat's."

The county of **ORKNEY and SHETLAND** consists of the groups of islands so called. Both groups are nearly, if not entirely, destitute of trees, and are bleak and barren, with the exception of a few fertile tracts ; but the scenery along the rocky coast is interesting, and in some places grand, while the climate, though humid, is comparatively mild, and snow seldom lies long in the Shetland Isles. Fishing is the chief industry.

KIRKWALL (3), situated on the largest of the Orkneys (called Pomona, or Mainland), is the county town. **Stromness** (2), 14 miles west of Kirkwall, has a good harbour. Between Kirkwall and Stromness are the "*Standing Stones of Stenness*"—ancient stone circles, like those at Stonehenge in England. **Lerwick** (4), the principal town in the Shetlands and the northernmost town in the British Islands, is on Bressay Sound, on the east coast of the Mainland, the largest island of the group.

IRELAND.

IRELAND,[1] the third largest island of Europe, lies to the west of Great Britain.

The shores of Ireland and Great Britain make the nearest approach to each other between Fair Head and the Mull of Cantire, where the channel is only 13 miles wide, and are furthest apart along the 54th parallel, between Dundalk Bay and Morecambe Bay, a distance of about 140 miles. St. David's Head, the most westerly point of Wales, is about 50 miles distant from Carnsore Point on the opposite Irish coast.

BOUNDARIES.—On three sides—the *north*, *west*, and *south*—Ireland is bounded by the **Atlantic Ocean** ; on the *east*, by the **North Channel, Irish Sea, and St. George's Channel.**

Ireland is divided from *England* by the Irish Sea ; from *Wales*, by St. George's Channel ; and from *Scotland*, by the North Channel.

EXTENT.—The superficial extent of Ireland is 32,530 square miles, or rather more than one-half the area of England and Wales.

The greatest length (from Malin Head to Mizen Head) is 290 miles, the greatest breadth (from Howth Head to Slyne Head) is 175 miles, while the least breadth (between Donegal and Belfast) is 90 miles.

---

1. Ireland (Gaelic, western isle. The native name is *Erin*. The Romans called it *Hibernia*. To the Phœnicians it was known as *Ierne*. It is often called the "Emerald Isle," on account of its verdure.

COASTS.—The western and south-western coasts of Ireland are more indented than the eastern side of the island. The entire length of coast-line, including the larger inlets, is about **2,200 miles,** or 1 mile of coast to every 15 square miles of area.

The East Coast of Ireland is, on the whole, flat and regular, and the approach from Great Britain is obstructed by numerous sandbanks and sunken rocks.

The West Coast, on the contrary, is high, rocky, and in parts very irregular, especially in the south-west. The scenery along the rugged and broken coasts of Donegal and Connaught is of surpassing grandeur, and the noble indentations between Loop Head and Mizen Head are undoubtedly, in point of scenery, the finest in the kingdom.

The North Coast is also bold and rocky, and is broken by two deep inlets—Lough Swilly and Lough Foyle. The Giant's Causeway, on the north coast of Antrim, is one of the chief natural wonders of the kingdom ; it is a vast assemblage of columns of basaltic rock, which line a part of the shore, and advance, by successive rows, into the sea. The "Causeway" is 2,000 feet long and about 1,000 feet broad. Similar basaltic columns form the so-called "Fingal's Cave" in the island of Staffa, on the west coast of Scotland.

The South Coast possesses several magnificent *natural harbours*, one of which, Cork Harbour, the port of call for the Atlantic liners, is one of the finest in the world. [1]

CAPES : The chief capes are the following :—

On the *North Coast*, Fair Head (or Benmore), Bengore Head, Malin Head, and Horn Head ; on the *West Coast*, Bloody Foreland, Rossan Point, Erris Head, Achil [2] Head, Slyne Head, Loop Head, Dunmore [3] Head ; on the *South Coast*, Mizen Head, Cape Clear, and Carnsore Point ; on the *East Coast*, Wicklow Head and Howth Head.

The most northerly point of Ireland is *Malin Head ;* the most westerly, *Dunmore Head ;* and the most southerly, *Mizen Head.* Cape Clear is the extreme point of a small island which lies off the south-west coast, and is the first land sighted, with the exception of Fastnet Lighthouse, in coming from America.

INLETS : The following are the principal inlets :—

On the *east coast*, Wexford Harbour, Dublin Bay, Dundalk Bay, Carlingford Lough, [4] Dundrum Bay, Strangford Lough, and Belfast Lough ; on the *north coast*, Lough Foyle and Lough Swilly ; on the *west coast*, Donegal Bay, Sligo Bay, Killala Bay, Blacksod Bay, Clew Bay, Galway Bay, the estuary of the Shannon, Tralee Bay, Dingle Bay, Kenmare Bay, and Bantry Bay ; on the *south coast*, Cork Harbour, Youghal Bay, Dungarvan Harbour, Tramore Bay, and Waterford Harbour.

Many of the numerous inlets on the southern and western shores of Ireland form splendid harbours. Both Bantry Bay and Cork Harbour could contain the entire British navy, and no less than twelve others could float the largest men-of-war. [5] Bantry Bay is so spacious, so deep and well sheltered, that all

---

1. Our remarks (pp. 149-50) anent the contrasts between the eastern and western coasts of Scotland and England apply equally to the dissimilarities in form of the east and west coasts of Ireland.
2. Achil, eagle.
3. Dunmore, *Dun*, a fort, and *mor*, great.
4. The term *lough*, in Ireland, is equivalent to *loch* in Scotland. It is given both to inland lakes and to the nearly land-enclosed inlets along the coast.
5. Blacksod Bay is the chief resort of ships of war on this coast; it affords safe anchorage in any weather.

the fleets in the world might safely anchor in it.[1]   On the east coast the only good harbour is Strangford Lough, and the entrance to that is somewhat dangerous.   Dublin Bay is partially protected by two granite sea-walls.

ISLANDS : Of the many islands along the coasts of Ireland, none are of any considerable magnitude.

The principal islands are :—On the *north coast*, **Rathlin** and **Tory** : on the *south coast*, **Clear, Spike**,[2] and **Saltee Islands** ; on the *east coast*, **Dalkey, Ireland's Eye** and **Lambay** ; on the *west coast*, **North Aran, Achil, Clare, Aran Islands**, and **Valentia**.[3]

MOUNTAINS.—Ireland is generally level in the interior, but moderately elevated highlands adjoin various portions of the coast. The highest mountains are in the south-west, within the county of Kerry, but there are nowhere any continuous chains.

The mountains and hills of Ireland may be arranged in four groups, including, respectively, the Northern, Eastern, Southern, and Western Highlands.

THE NORTHERN HIGHLANDS include the Mountains of Donegal, which culminate in *Mount Errigal*, 2,466 feet in height, and the Mountains of Antrim, 2,400 feet in height.

THE EASTERN HIGHLANDS include the granitic masses of the Mourne Mountains, between Dundrum and Dundalk Bays, with *Slieve*[4] *Donard*, 2,796 feet high, and the Wicklow Hills, famed for their scenery, and rising in *Lugnaquilla* to an elevation of 2,039 feet above the sea.

THE SOUTHERN HIGHLANDS include the less detached ranges of the Slieve Bloom, 1,733 feet ; Silvermine, 2,278 feet ; Galty Mountains, 3,015 feet ; and Knockmeildown Mountains, 2,609 feet.

THE WESTERN HIGHLANDS include the Kerry Mountains, in the south-west, a series of parallel ranges, separated by Dingle Bay and other inlets, and rising in *Carrantuohill*, the highest point in Macgillicuddy Reeks, to 3,414 feet above the sea. The Mountains of Connaught include the Nephin Beg, 2,065 feet, Croagh Patrick, 2,510 feet, and Muilrea, 2,688 feet, in the county of Mayo ; and the Mountains of Connemara, the Twelve Pins group, the Mamturk Range and the Slieve Aughty, in the county of Galway.

PLAINS.—A great limestone plain extends across the middle part of the island, from Dublin Bay on the east, to Galway Bay on the west, on either side of which are several minor plains.

In some parts of this Great Central Plain, and also in the various mountain regions, there are extensive bogs,[5] which cover perhaps *one-seventh* of the entire area of the island. These bogs, of which the Bog of Allen, in Leinster, is the largest, furnish abundance of peat (used as fuel), and are capable, when drained, of being brought under cultivation.[6]

---

1. It was of this magnificent expanse that Thackeray said, "Were such a bay lying along English shores, it would be a world's wonder ; perhaps if it were on the Mediterranean or the Baltic, English travellers would flock to it in hundreds."

2 Spike Island is in Cork Harbour, and is fortified.

3 Valentia is an important telegraph station, being the western terminus of the submarine cable to Newfoundland.

4. Slieve, Irish, Sliabh, a mountain.

5. Organic substances buried in the bog are curiously preserved from corruption ; remains of large forests not unfrequently are discovered firm and sound, and *bog-oak* is a recognized material for making trinkets ; deer that perished centuries ago are found as if but recently slain ; and at times a human body, belonging to generations long forgotten, is strangely brought back to the light of day.

6. Of the bogs of Ireland, the *Black Bog* is most valuable for fuel, but is not so capable of being reclaimed as the *Red* and *Brown Bogs*.

RIVERS.—Ireland abounds in inland waters, but the rivers, with one exception, the Shannon, have short courses, and, though navigable, are commercially unimportant. The principal rivers are :—

On the north, the Bann (100 miles long), draining Lough Neagh ; and the Foyle, flowing into Lough Foyle.

On the east, the Lagan (42 miles), flowing into Belfast Lough ; the Boyne[1] (80 miles), which is navigable to Navan ; the Liffey (75 miles), which has Dublin, the metropolis of Ireland, on its banks ; the Slaney (70 miles), flowing into Wexford Harbour.

On the south, the Barrow (114 miles long), which rises in the Slieve Bloom Mountains, and is navigable to Athy, 60 miles from the sea ; the Suir and Nore, tributaries of the Barrow ; the Blackwater (90 miles), which rises in the Kerry Mountains and falls into Youghal Harbour ; the Lee (60 miles), whose estuary forms the splendid harbour of Cork ; and the Bandon (40 miles), which flows into Kinsale Harbour.

On the west, the Shannon (224 miles), the longest river in Ireland, which flows from a small pond in Cavan, through Loughs Allen, Ree, and Derg, entering the Atlantic by a broad and deep estuary 60 miles long. The Shannon is navigable to Lough Allen, 213 miles from the sea ; and, as its *fall* is only about 9 inches per mile, its current is very sluggish.

LAKES abound in Ireland : the largest of them, **Lough Neagh,** is in Ulster ; the most beautiful, the **Lakes of Killarney,** are in Munster ; but the greater number are in Connaught.

Lough Neagh, the largest lake in Ireland, is 150 square miles in extent, and is thus larger than any other lake in the British Isles, being more than 3 times the size of Loch Lomond in Scotland, and 15 times larger than Windermere in England. Lough Erne is singularly beautiful, its surface being studded with small wooded islands, which are said to equal in number the days in the year. Lough Mask and Lough Corrib, which are connected together by a subterranean channel, are both partly in Mayo and partly in Galway.

Lough Allen, Lough Ree, and Lough Derg are within the course of the River Shannon, and may be regarded as expansions of that river. The Lakes of Killarney, in Kerry, three in number, and altogether less than 10 square miles in extent, are celebrated for the contrasts afforded by their scenery, from the "soft, verdant, and beautiful, to the wild, rugged, and sublime." The highest mountains in Ireland rise immediately above their western shore.

CLIMATE.—The climate of Ireland is moister than that of England. The winters are nearly always mild, and the prevalent winds, which are from the west, are laden with the warm and moist vapours derived from the waters of the Atlantic.

Vegetation : The moist climate of Ireland preserves a more constant verdure to the fields, and a superior freshness and brightness of colour to its general vegetation, so that the island is most appropriately named the " Emerald Isle." The vegetation native to the coasts of Kerry (the south-westernmost county) is especially distinguished for its rich luxuriance.

Ireland is at all times much more humid than England, and is, indeed, the rainiest country in Europe, and more rain falls on its western and southern than

1. The " Battle of the Boyne" was fought on the 1st of July, 1690.

on its eastern coasts. Thus, the average annual *rainfall* at Cork is 40 inches, but at Dublin only 31 inches.

MINERALS.—As regards mineral productions, Ireland is inferior to England and Scotland in one essential particular—coal—the paucity of which affects injuriously its manufacturing industry.

The coalfields of Ireland—diffused, at wide distances apart, through the North-Eastern, Midland, and South-Western counties—are of limited extent compared to those of Great Britain, and their produce small in amount. Peat is the fuel most generally consumed, but coal is imported from England and Scotland.

Ireland is rich in iron ore, and there are small deposits of copper, lead, and silver. In many parts of the country there is a great variety of marbles and building-stones.

INHABITANTS.—Ireland contains a population of nearly 4¾ millions, and has thus half a million more inhabitants than Scotland, but scarcely a sixth of the population of England.

Ireland contained, in 1891, a population of 4,706,162, or fewer by nearly 1¼ millions than had belonged to it eighty years earlier, and little more than one-half the amount of its population in 1841, when it amounted to 8,175,124. But vast numbers of the Irish people emigrated to other lands during the intervening period ; and famine, with its attendant sickness and suffering, contributed to thin the population. According to the Census Returns of 1891, the population had declined to an average of only 145 per square mile, or less than one-third of the density of England and Wales.

Race and Language.—The great majority of the Irish population belong to the Celtic race—the same that peoples the Highlands of Scotland and the mountain-region of Wales. It is chiefly in the province of Ulster (the north-eastern part of the island) that the Anglo-Saxon race is found settled on Irish soil. The people of Ulster are the descendants of immigrants from the Scottish Lowlands, and preserve the social habits and industry of Scotland. People of English descent are numerous in the neighbourhood of Dublin, and are also scattered over every portion of the island. The native language of Ireland, called *Erse*, a Celtic dialect, is rapidly becoming superseded by the English tongue, and nearly all who speak it also understand and speak English.

INDUSTRIES.—Some manufactures, such as *linen*, *lace*, and *poplins*, are carried on on a large scale, but Ireland is chiefly an agricultural country.

Agriculture.—About two-thirds of the surface of Ireland is arable, but a very large portion of the land is in pasture. *Cattle*, *sheep*, and *pigs*, with various *farm-produce*, constitute (over by far the greater part of the island) its chief industrial wealth.

Manufactures.—These flourish principally in Ulster, where the *linen* manufacture is pursued on a scale of great extent. *Woollen* and *cotton* goods are also made, but in smaller quantities.

Commerce.—A great part of the commerce of Ireland consists in the *export* of its agricultural produce to the English markets, and in the *import* of coal, with various articles of British manufacture and foreign produce.

Ports.—The principal ports are Dublin, Belfast, Cork, Waterford, Limerick, Galway, and Londonderry. The greater part of the trade with Great Britain is carried on between these ports and Glasgow, Liverpool, and Bristol.

The chief passenger and mail route between England and Ireland is from Holyhead to Kingstown (64 miles distant), the outport of Dublin.

Internal communication is facilitated by excellent turnpike roads, and over 2,700 miles of railways, connecting Dublin and Belfast with all the chief centres of population. Cheap water-carriage is provided by several canals and numerous navigable rivers.

TOWNS.—Two only of the towns of Ireland contain over 200,000 inhabitants. These are the city of **Dublin**, the capital, and **Belfast**, the chief manufacturing and commercial city of the island.

Dublin has over 350,000 inhabitants; Belfast, 273,000; Cork, 98,000; Limerick, 37,000; Londonderry, 33,000; Waterford, 22,000; and Galway, 14,000.

PROVINCES AND COUNTIES.—Ireland is divided into four **Provinces**, which are sub-divided into thirty-two **Counties.** The provinces are, *Leinster* in the east, *Ulster* in the north, *Connaught* in the west, and *Munster* in the south.

LEINSTER contains 12 counties—**Five Maritime**—Dublin, Wicklow, Wexford, Meath, and Louth; and **Seven Inland**—Kilkenny, Carlow, Kildare, Queen's County, King's County, Westmeath, and Longford.

ULSTER contains 9 counties:—**Four Maritime**—Antrim, Down, Londonderry, Donegal; and **Five Inland**—Armagh, Tyrone, Fermanagh, Monaghan, and Cavan.

CONNAUGHT contains 5 counties:—**Four Maritime**—Leitrim, Sligo, Mayo, and Galway; and **One Inland**—Roscommon.

MUNSTER contains 6 counties:—**Four Maritime**—Waterford, Cork, Kerry, and Clare; and **Two Inland**—Limerick and Tipperary.

I.—THE PROVINCE OF LEINSTER.[1]

DUBLIN, the metropolitan county, is hilly in the south; the rest of the county is a rich, level, and well-cultivated plain.

DUBLIN (353), the capital of Ireland, stands at the mouth of the River Liffey. It is less populous than either Manchester, Liverpool, or Glasgow, but it is a great seat of trade, and has numerous fine public buildings. Dublin has the rank of an archiepiscopal city, and possesses two cathedrals. It is also the seat of several universities. **Kingstown** (17), on the south side of Dublin Bay, has a fine artificial harbour, and is the mail-packet station between Dublin and England, and for steamers plying to Holyhead and Liverpool.

---

1. The areas and population of the Counties of Leinster are as follows:—
(1.) Dublin, 354 sq. m., pop. 429,111.
(2.) Wicklow, 781 sq. m., pop. 65,935.
(3.) Wexford, 901 sq. m., pop. 131,534.
(4.) Meath, 906 sq. m., pop. 76,616.
(5.) Louth, 315 sq. m., pop. 70,852.
(6.) Kilkenny, 796 sq. m., pop. 87,154.
(7.) Carlow, 346 sq. m., pop. 40,706.
(8.) Kildare, 654 sq. m., pop. 69,928.
(9.) Queen's County, 664 sq. m., pop. 64,639.
(10.) King's County, 772 sq. m., pop. 65,408.
(11.) Westmeath, 708 sq. m., pop. 62,026.
(12.) Longford, 421 sq. m., pop. 52,354.

A short distance to the south of Dublin begins the romantic district of the Wicklow Mountains. The small seaport of **Balbriggan** is famous for its hosiery. At *Clontarf* a famous battle was fought, in which King Brian Boru defeated the Danes in A.D. 1014.

WICKLOW has a precipitous coast-line, and the interior is a mass of mountains ; the wooded valleys of the Avoca and its tributaries are extremely beautiful. There are **copper** and **lead mines**, and some *gold* has been found.

WICKLOW (4), the county town, **Bray** (6), a beautifully situated wateringplace, and **Arklow** (5), a port and fishing station at the mouth of the Avoca, are the largest towns, and are all on the coast.

WEXFORD is for the most part a level **plain**, fringed on the north-west by offshoots of the Wicklow Hills. Agriculture, **dairy** farming, and fishing are the chief industries.

WEXFORD (12), the county town, exports large quantities of provisions and fish. **New Ross** (7), is an important river-port on the Barrow. Near **Enniscorthy** (6) is *Vinegar Hill*, where the Irish rebels were defeated in 1798. **Tuskar Rock** Lighthouse, on a dangerous rock, lies 5 miles E.S.E. of Greenore Point.

MEATH is nearly all level, and the soil **fertile** and **well cultivated.**

The county is watered by the **Boyne**, on which stands **TRIM** (2), the county town. **Kells** and **Navan** are small inland towns. **Tara Hill** (507 feet high), on which the ancient kings of Ireland held their Councils, has ancient earthworks and other antiquities.

LOUTH is the **smallest** county in Ireland.   The peninsular portion contains a group of picturesque granite hills—the rest of the county is **level** and **fertile.**

The county town, **DUNDALK** (13), has considerable trade and manufactures.   **Drogheda**, built on both sides of the Boyne, 4 miles from its mouth, is a flourishing port ; Drogheda was besieged and taken by Oliver Cromwell in 1649.   The *Battle of the Boyne*, in 1690, between the armies of William III. and James II., was fought on the banks of the river a short distance above the town.

KILKENNY is mainly a fertile plain, diversified with gentle undulations.   **Anthracite coal** and **black marble** are found.

KILKENNY (11), on the **Nore**, a tributary of the Barrow, is the county town and the second among the towns of Leinster in point of population, and also the largest inland town in Ireland.

CARLOW forms part of the great central plain, and includes a large area of bog-land.

CARLOW, the county town, is on the Barrow.   Fine *granite* is quarried at **Bagenalstown**, on the Barrow, 10 miles south-west of Carlow.

KILDARE includes part of the great Bog of Allen—the rest of the county has a rich and fertile soil.

ATHY (4), on the Barrow, is the county town. Near the ancient town of Kildare is the famous *Curragh*, a military camp and race-course—the finest in the world.

QUEEN'S COUNTY is mountainous in the north-west—the rest is level and fairly fertile, but there are some large bogs.

MARYBOROUGH (3) is the county town. Mountmellick and Portarlington are small manufacturing towns.

KING'S COUNTY is flat and in great part boggy, except in the south-west, where it is bordered by the Slieve Bloom Mountains.

TULLAMORE (5), the county town, is on the Grand Canal. Sixteen miles north of Birr or Parsonstown (where Lord Rosse's great astronomical telescope is erected), on the Shannon, are the ruined churches, round towers, ancient crosses, and tombs of *Clonmacnoise*.

WESTMEATH, though level, is beautifully diversified with fine woods and numerous lakes, studded with pretty islets. The arable land is very fertile.

MULLINGAR (5), the county town, and Athlone, on the Shannon, are important military stations, and carry on a large trade in cattle and dairy produce.

LONGFORD is a pastoral county, level and fertile, except in the bog areas.

The county town, LONGFORD (4), communicates by canal and rail with Dublin. At Pallas, a village in the south of the county, Oliver Goldsmith was born in 1728.

II.—THE PROVINCE OF ULSTER.[1]

ANTRIM is distinguished, commercially, as the chief manufacturing county in Ireland, and physically, for the long and narrow plateau which extends along the coast from Belfast Lough to the basaltic cliffs of Fair Head and the still more wonderful **Giant's Causeway**.

The county town of Antrim, BELFAST (273), is also the capital of Ulster and the commercial capital of Ireland. It is the chief seat of the great *linen* industry of the province, and also has cotton and muslin factories, large shipbuilding yards, foundries, glass and chemical works, &c. The import and export trade of Belfast is larger than that of any other town in Ireland. Carrickfergus (5) is on the north side of Belfast Lough; here William III. landed, in 1690, previous to the battle of the Boyne. Larne is a beautiful place at the mouth of Lough Larne, and a port of call for the Clyde steamers.

1. The areas and population of the Counties of Ulster are as follows :—

(1.) Antrim, 1,100 sq. m., pop. 427,698.
(2.) Down, 957 sq. m., pop. 266,863.
(3.) Londonderry, 816 sq. m., pop. 151,660.
(4.) Donegal, 1,870 sq. m., pop. 185,211.

(5.) Armagh, 513 sq. m., pop. 143,055.
(6.) Tyrone, 1,260 sq. m., pop. 171,273.
(7.) Fermanagh, 714 sq. m., pop. 74,727.
(8.) Monaghan, 500 sq. m., pop. 102,099.
(9.) Cavan, 746 sq. m., pop. 111,079.

The whole coast from Larne to Portrush—a frequented watering-place on
the north-west coast, and the station for the Giant's Causeway (passengers to
which are conveyed by an Electric Tramway)—is most picturesque.   **Lisburn**
(10), on the Lagan, above Belfast, **Ballymena** (9), in the centre of the county,
near the north-eastern shores of Lough Neagh, and other towns, are all
engaged in the *linen* trade and manufacture.

DOWN presents an endless succession of cultivated hills,
valleys, and small plains, except in the south, where the grandly
picturesque Mourne Mountains rise direct from the sea-board.
Cereals and flax are largely grown; linen is the staple manufacture,
and fishing is also an important industry on the coast.

The county town is **DOWNPATRICK**, near the southern shores of Strangford Lough, and near the north end of the same lough is **Newtownards**, a muslin
weaving town.   Both are much less populous than **Newry (13)**, a large manufacturing and trading town on the Newry Canal, about 6 miles above **Warren-
point**, its outport at the head of Carlingford Lough.   Bangor is a favourite
watering-place on Belfast Lough.   **Donaghadee** is the nearest port to Scotland,
the distance from Portpatrick, on the opposite coast of Galloway, being 21½
miles.

LONDONDERRY is level and fertile in the north and centre :
on the southern border are the Sperrin Mountains and other ranges.
Linen is the staple manufacture, but agriculture and cattle-rearing
form the chief industry.

The picturesque port of **LONDONDERRY** (33), on Lough Foyle, is the
county town.   On the old walls are the cannon used during its famous siege in
1688-1689.   Londonderry has a large coasting trade.   **Coleraine** (6), on the
Bann, is another important port and manufacturing town.

DONEGAL has magnificent coast scenery.   From Malin Head
to the mouth of the Erne is a bewildering succession of beautiful
bays, high headlands, grand cliffs, and innumerable islands. Inland
are high mountains, bleak moorlands, long valleys, large bogs,
and numerous fine lakes and salmon rivers.   Half the county is
irreclaimable bog and waste land.

The county town, **CLIFFORD**, is a village ; only one town, **Ballyshannon**,
famous for its salmon fishery, at the mouth of the Erne, has above 2,000 inhabitants.   Moville, on the western side of Lough Foyle, is a sea-bathing resort
and a port of call for the Atlantic "liners" to and from Glasgow and Liverpool.

ARMAGH is flat and boggy in the north, and hilly in the south-
east.   The rest of the county is gently undulating and well cultivated.   Linen is the chief manufacture, and good marble is quarried
near the city of Armagh.

ARMAGH, the county town, is also the northern ecclesiastical metropolis of
Ireland.   **Lurgan (11)** and **Portadown** (8) are busy manufacturing towns *(linen,
muslin, &c.)*   Part of **Newry** is in this county.

TYRONE is mountainous in the north, hilly in the west and south, sloping gently on the east towards the shores of Lough Neagh. On the whole, Tyrone is a county of "gentle hills, fruitful valleys, pretty glens, and small plains." Agriculture and the manufacture of linen and woollen goods are the staple industries.

The county town, OMAGH (4), stands on a hill'in the centre of the county. Strabane (4), and Dungannon (4), are engaged in the linen trade.

FERMANAGH consists of a long and fertile valley, in which lie the beautiful Loughs of the Erne, enclosed on either side by high and bleak uplands.

ENNISKILLEN (6), the county town, is beautifully situated on an island in the River Erne, between the two loughs.

MONAGHAN is hilly, with many bogs and numerous small lakes.

The three largest towns are MONAGHAN (3), the county town ; Clones, an ancient town, occupying the summit of one of those round hills so numerous in that part of the limestone plain ; and Carrickmacross (2).

CAVAN, like Monaghan, is hilly, with many lakes and bogs. There are deposits of coal and iron, copper and lead ores ; but agriculture and flax-growing are the chief industries.

CAVAN (3), the county town, Cootehill and Belturbet (each 2), are small towns with some trade in corn and flax.

THE PROVINCE OF CONNAUGHT.[1]

LEITRIM is cut in two by Lough Allen and the Shannon—both divisions are hilly, with much bog and waste land. There are several coal-pits near Lough Allen.

The county town, CARRICK-ON-SHANNON, has some trade in *grain* and *provisions* by the river.

SLIGO is hilly in the north-west ; the rest of the county—north and south of the Ox Mountains—is low, and in some parts moderately fertile. Agriculture and fishing chiefly occupy the people.

The county town, SLIGO (10), a port at the head of Sligo Bay, has considerable trade.

MAYO is level in the north and east, but Clew Bay is surrounded by a mass of wild and rugged mountains and bleak uplands.—"The great conical mass of Nephin Beg is a conspicuous landmark all over the west of Ireland. Sheep and cattle rearing and fishing are the chief industries."

CASTLEBAR (4), in the centre of the county, is the chief town. Ballina (5) is a port at the mouth of the Moy. Killala, on Killala Bay, has exports of

1. The areas and population of the Counties of Connaught are as follows :—

(1.) Leitrim, 613 sq. m., pop. 78,379.
(2.) Sligo, 721 sq. m., pop. 98,338.
(3.) Mayo, 2,116 sq. m., pop. 218,405.
(4.) Galway, 2,452 sq. m., pop. 214,256.
(5.) Roscommon, 949 sq. m., pop. 114,194.

grain and *provisions.* The French, under General Humbert, landed near the town in 1798. The port of Westport, on Clew Bay, is the prettiest town in all Ireland.

GALWAY is divided into two unequal and strongly contrasted portions by Lough Corrib. The larger eastern division is level, and comparatively fertile ; the western portion is one of the wildest districts in Ireland, and includes the romantic tract of country known as *Connemara*, a region of alternate hills and valleys, with enclosed lakes and mountain streams.

GALWAY (14), the county town, is a seaport and railway terminus, near the head of Galway Bay. At one time it was expected that Galway, with its fine harbour and floating dock, would become important as a mail packet station for America. Ballinasloe, on the Suck, is noted for its great horse, sheep, and cattle fairs. Aughrim, a village in the eastern part of the county (a few miles distant from the right bank of the Suck), witnessed a decisive victory gained by William III. over the troops of James II. in 1691. **Tuam** (3) and **Athenry** are ancient towns in the interior.

ROSCOMMON, a pastoral county, includes the fine and rich grazing lands west of the Shannon and its Loughs, and extending to the Suck in the south, and the Curlew Hills (863 feet) in the north.

The county town, **ROSCOMMON** (2), has some manufactures of *linen, woollen,* and *pottery* goods, and contains the ruins of a fine old abbey, and a beautiful Anglo-Norman castle, both built in the 13th century.

THE PROVINCE OF MUNSTER.[1]

WATERFORD contains long ridges running across the entire country from east to west, between which are deep and fertile valleys. The valley of the Blackwater, from Cappoquin to the sea, is the finest in Ireland.

WATERFORD (22), the county town, exports large quantities of dairy produce to Liverpool and Bristol. Lismore, on the Blackwater, has a large salmon fishery. Dungarvan (6) is a fishing port, and also has some trade in *agricultural produce.* There are valuable quarries of *marble* near Cappoquin.

CORK is the largest and most southerly county in Ireland. Its surface is broken by long ridges and hilly woodlands, between which are rich, fertile valleys. The coast is broken by numberless bays and inlets—one of which, Cork Harbour, is one of the finest in the kingdom.

The county town, **CORK** (97), the chief manufacturing and trading city of the south of Ireland, is picturesquely situated on the Lee, 10 miles above its entrance into Cork Harbour. Cork butter, and other dairy produce, are famous.

1. The areas and population of the Counties of Munster are as follows:—
(1.) Waterford, 721 sq. m., pop. 98,130.
(2.) Cork, 2,890 sq. m., pop. 438,641.

(3.) Kerry, 1,850 sq. m., pop. 178,919.
(4.) Clare, 1,294 sq. m., pop. 123,839.
(5.) Limerick, 1,064 sq. m., pop. 158,563.
(6.) Tipperary, 1,659 sq. m., pop. 172,863.

On Great Island, in Cork Harbour, is **Queenstown**, the port of call for the American "liners" to and from Liverpool. The American mails are landed here, and taken by train to Dublin, and thence by steamer to Holyhead. Youghal (6), at the mouth of the Blackwater, and **Kinsale** (4), at the mouth of the Bandon, are thriving seaports.

KERRY, the most westerly county of Ireland, is the most deeply indented and mountainous portion of the island. The coast scenery is unsurpassed in variety and grandeur. The Kerry mountain region combines the grandest mountain, valley, and lake scenery in the British Isles, and this combination of the wild, the beautiful, and the sublime is seen in perfection at Killarney.

TRALEE (10), the chief town, is a small port near the head of Tralee Bay. Killarney (6), a mile east of the Lower Lake, is a favourite tourist resort. The splendid mountain and lake scenery attracts thousands of visitors every year to this part of Ireland. Listowel (2), on the River Feale, is in the northern part of the county. Valentia, an island at the southern entrance of the fine inlet of Dingle Bay, is the terminus of the Atlantic Cables.

CLARE is hilly in the east ; the middle and south form a broad and fertile plain, dotted with picturesque lakes. Agriculture, the manufacture of linens and **woollens,** and fishing, are the chief industries.

ENNIS (6), on the River Fergus, in the centre of the county, is the county town. Kilrush, on the estuary of the Shannon, is a flourishing port, and a harbour of refuge. Killaloe, on the Shannon, has *marble* and *slate* quarries in the vicinity.

LIMERICK is extremely fertile in the centre and north ; from its richness, the plain east of the city of Limerick is called the "Golden Vale." Agriculture, and **cattle** and **sheep rearing,** are the chief industries.

LIMERICK (37), the county town, stands on either side of the Shannon, a short way above the estuary, and has a large export trade in Irish produce, and is famous for its manufactures of *lace, gloves,* and *fish hooks.* The produce of a large extent of agricultural country is brought down the Shannon for shipment at Limerick. The city is rich in historic associations, and sustained two memorable sieges on behalf of James II. in 1690-91.

TIPPERARY is mountainous in the west and boggy in the north, but the central portion of the county is a magnificent and extremely fertile plain, watered by the River Suir and its tributaries. As in Limerick, **agriculture** is the chief industry, and much **cattle** and **dairy produce** are exported.

The county town, CLONMEL (10), is beautifully situated on the Suir ; lower down the river is the river-port of Carrick-on-Suir (6). Cashel (4) is a cathedral city, in the "Golden Vale." The "Rock of Cashel" (300 feet high) is crowned with the ruins of a round tower, a chapel, and a cathedral. Thurles (5) is a flourishing town, with interesting ruins.

THE UNITED KINGDOM.

Though comparatively small in *area*, and surpassed by many other countries in point of *population*, the United Kingdom of Great Britain and Ireland is the greatest commercial and manufacturing country in the world.

The total area of the United Kingdom is a little over 120,000 square miles, or rather more than *one-half* the area of France or Germany, and scarcely *one-seventeenth* part of the size of the Russian territories in Europe.

As regards number of inhabitants, Great Britain and Ireland contain a few thousand more people than France, but 10 millions less than Germany, about 24 millions less than the United States, and 57 millions less than Russia, the most populous of European countries.

According to the Census Returns for April, 1891, and inclusive of the islands in the British Seas, the population of the *United Kingdom*, in 1890, amounted to 38 millions, of which *England* and *Wales* contained 29 millions, *Scotland* a little over 4 millions, and *Ireland* nearly 4¾ millions.

The population of Great Britain has nearly trebled since 1801, but that of Ireland, which increased by 3 millions in 40 years (1801-41), decreased also by 3 millions during the next 40 years (1841-81), and now the island contains fewer people than it did ninety years ago.

The density of population in the United Kingdom is *greatest* in England and Wales, where, according to the latest returns, it amounts to 498 per square mile, and *least* in Scotland—134 per square mile. In Ireland, the density amounts to 145 per square mile. In 1891, England itself had 540 persons to each square mile, as against 206 in Wales, 134 in Scotland, and 145 in Ireland. In the same year, the average density of population in the British Isles, as a whole, was 316 per square mile.

Emigration from the United Kingdom, and especially from Ireland, has been very active during the present century, and between 1815 and 1889 upwards of 10¼ million persons of British origin emigrated, the majority of them settling in the United States of North America. During the last 35 years, nearly 7 million British emigrants left their native land; 4½ millions going to the United States—of these nearly 2 millions were English, 360,000 Scotch, and over 2¼ millions Irish—the rest settling chiefly in Canada, Australasia, and South Africa.

The industrial and commercial supremacy of the United Kingdom is due to its splendid *geographical position* and almost unparalleled *development of coast-line*, a favourable *climate* and an *abundant supply of coal*, the most essential of all minerals, and *iron*, the most useful of all metals, the *energy* and *enterprise* of the people, ample *capital* and efficient *labour*, unrivalled *facilities for carrying on industrial operations* on a vast scale, *colonies* and *dependencies* in all parts of the globe, and a *mercantile marine* larger than the merchant navies of all other countries taken together.

The geographical position of the British Isles is undoubtedly the best in the world, because, although "detached from the great continental masses, they are peculiarly and influentially situated with reference to them," occupying, as they do, a central position among the countries of the world—the exact centre of the land hemisphere being within a few miles of Falmouth—and also most favourably placed with regard to the chief industrial and trading countries, with which they easily communicate by what Professor Seeley happily terms "an incomparable road-system," namely, the sea.

The development of coast-line is such that no part of the country is far from tidal waters, and products are thus readily received and exchanged with all parts of the world—this interchange being all the more easily carried on because of the abundance of seaports,[1] and the unrivalled facilities for internal communication by roads, railways, and navigable rivers.

Another peculiarity is the remarkable contraction in breadth of both Great Britain and Ireland at various points by deep indentations on opposite coasts. Thus, between the tidal waters of the Thames and the Severn, the distance is under 100 miles ; while Hull, on the Humber, is 113 miles from Liverpool, on the Mersey. The mouth of the Tyne is only 60 miles from the head of the Solway Firth, and the Firths of Forth and Clyde are divided only by 32 miles of land. In Ireland, there are similar contractions between Dundalk and Sligo and Dublin and Galway.

The numerous navigable rivers, with their broad and deep estuaries, into which the tide penetrates very far, together with an admirable, and so far complete, system of canals, which are much used for the conveyance of minerals and heavy goods, are all favourable to the activity and growth of British commerce.

The entire extent of the coast-line of Great Britain and Ireland, including the larger indentations, is vastly in excess of the direct distance between the extreme points of the land, while the true salt-water coast-line has a far greater extent, and the line of tidal influence is still more enormous in proportion to the area of the country. Further, the British Isles enjoy the advantage of higher tides than most other countries, which enable vessels of considerable burden to penetrate almost to the heart of the country.

The climate of the British Isles is remarkably mild and equable, as compared with that of continental countries under the same parallels of latitude. The variation of temperature is comparatively slight, and in no other part of the world do the isotherms of 46° and 50° F. reach so far north as in this country, the truly Oceanic climate of which is the most favourable for industry and trade in the world. Out-door work and railway traffic are carried on with little interruption all the year round, and the tidal estuaries and navigable rivers are never frozen over even in the severest winters.

Of the energy and enterprise of the British people, industries on a vast scale and a world-wide commerce are the best evidence. The national characteristics of the British people are, according to a German writer,[2] solidity, energy, endurance, enterprise, strict respect for the law, and great industry ; while a French *savant*[3] declares that the British race is extraordinarily vigorous, and that in physical strength, practical intelligence, mental soundness and tenacity of purpose, it is the equal of any on the globe. The British, he adds, possess inventive genius, the love of adventure, the innate instinct of trade, a passion for success, and imperturbable courage.

---

1. " In the British Isles there are more than 20 seaports with a depth of at least 25 feet at high water, and most of these are situated in the vicinity of the great seats of production. In view of the increasing size of the shipping of the present day, this large number of deep harbours is a matter of peculiar importance."—(Hand-Book of Commercial Geography—Chisholm.)
2. Carl Zehden.
3. Onésime Reclus.

The wide extension of the English language, which is assuming more and more the character of a universal language, is an important factor in the development of British trade and industry. **Ample capital**, for carrying on industrial and commercial operations on a vast scale, is readily drawn from the enormous wealth of the country, while an abundant supply of efficient labour is secured by a dense and prolific population.

Between the United Kingdom and the British Colonies and Dependencies in all parts of the world, an interchange of productions naturally arises, and the bulk of the trade of Great Britain is done with them and the kindred English-speaking States of North America.

The **British mercantile marine** is larger than that of any other country, and, indeed, exceeds those of all other countries taken together. England possesses more than half the merchant service of the world, and her vessels have become the "ocean-carriers of the world."

INDUSTRIES.—Agriculture and stock-raising are important industries, and the fisheries are a great source of wealth, but mining, especially for coal and iron, manufacturing and commerce, occupy and support most of the people of the United Kingdom.

In England, the *industrial* class is nearly 5 times larger than the *agricultural* class, and the number of workers employed in mining and manufacturing is nearly 7 times the number engaged in *commercial* pursuits. In Scotland, there are nearly 4 times as many people engaged in mining and manufacturing as in agricultural occupations; but in Ireland the agricultural population is half again as numerous as the industrial class, and nearly 14 times the number occupied in carrying on the trade and commerce of the island. Great Britain, therefore, is mainly industrial and commercial, while Ireland is chiefly agricultural.

AGRICULTURE.—England is the most highly-cultivated country in the world, but most of the land in Scotland, Ireland, and Wales is in pasture.

The cultivable area of the United Kingdom is a little more than half the total area (58·5 %), while *woods* cover 3½ % of the surface, and *mountain-land, heath*, &c. nearly 40 %.   In England no less than 77 % of the total area is under cultivation or in permanent pasture; scarcely 5 % of the country being occupied by *woods*, and about 18 % consisting of *mountain-land, heath, water*, &c. In Wales the cultivable area is 60 % of the whole, while in Scotland, the productive land only amounts to a fourth of the total area.   In Ireland, the proportion of cultivable and pasture land is nearly the same as in England.

In Great Britain, a little more than one-half the cultivable area (16½ million acres out of 32½ million acres) is *under crops* or *sown grass*—the other half being in *permanent pasture*.   In Ireland, out of 15 million acres of cultivable land, over 12¾ million acres are in *pasture*, and only 2¼ million acres *actually tilled*.

The chief objects of culture are *wheat, barley, oats, potatoes, turnips*, and other *root crops*.   Wheat is principally grown in the eastern counties of England and the southern division of Scotland, but scarcely one-third of the quantity required for home consumption is grown within the country; enormous quantities are consequently imported from the United States, India, Australia, Russia, and other countries, and the proportion of foreign to home-grown wheat consumed is steadily increasing—the cheapening of the means of transit, and such advantages as are derived from cheap land in America and Australia, and cheap labour in India, rendering wheat-growing, at a profit, almost impossible for the British farmer, who is thus forced to devote more attention to dairy-farming, fruit-growing, &c. That foreign competition has radically affected the agricultural industry in this country, is evident from the fact that the wheat area has fallen from 4 million acres in 1867 to less than two million acres in 1893. There is also a slight decrease in the area under barley, while the acreage under oats, potatoes, and other root crops, is practically the same now as 20 years ago.

Total production of the chief crops in 1892 :—

	Wheat, 1,000 bush.	Barley, 1,000 bush.	Oats, 1,000 bush.	Potatoes, 1,000 tons.	Turnips, &c. 1,000 tons.
Great Britain .	58,561	70,485	116,295	3,049	27,348
Ireland. .	2,214	6,454	51,886	2,585	4,071
United Kingdom	60,775	76,939	168,181	5,634	31,419

The soil of the United Kingdom is in fewer hands than that of any other country in Europe,[1] and in both Great Britain and Ireland comparatively few of the farmers own the land they till.

In England and Scotland, however, the farms are, on an average, much larger than in Ireland, where the "division of the soil among tenants and sub-tenants is carried so far that in some cases barely *one* acre is cultivated by the farmer, a fact which explains much of the misery of the Irish peasantry, while 15 acres are considered the minimum for a successful farm in England."[2]

The 32½ million acres of cultivable and pasture land in Great Britain are parcelled out into half-a-million holdings or farms, one half of them under, and one half above, 20 acres in extent. In Ireland, a cultivable area of 15 million acres, less than half that of Great Britain, is yet divided among nearly the same number of occupiers.[3]

The domestic animals of Great Britain and Ireland are among the finest in the world, and famous breeds of *horses, cattle,* and *sheep* are fed on the magnificent pasture lands.

Nearly two-thirds of the cultivable area (32½ million acres) of Great Britain are in permanent pasture (16 million acres), or under clover and mature grass (4·8 million acres) ; and in Ireland four-fifths of the cultivable land are in pasture. Great attention is paid to the *breeding of horses and cattle,* and English horses are unrivalled for speed, endurance, and strength ; while English cattle are rich milk-givers, and English sheep furnish wool and meat of the best quality. But in addition to the home supply, the enormous demands of this "flesh-eating nation" necessitate an annual importation of live animals and dead meat to the value of no less than thirty millions sterling.

Live Stock (1,000 head) in the United Kingdom in 1893 :—

	Horses.	Cattle.	Sheep.	Pigs.
England . .	1,173	4,744	16,805	1,793
Wales . . .	147	738	3,101	200
Scotland . .	203	1,218	7,373	119
Ireland . . .	545	4,464	4,421	1,152
United Kingdom	2,068	11,164	31,700	3,264

1. For full details, see the "Statesman's Year Book" for 1894, p. 69.
2. See further Zehden's *Commercial Geography.*
3. 525,000 of whom 50,000 hold less than an acre each, and only 32,000 have farms of over 100 acres.

In Great Britain there are only 23,000 holdings of less than an acre, but 280,000 farms of 1 to 20 acres, while nearly 100,000 farmers have above 100 acres of land each.

FISHING is an important industry, especially along the eastern coasts of Scotland and England.

Upwards of 27,000 boats and 120,000 men are employed in the British sea fisheries, and the value of the fish landed every year is about eight millions sterling. Of the total quantity caught in 1893, 328,000 tons were landed on the English and Welsh coasts (255,000 tons, or five-sixths of the whole on the east coast of England), 309,000 tons on the Scottish, and barely 35,000 tons on the Irish coasts. About half the fish landed is conveyed inland by rail, the rest is carried by fast steamers and welled smacks to London, Grimsby, Harwich, and other ports, and thence distributed all over the country. *Billingsgate*, in London, is the largest fish market in the world, and the catch of the Hull, Grimsby, and Yarmouth trawlers, on the Dogger Bank and other fishing grounds in the North Sea, is brought to it by fast steamers, which run regularly between the fishing fleet and the Thames.

The herring-fishery is actively prosecuted on the Scottish coasts and along the east coast of England—the great centres being Wick in Scotland and Yarmouth in England. The cod fishery round the Shetland Islands, and the pilchard and mackerel fishery off the Cornish coasts, rank next in value. Considerably over half a million pounds' worth of salmon is caught and landed in Scotland and Ireland every year. The coasts of Ireland swarm with fish of all kinds, but the catch is scarcely one-seventh of that along the coasts of Scotland.

MINING is one of the most important of British industries. Great Britain contains vast stores of mineral wealth—in fact, no other country in the world possesses in such variety and abundance the material elements of prosperity.

The mining portion of England lies north-west of a line drawn from the Humber to the Exe—to the east and south of this line are the purely agricultural districts. In Wales, the coal and other mines are found in the south-eastern and north-eastern portions of the country—the former are among the richest and most productive in the world. In Scotland, the mining and manufacturing districts are limited to the Lowland region, within the basins of the Forth and Clyde. Ireland is deficient in minerals; there are extensive deposits of excellent iron-ore, but Irish coal is inferior in quality, and small in amount.

Coal and iron-ore are the chief objects of mining industry, but tin, copper, lead and zinc ores are also worked; slates, limestone, sandstone, and granite are largely quarried, while the salt mines of Cheshire yield enormous supplies of rock-salt.

The manufacturing and commercial supremacy of Great Britain is principally due to the enormous wealth of the country in coal and iron.

Coal and Iron are the essential elements of modern industry and commerce, and it is to the abundance and juxtaposition of these great resources that the manufacturing superiority of our country is directly due. Had we not, as Mr. Keltie justly remarks,[1] possessed these resources within our own country, we might never have been able fully to avail ourselves of our undoubtedly great geographical advantages. "Agriculture and stock-raising are, no doubt, impor-

---

1. See further the admirable review of the commercial position of the British Empire in Mr. | Keltie's work, *Applied Geography*, published by George Philip & Son.

tant British industries; in England and the lowlands of Scotland, the country is admirably adapted for such operations. But, had we been nothing more than an agricultural nation, it is to be feared that we should have remained far behind in the race of nations. A purely agricultural country can never support a very dense population, and in so small a country as ours, we could never have had much surplus capital for great enterprises, nor surplus inhabitants for purposes of colonization. Our coal and our iron have, to a great extent, been the making of us, and have enabled us to avail ourselves of our geographical advantages. Though agricultural products, wheat, and other cereals, animals and animal food, figure largely in our imports, they are almost nowhere in our exports. We have to import most of our food, and if we were fools enough to let our ports be blockaded, we should assuredly be starved out in case of war. We could never, it is obvious, have risen to our present position on the cultivated products of our soil. With the coal, itself a great export, we have been able to manufacture iron for ourselves and for export to other countries, and to turn to profitable account the raw materials, the cotton and the wool and the other textiles which our ships bring so plentifully to our shores. We have been able with our coal and iron, infinitely more valuable than all the gold mines of the world, to build our ships, our railways, our machinery ; to become, in short, a great manufacturing as well as a great carrying nation, and to nurture a succession of inventors to devise fresh methods of making the most of our exceptional advantages. All this manufacturing activity, moreover, encouraged a rapid increase of population, an increase so great in course of time, that, busy as we were, there was not work for all, and thousands swarmed off from the central hive to found Britains elsewhere, and so to lay the foundation of that empire beyond the seas which has reached such enormous dimensions at the present day."

COAL : The annual production of coal in the United Kingdom, for manufacturing purposes, househo'd use, and for export, amounts to the enormous quantity of over 180 million tons, or more than one-half the entire output of all other countries taken together. Coal-mining and the coal trade employ directly about six hundred thousand men, and the export trade in coal alone employs a very large amount of shipping.

Both the production and the export of coal have increased enormously within the last half century. In 1855, sixty-four million tons of "black diamonds" were raised. Thirty years later, in 1885, the output exceeded this by nearly 100 million tons, and now the total annual production has almost reached the enormous amount of fully 180 million tons !

The export of coal shows proportionately a very much larger increase. In 1851, only 3¾ million tons were exported. Thirty years later, 19½ million tons were shipped abroad, and at present the total export is upwards of 30 million tons a year.

The coalfields of England and Wales, and southern Scotland, are by far the largest and richest in Europe, and the most productive in the world.

The coalfields of England lie north of a line drawn between the Wash and the Severn, those of Scotland are all in the Forth and Clyde district. The principal coalfields of Ireland are at some distance from the coast, and are, besides, of small extent, and but little worked.   The rich coal-mines of Wales lie in the extreme south-eastern and north-eastern part of the principality.

The Northumberland and Durham Coalfield annually produces upwards of 30 million tons of coal, most of which is used in the great engineering and chemical works, shipbuilding yards, and other industries of the district.   The coal trade of NEWCASTLE and SUNDERLAND, NORTH and SOUTH SHIELDS, and the HARTLEPOOLS, is enormous, and the ironworks of MIDDLESBOROUGH, supplied with fuel from this coalfield, produce more than one-third of the iron smelted in Europe.   On the opposite side of the country is a much smaller but still important coalfield, namely,

The Cumberland Coalfield, which extends along the west coast to the north of St. Bees Head, and of which the produce is shipped from WHITEHAVEN and MARYPORT.   Further south lies the great

South Lancashire Coalfield, the second in importance in England, with an annual output of 22 million tons, nearly all required for the great textile industries of the district, the busiest and most populous portion of the "world's workshop."   From this field the coal is supplied to the innumerable cotton and other factories of MANCHESTER and SALFORD, OLDHAM, BOLTON, BURY, ROCHDALE, BLACKBURN, and PRESTON, the glass and chemical works of WIGAN, ST. HELENS, the shipbuilding yards, engineering and other works, and the shipping of LIVERPOOL and BIRKENHEAD.   There is a denser population on this coalfield than in any part of England, except the metropolitan district.   On the eastern side of the Pennine Range extends an equally important coal area, namely,

The Yorkshire Coalfield, which yields about the same amount—23 million tons a year, and the produce of which, like that of South Lancashire, is consumed mainly in the industries of the district.   On this coalfield are all the great centres of the woollen and clothing manufactures—LEEDS, BRADFORD, HUDDERSFIELD, HALIFAX, WAKEFIELD, DEWSBURY, BATLEY, and SALTAIRE—and also the chief centre of one branch of the hardware trade, SHEFFIELD, the great cutlery town.

The Derbyshire Coalfield—annual output, 11 million tons—and the Nottingham Coalfield—annual output, 7 million tons—are the southern extension of the larger Yorkshire coalfield ; in fact, the three are often regarded as one great coal area, stretching from the Wharfe to the Derwent, and thus, with its annual production of more than 40 million tons, is not only the most extensive, but also the most productive of British coalfields.   ROTHERHAM is the centre of the iron-smelting industry on this coalfield.

The North Staffordshire Coalfield supplies the large earthenware industries of the POTTERIES, while the South Staffordshire Coalfield sustains the huge hardware industries of WOLVERHAMPTON, WALSALL, WEST BROMWICH, DUDLEY, WEDNESBURY, and BILSTON ; the ribbon and cycle manufactures of COVENTRY ; and the multifarious metal industries of BIRMINGHAM—the industrial metropolis of the Midlands.

The Shropshire Coalfield is small in extent, but on it are the ironworks of COALBROOKDALE, WELLINGTON, and BRIDGNORTH.

WALES has two coalfields, one in the north-east, and another, and far more important, in the south-east.   On the former—the Flint and Denbigh Coalfield—are some important chemical and lead works ; FLINT, MOSTYN, HOLYWELL, WREXHAM, and RUABON being the chief mining centres.

The South Wales Coalfield has an area of 1,000 square miles, and annually produces upwards of 23 million tons of coal and anthracite. One-third of the entire British export of coal is shipped from CARDIFF, the chief outlet of the great mining and metal centres of MERTHYR TYDVIL, ABERDARE, DOWLAIS, PONTYPRIDD, &c., and from SWANSEA (the outlet for the western division of this rich coalfield), which also has important industries—copper-smelting works, &c.—of its own.

The Bristol Coalfield is small, and the deposits are difficult to work. There is also a small coalfield in the Forest of Dean, in Gloucestershire, and isolated mines are worked in other parts of England.

The Scottish Coalfields have a total production of about 27 million tons a year.

The mines of Lanark and Ayrshire supply the ironworks of HAMILTON, AIRDRIE, BATHGATE, FALKIRK, MOTHERWELL, and COATBRIDGE, and the great shipbuilding and engineering establishments of GLASGOW and GREENOCK, the cotton and thread factories of PAISLEY, and the woollen and carpet factories and iron-foundries of KILMARNOCK. The mines of the Lothians supply EDINBURGH and LEITH, while those of Fife and Forfar sustain the linen and jute manufacture of DUNDEE, DUNFERMLINE, and other manufacturing towns on the east coast.

The Irish Coalfields are much inferior in extent and amount of production to those of Great Britain.

The only important mines are those of KILKENNY, TIPPERARY, and TYRONE; but Irish coal is of inferior quality, and the total output scarcely exceeds 100,000 tons a year. Hull, in his " Physical Geology and Geography of Ireland," says that the coal measures once overspread all the area now occupied by carboniferous limestone, that is, all the central limestone plain of Ireland, and that then the surface of the Irish area remained in a state of dry land, while that of England was submerged beneath the waters of the sea. Little by little the carboniferous strata were swept by sub-aerial waters into the adjoining ocean, " to form, perhaps, some of the strata which were being piled up over the ocean-bed of the British area. At this time Ireland contributed to the future mineral wealth of England; she stript herself to clothe her sister, and to supply materials for protecting from atmospheric waste her vast stores of coal, upon which her greatness and prosperity now so largely depend."

" Of the upper carboniferous beds," states another writer. " which, at one time, overspread the central plain of Ireland, only small patches remain in isolated spots, serving chiefly as an indication of the immense loss that has been sustained in an important element of material prosperity."

IRON.—Iron-ore, by far the most valuable of all metallic ores, occurs abundantly within and near the coal areas of England and Wales and southern Scotland, and there are also rich deposits of this ore in Ireland.

In Ireland not only are the few coal mines situated at a considerable distance from the coast, but also from the iron-ore districts. In Great Britain, on the contrary, practically inexhaustible deposits of iron-ore are found not only within or close to the coal areas, but often in the same mines, and it is this abundance and juxtaposition of the iron-ore and of the coal to smelt it, that has given Great

Britain the lead among the industrial nations of the world.   Iron-working, in
England and Wales and Southern Scotland, is on this account so cheaply
carried on, that few other countries can compete with us in the iron markets of
the world.   Rather less iron-ore is now mined in South Wales than formerly,
owing to the enormous and rapidly-increasing import of cheap ores from Spain
and other countries.   Spain alone now sends us upwards of $3\frac{1}{2}$ million tons of
iron-ore, out of a total import of 4 million tons.

The annual production of iron-ore in the United Kingdom is
about 11 million tons, from which nearly 4 million tons of metal
are produced.

The chief iron-mines are in *South Wales, South Staffordshire, Yorkshire*, and
Southern Scotland.   In Ireland, the richest deposits of iron-ore are in county
Antrim.   From the clay-band ores of the *Cleveland* hills around MIDDLES-
BOROUGH, one-third of the iron smelted in England is produced, while the red
hematite ores of North Lancashire and Cumberland supply the great steel
works of BARROW-IN-FURNESS.

Besides the vast deposits of coal and iron, there are productive
ores of tin, lead, copper and zinc, and some gold and silver are also
produced.   Slate, clay, salt, and other minerals are found in abundance.

Rich deposits of tin and copper ore are found in *Cornwall* and *Devon*, but
more tin, and much more copper (either in the form of ore or partly refined
metal) is now imported into, than is produced in, the country.   Ores of lead,
some of them containing silver, are found and worked on both sides of the
Pennine Range and among the Cumbrian and Welsh mountains, the Wicklow
Hills in Ireland, and at Leadhills in the south of Scotland.   Zinc ores are found
in the *Isle of Man*, and in Wales and Northumberland.   A little gold is produced from gold-ores worked near DOLGELLY, in Merionethshire, and a considerable quantity of silver is produced by the *desilverisation* of lead and copper
ores.

Slates are extensively quarried in *Wales*, chiefly at BETHESDA, LLANBERIS,
and FESTINIOG; building stones, granites, and marbles are largely quarried in
various parts of the country; in the eastern and south-eastern divisions of
England, the clay in which they abound supplies the chief building material—
brick; and great quantities of china clay are sent from the south-western counties
of England to the potteries of Staffordshire.   More salt is produced in Great
Britain than in any other country in the world; it is chiefly derived from the
rock-salt mines of *Cheshire* and *Worcestershire*.

Summary of the mineral produce of the United Kingdom.
(1892.)
I.—METALLIC MINERALS.

	Iron.	Lead.	Tin.	Copper.	Zinc.	Gold.[1]
Ores raised (in 1,000 tons) .	11,312	40	14	6	26	10
Value in £1,000 . .	2,970	296	734	12	104	9

1. 271,259 oz. of silver, of the value of £44,998, were produced in 1893, chiefly from lead ores.

II.—NON-METALLIC MINERALS.

	Coal.	Stone.	Slates.	Clays.	Salt.	Oil Shale.
Amount in 1,000 tons,	181,786	—	1,025	3,103	1,956	2,089
Value in £1,000 .	66,050	8,667	1,025	889	861	522

MANUFACTURES : The United Kingdom is the chief manufacturing country in the world, and in England and Wales, and Lowland Scotland, more people are engaged in manufacturing pursuits than in any other branch of industry.

The most important manufacturing industries of the United Kingdom are the great textile manufactures and metal industries, with the chemical industries, and the leather manufacture. The making of earthenware, glass, paper, watches and clocks, &c., are all important industries, but none of them are upon a scale of such magnitude as the textile fabrics and metal wares, chemicals and leather goods, which form the great staples of British manufacturing industry.

The textile manufactures of the United Kingdom are the most extensive in the world. British textile factories employ over one million people, and at least five millions depend for their support directly upon these industries.

"A century ago," says Mr. Ellison, of Liverpool, "the value of cotton, woollen, and linen yarns and piece goods produced in Great Britain and Ireland was about £22,000,000—say, woollen £17,000,000, linen £4,000,000, and cotton £1,000,000. Of recent years the value has been about £170,000,000 —say, cotton £100,000,000, woollen £50,000,000, and linen £20,000,000. The total amount of capital employed is about £200,000,000, and at least 5,000,000 people—men, women, and children—are dependent upon these industries for their livelihood. Moreover, one-half of the value of British and Irish products exported consists of textiles."—(The *Statesman's Year-Book*, 1894, p. 77).

In 1890, the 7,190 textile factories—6,180 in England and Wales, 747 in Scotland, and 263 in Ireland—contained upwards of 53½ million spindles and 822,000 power-looms, and employed more than 1 million hands.

The Cotton Manufacture is by far the most important of British Industries, and the cotton factories of Lancashire and Lanarkshire produce more than one-half of the cotton goods of the world.

The cotton factories (2,500 in number) of the United Kingdom employ over half-a-million operatives, and annually consume about 1,700 million lbs. of raw cotton, and produce 2 million miles of cotton cloth for export, over and above the large quantities required for home consumption. Nearly 2,000 million lbs. of raw cotton are imported every year, and almost exclusively through LIVERPOOL.—the greatest cotton market of the world—and thence distributed to the great centres of the cotton industry.

The Chief Centres of the Cotton Industry are : MANCHESTER, the commercial centre of the densely-populated coal area of South Lancashire, which contains over 300 towns and villages, all actively employed in spinning or weaving cotton. The larger towns thus engaged are : BLACKBURN, OLDHAM, PRESTON, BOLTON, BURY, ROCHDALE, BURNLEY, ACCRINGTON, CHORLEY, and WIGAN, in Lancashire ; STOCKPORT and HYDE, in Cheshire ; GLOSSOP, in Derbyshire. Oldham and Bolton are chiefly engaged in cotton-spinning ;

and Preston, Blackburn, Accrington, and Burnley in cotton-weaving. NOTTING-HAM is also actively engaged in certain branches of the cotton trade—the staple industry being the making of *cotton hosiery*, *net-work*, and *lace*. In Scotland, *cotton* goods are chiefly made at GLASGOW and PAISLEY—the latter town is famous for its *thread*. There are also large cotton factories in Ireland, at BELFAST.

The Woollen Manufacture is the second great industry of the United Kingdom, and the *woollen, worsted, and shoddy* factories of the country, 2,700 in number, employ over 280,000 people, and annually use up over 770 million lbs. of foreign wool, in addition to 153 million lbs. produced at home.

The woollen manufacture is the most ancient of our textile industries, and British woollen goods have for centuries enjoyed a high reputation for excellence and finish, and at the present time Great Britain surpasses all other countries in this branch of industry (Zehden). In 1890, the 2,700 British factories—1,793 woollen, 753 worsted, and 125 shoddy—contained nearly 6¼ million spindles, and 140,000 power-looms.

The Chief Centre of the Woollen Industry is the West Riding of Yorkshire, where LEEDS, the great centre of the *woollen cloth* trade, occupies a geograph-ical and industrial position analogous to that of Manchester in the cotton trade and manufacture. BRADFORD, another large town on the Yorkshire coalfield, is the chief centre of the *worsted* manufacture, and near it stands the model industrial town of SALTAIRE, which has grown up around Sir Titus Salt's great *alpaca* works. HUDDERSFIELD is famous for its high-class plain and fancy *woollen fabrics*, and HALIFAX for its *carpets* and *baizes*. Heavier fabrics are made in the *shoddy* and *blanket* works of DEWSBURY and BATLEY. Woollen goods of various kinds, or yarn, are, in fact, made in almost every town and village on the eastern slope of the Pennines, and even on the western side; some of the great cotton towns, such as ROCHDALE, BURY, ASHTON, and GLOSSOP, have also a considerable woollen manufacture.

Three counties in the west of England are an old and still famous centre for the finer broadcloths. These "*West of England cloths*" are made at TROW-BRIDGE, BRADFORD, and WESTBURY, in the west of Wiltshire; at STROUD, in Gloucestershire; and at BATH and FROME, in Somersetshire. Of outlying towns, KENDAL, in the north-west, and NORWICH, in the east of England, which produced woollen goods five hundred years ago, still retain the industry to some extent; but other places in the south of England, such as NEWBURY, in Berkshire, once famous for their cloth trade, are now entirely agricultural.

In Scotland, the woollen manufacture is carried on chiefly in the valley of the Tweed—GALASHIELS, SELKIRK, HAWICK, and JEDBURGH producing the famous *tweed* cloth. Tweeds and woollen hosiery are also made at DUMFRIES, plaids and tartans at STIRLING, tweeds at BANNOCKBURN, and other woollen goods at KILMARNOCK and AYR.

In Ireland, the industry is practically confined to the making of some coarse *woollen goods* in LEINSTER.

Flannels are largely made at WELSHPOOL and DOLGELLY, in Wales; and also at ROCHDALE, HALIFAX, and other English towns. *Blankets*, first made at Bristol, by Thomas Blanket, in 1340, are now made chiefly at DEWSBURY, and other places in the West Riding of Yorkshire, and at WITNEY, in Oxford-shire. The so-called "Brussels" *carpets* are made at KIDDERMINSTER, while the " Kidderminster" carpets are made at KILMARNOCK, in Scotland, and at HALIFAX, in Yorkshire. *Woollen hosiery* and *elastic webbing* are the staple textile industries of LEICESTER.

The Linen Manufacture is almost confined to a few towns in *Ulster*, in Ireland ; *Fife* and *Forfar*, in Scotland ; and the *West Riding of Yorkshire*, in England.

The Linen Trade of Ireland employs about 60,000 workers, and centres at BELFAST. The linen and jute works of DUNDEE, ARBROATH, MONTROSE, DUNFERMLINE, and other towns on the eastern coast of Scotland, employ upwards of 80,000 people, or more than four times the number employed in the linen mills of BARNSLEY, LEEDS, and other towns in the West-Riding of Yorkshire. Sail-cloth is largely made at SUNDERLAND, STOCKTON, LIVERPOOL, and other seaport towns. Over 10 million pounds' worth of flax, hemp, and jute is imported into the United Kingdom every year, in addition to the large quantities of flax and tow produced at home. Most of the jute imported is sent to Dundee to be made into sacking and cordage, but there are large jute works in Belfast and London.

The Silk Industry, in which France surpasses all other countries, has never been developed to any extent in England.

The silk industry proper, *i.e.*, the spinning and weaving of *thrown silk*, has declined within recent years, and is still decreasing, but the making of silk plushes and other fabrics from *spun silk* (prepared from silk waste, in the same way as cotton or woollen yarn) is increasing. COVENTRY was formerly noted for its ribbons ; velvets and silk plushes are made at BRADFORD ; LEEK has silk thread and silk dyeing works, and there are numerous silk factories at MACCLESFIELD, CONGLETON, DERBY, and NORWICH. Bethnal Green and Spitalfields, in London, were formerly famous for their silk manufactures.

The Metal Industries of the United Kingdom are by far the most important and extensive in the world.

The British Metallic industries include the preparation of the "raw material' —the smelting of the iron, lead, tin, copper, and other metallic ores—and the manufacture therefrom of almost everything that can be made in metal, from tiny needles or delicate hair-springs, to huge anchors and the most powerful steam engines. And, in spite of the keen and rapidly increasing competition of foreign countries, especially the United States and Germany, British metal goods are yet unsurpassed in quality, quantity, and cheapness. One inestimable advantage, as we have already pointed out, is the possession of such vast stores of coal and iron in close proximity ; but the excellence and cheapness of English manufactures are chiefly due to the localization of the various industries—different manufactures being carried on in different localities, and these, generally speaking, admirably suited for the purpose, especially when, as in the case of the great textile and iron manufactures, the goods must be produced on a very large scale.

Certain districts and towns are thus intimately associated with certain industries, which are therein brought to the highest possible excellence at the least possible cost. The greatest of British textile industries—the cotton manufacture—is thus localized on the rich coalfield to the west of the Pennine chain ; while the second great industry—the woollen manufacture—is likewise carried on on a large scale on the Yorkshire coalfield, on the eastern side of the same range; and, similarly, the third great industry of England—iron and hardware—has been chiefly developed on the Staffordshire coalfield, to the south of the Pennines. This concentration of particular industries in particular localities is still more striking in the case of towns such as Manchester, Leeds, Birmingham, Glasgow, &c., so closely associated with cotton, woollen, iron goods, and shipbuilding respectively. And in regard to the metal industries, these have, for centuries, been concentrated in and around Birmingham—the metropolis of the English iron and metal trades.

The Iron Trade includes the smelting of the iron-ore and the production of *pig-iron*, and the manufacture of *iron and steel goods* of every description.

The Smelting of Iron-ore is carried on in the *Cleveland* district, in the North Riding of Yorkshire; in the *Furness* district, in North Lancashire; on the coalfields of *South Wales, South Staffordshire, Yorkshire, Shropshire,* and *Cumberland;* and in Scotland, on the rich coalfield of the Lowland Plain, chiefly in *Lanarkshire* and *Ayrshire.*

There are over 1,000 blast furnaces for smelting iron in Great Britain, but only 362 of these were in blast in 1892. There were, besides, in the same year, over 3,000 puddling furnaces for the manufacture of puddled bar iron, about 80 Bessemer steel converters, and 250 open-hearth steel furnaces. These numbers show the magnitude of the iron and steel industry, and the enormous capital and labour that must be employed in the trade.

To supply the smelting-furnaces, 11½ million tons of iron-ore are raised at home every year, in addition to an import of 4 million tons, chiefly from Spain.

From the 15 million tons of iron-ore annually obtained, about 7 million tons of pig-iron are produced, and considerably more than one-third of this is converted into steel, principally by the Bessemer process. Steel is rapidly superseding iron for many purposes—bridges, railways, ships, armour plates, &c.—and Great Britain now produces 3½ million tons of steel, or much more than half the entire steel production of the world, every year.

More than one-third of the iron-ore raised in England is smelted in the numerous furnaces at and around MIDDLESBOROUGH, in Yorkshire, and those in the south of *Durham;* the rich, red hematite ores of North Lancashire and South Cumberland are chiefly smelted at WORKINGTON and BARROW. (Barrow has the largest steel works in the kingdom).

Iron-smelting and coal-mining are also the great industries of *Glamorganshire* and *Monmouthshire,* where the furnaces cluster thickly round MERTHYR TYDVIL, ABERDARE, SWANSEA, NEWPORT, and other towns. The ore now used in the Welsh iron-works is chiefly that imported from Spain. More than one-third of the total British import of ore is landed at NEWPORT, CARDIFF, and SWANSEA.

The South Staffordshire Coalfield is another great iron-smelting region, and at night the dreary expanse of the "Black Country" is illuminated by the flames of hundreds of furnaces at and around DUDLEY, BILSTON, WEDNESBURY, WALSALL, WEST BROMWICH, TIPTON, and WOLVERHAMPTON; while further north, on the Yorkshire coalfield, are the iron-works of ROTHERHAM,

&c. On the Scottish coalfield, iron-smelting on a large scale is carried on at AIRDRIE, COATBRIDGE, HAMILTON, and other places in Lanarkshire and Ayrshire.

One hundred and sixty years ago, the iron and hardware manufactures of BIRMINGHAM employed and supported upwards of 50,000 people, and at the present day, the manufacture of all kinds of *iron* and *steel* goods and other metal wares, from *needles, pins,* and *pens*, to *steam-engines, machinery,* and *cannons,* supports ten times the number in the town itself, and also occupies the numerous towns and villages all crowded together on the adjoining coalfield, each of them actively engaged in one or more of the multifarious branches of the metal trade and industry. WOLVERHAMPTON is known everywhere for its locks; DUDLEY and BROMSGROVE for their nails and hinges; CRADLEY HEATH for its chains; and SOHO, a suburb of Birmingham, for its machinery, &c.

The machinery for textile manufactures is made principally in the district in which it is used; thus, cotton-spinning and weaving machines are made at OLD-HAM, BOLTON, MANCHESTER; woollen and worsted machinery at KEIGHLEY, LEEDS, and other places in the woollen-manufacturing district; elastic-webbing machinery at LEICESTER. Similarly, agricultural implements are chiefly made in the great centres of agricultural industry, such as GRANTHAM, BEDFORD, LINCOLN, NORWICH, IPSWICH, and GAINSBOROUGH. Enormous quantities of tin-plate are manufactured at SWANSEA, LLANELLY, CARDIFF, NEATH, NEW-PORT, and other towns on the South Wales coalfield, and form a large export to the United States for the fruit and fish canning industries.

Marine engines and locomotives are made at many of the larger towns—such as Manchester, Birmingham, Newcastle, Darlington, and Glasgow—and all the great railway companies have engine and carriage works of their own. The L. and N. W. R. Co's. engine works are at CREWE, and carriage works at WOLVERTON; those of the Midland are at DERBY; the Great Western Railway, at SWINDON; the Great Northern, at PETERBOROUGH; the Cambrian, at OSWESTRY; the North British, at Cowlairs, near Glasgow, and St. Margaret's, near Edinburgh; and the Glasgow and South-Western, at KILMARNOCK. Railway trucks are largely made at Birmingham, Manchester, Shrewsbury, and numerous other industrial centres.

Cutlery and Tools : One branch of the hardware trade, the making of cutlery and tools, has its chief centre at SHEFFIELD, in the south of Yorkshire. Swedish iron is used for making the high-class cutlery and tools for which Sheffield is famous all over the world. Continental competition, principally German, forces Sheffield to produce enormous quantities of the cheaper articles, and the United States is a formidable competitor in higher-class goods; but for the best kinds of cutlery and tools, the cutlers and tool-makers of Sheffield are absolutely unrivalled.

Besides its special industry, Sheffield has also large steel works, which are chiefly engaged in the production of *armour plates* and *steel rails.* Steel rails are also very largely made at BARROW and MIDDLESBOROUGH. Iron wire of all kinds is made at WARRINGTON; needle-making forms the special industry of REDDITCH in Worcester; pins and steel pens are important articles of manufacture at BIRMINGHAM. A million steel pens are made every year at Birmingham alone. British metal work also includes a vast number of articles in copper, lead, zinc, gold, silver, and especially brass, the latter chiefly in Birmingham; and clocks, watches and scientific instruments, of unrivalled excellence and finish, are made in the larger towns, chiefly in London, Birmingham, Sheffield, and Liverpool.

The **shipbuilding industry** of the United Kingdom is by far the most extensive in the world.

More vessels have been, and are being, built in the ship-yards on the **Clyde**, the **Tyne**, the **Tees**, the **Wear**, the **Mersey**, and the **Thames**, than in those of all other countries taken together.   Indeed, the Clyde shipbuilding yards alone turn out, in some years, a larger tonnage than the rest of the world.

More than half the merchant service of the world is British-owned, and also British-built.   The greater number of the steel and iron steamers belonging to foreign countries, and sailing under foreign flags, except those of the United States, were built in British yards; and now that steel has all but supplanted iron and wood in the construction of both sailing and steam vessels, a larger proportion than ever of the world's shipbuilding will be done in British yards.

The chief centre of the shipbuilding industry of the United Kingdom, and, in fact, of the world, is the Lower Clyde. Both banks of the Clyde, from Glasgow to Greenock, are lined by immense shipbuilding yards and engineering works; every branch of the trade being actively carried on in one or other of the Clyde ports—GLASGOW, PORT-GLASGOW, GREENOCK, DUMBARTON, &c.   The Clyde shipbuilders have built and fitted-up some of the largest ocean-steamers and the most powerful ironclads afloat, and at the present time there are about twice the number of vessels (four times the tonnage) being built in the Clyde, than in all the ship-yards of the United States taken together.

Next in importance is the **Tyne** district, the shipbuilding yards of NEWCASTLE, JARROW, SOUTH SHIELDS, and SUNDERLAND having, in 1890, as many vessels under construction as those on the Clyde; the **Tees** district, which includes MIDDLESBOROUGH, STOCKTON, and the HARTLEPOOLS, being third, followed by the **Wear** district, BELFAST and LONDONDERRY, and the **Mersey** (LIVERPOOL and BIRKENHEAD).   Splendid ocean and coasting steamers have also been built at BARROW and on the Thames; while considerable numbers of the smaller wooden vessels and boats have been constructed in the various seaports along the coast.

The chief **Government Dockyards** at PORTSMOUTH and DEVONPORT, CHATHAM, SHEERNESS, and PEMBROKE, are all more or less actively engaged in building ships for the navy, and in repairing and refitting our *ironclads* and *armed cruisers*, &c.   Ironclads, cruisers, and torpedo boats are also built in the private yards on the Clyde, the Thames, &c.

Judging from the returns showing the average unemployed labour in the shipbuilding trade, "none of the industries of the country exhibit so rapid and marvellous a change in one decade as the shipbuilding and boiler-making industries."   In 1884, nearly one-fourth of the available labour was unemployed; in 1888, it amounted to 9¼ per cent.; and in 1890, it fell to less than 1 per cent.   If we compare the vessels now built with those constructed 15 years ago, we find that the average size of the new vessels is considerably more than twice that of the ships then built; that steel was not then used for shipbuilding purposes, while now it has all but supplanted iron; and that then the tonnage in hand was equally divided between steamers and sailing vessels, while now the proportion of steam to sailing tonnage is about nine to one.[1]

In 1892, 681 mercantile vessels, of 1,109,950 gross tonnage, were launched in the United Kingdom, and 30 warships, of 151,157 tons displacement. "Of the merchant steamers (512 in all) one—the Cunard liner *Campania*—was of

1. For tables and statistics, see *Whitaker's Almanac*, 1894, p. 674.

12,950 tons ; 36 were between 4,000 and 7,000 tons each ; and 234 were between 1,000 and 4,000 tons. Of sailing vessels (169 in all) the largest was the " *Somali*," of 3,537 tons ; six others were between 3,000 and 4,000 tons ; and 117 were between 1,000 and 3,000 tons." The merchant and other vessels (not warships) built abroad during the same year were 165 steamers, of 132,323 tons, and 252 sailing vessels of 123,832 tons.

The **Earthenware** and **Porcelain** manufacture is the leading industry in the "*Potteries*" district in North Staffordshire, and is also carried on at *Lambeth*, in London, and other places.

Most of the enormous quantities both of earthenware and porcelain required for home use and for export, chiefly to the Colonies and America, is made in the towns and villages of the "*Potteries*" district—BURSLEM, STOKE-UPON-TRENT, HANLEY, NEWCASTLE-UNDER-LYME, ETRURIA, LONGPORT, &c.—but the finest porcelain ware is produced at WORCESTER, DERBY, and COALPORT, while crucibles and firebricks are made at STOURBRIDGE. In this industry, as in almost all others, the abundance of the raw material and of the necessary fuel in close proximity, is the reason why the making of earthenware on a large scale is confined to the North Staffordshire coalfield.

The **Glass** manufacture is chiefly carried on at *St. Helens* in Lancashire, and also at *Birmingham, Dudley* and *Stourbridge, Newcastle* and *South Shields, Glasgow* and *London*.

Common glass-ware is largely imported from Belgium, Germany, and Bohemia ; but the finer British-made plate-glass (mirrors, &c.) is as largely exported, chiefly to the Colonies and the United States.

The **Chemical Works** of the United Kingdom are the largest in the world, and produce enormous quantities of the *alkalies* used in the manufacture of glass and soap, *sulphuric acid*, various *dyes*, &c.

The largest **Alkali Works** are at WIDNES and RUNCORN, in the Mersey basin, JARROW on the Tyne, and FLINT on the Dee. *Sulphuric acid* is made principally at SWANSEA and NEWCASTLE. The extraction of *aniline dyes* from coal-tar, although discovered and even yet largely worked in Great Britain, is now chiefly carried on in Germany.

The British **Leather Trade** is very important, the preparation of the *leather* and the making of *leather goods* employing directly about 500,000 people.

There are 800 tanneries in the kingdom, and enormous quantities of hides and skins, imported for tanning or already tanned, are required for the great boot and shoe factories of NORTHAMPTON and neighbouring towns, LEICESTER, STAFFORD, LONDON, &c., and for the saddlery and harness manufactures of WALSALL, BIRMINGHAM, LONDON, &c.

The best **gloves** are made at WORCESTER and WOODSTOCK, but the greater number of gloves sold in this country is made abroad, chiefly in France.

A larger quantity of paper of all kinds is made in British mills than in those of any other country. The largest *paper mills* are in *Kent* (Maidstone, &c.), *Lancashire, Hertfordshire*, and *Midlothian* (Penicuik, &c.) *Rags* are imported from the Continent and the Colonies, and very large supplies of the principal substitute used in paper-making, *i.e., alfa* or esparto grass, are imported from Spain and Algeria.

A great number of other industries, most of them of more or less local, and some even of national importance, afford employment to skilled workers in towns, and industrious peasants in the country districts. Thus clocks and watches are made in the larger towns, such as London, Liverpool, Birmingham, &c., while straw-plaiting is a peasant industry in Bedfordshire and Buckinghamshire, as is also the making of hand-made lace in Devonshire and Ireland. Then there are the enormous drink manufactures—chiefly beer and ale in England, and whisky in Scotland and Ireland. The 27,000 English breweries annually produce the enormous quantity of 11 hundred million gallons of malt liquor, and the capital invested in the English brewing trade alone is equal to the entire capital employed in the three great textile industries of the country—the cotton, woollen, and linen manufactures. The largest breweries in the world are at Burton-on-Trent, and several of the London breweries produce from 10 to 20 million gallons a year each. Dublin possesses the largest whisky distilleries, and there are also numerous distilleries in Scotland.

TRADE AND COMMERCE : In trade and commerce, as well as in mining and manufactures, the United Kingdom surpasses all other countries, its enormous internal trade, merging into a gigantic foreign trade, equalling in value one-fifth of the entire trade of the world.

The various causes which have so powerfully contributed to the development of British trade and industry have been already noticed (*vide ante*, p. 176-8). Commerce, being practically an exchange of commodities, and the basis of British commerce being the import of raw materials and produce, and the export of manufactured goods (or, in other words, the *raw materials* which flow into this country from all parts of the globe, are paid for chiefly with *manufactured goods*), the fact that the United Kingdom is the greatest manufacturing country, as well as the leading commercial power in the world, gives British merchants immense advantages over foreign traders in the markets of the world ; and thus, in spite of the keenest competition on the part of Germany, France, and the United States, Great Britain still holds the first place as regards both the quantity of her manufactures and the value of her trade. There is certainly a relative decline in the great textile industries, but this simply means, not that we have fallen behind, but that other countries have advanced in a relatively greater degree.

INTERNAL TRADE : The internal trade of the United Kingdom is very large, and the transport of goods and produce from place to place is quickly and easily effected by means of splendidly constructed railways, excellent roads, numerous canals, and navigable rivers.

In addition to the facilities for transit by *road, rail, river*, and *canal*, the unparalleled development of the coast-line, richly supplied with safe and commodious harbours, has given rise to a very large coasting trade, and thousands of coasting steamers and sailing vessels carry enormous quantities of goods and produce from port to port at a much cheaper rate than they could be transported by rail. Coasting vessels, therefore, act as auxiliaries and competitors to the railways, and help to keep down the *cost* of carriage to and from the great centres of the foreign trade of the country.

ROADS : Excellent turnpike and good cross roads traverse the country in all directions.

England alone has above 25,000 miles of well-made high roads, and over 100,000 miles of cross roads. Scotland has over 4,000 miles of excellent turnpike roads ; Ireland is also well supplied with good roads.

RAILWAYS have done more than anything else to **develop the internal trade and increase the foreign trade of the United Kingdom.**

The total length of British railways is over 20,000 miles, of which 14,400 miles are in England and Wales, 3,200 miles in Scotland, and 3,000 miles in Ireland.

Over 200 years ago coals were conveyed from the mine to the banks of the Tyne at Newcastle by "laying rails of timber exactly straight and parallel, the bulky carts having four rollers fitting those rails, whereby the carriage was made so easy that one horse would draw four or five chaldrons." In 1776, an iron railroad was made at the Sheffield Colliery. "Railways or tramways of wood, upon which waggons were propelled by animal power, were thus in use as early as the 17th century, but it was not until near the beginning of the present century that iron was substituted for wood. James Watt first conceived the idea of utilizing steam for locomotion. This was probably about 1780. George Stephenson, however, was the first to introduce steam locomotive power into practical use." This was on the Stockton and Darlington line in 1825, but the importance of the new mode of locomotion was not generally recognized until the opening of the Liverpool and Manchester line in 1830. Eighteen years later, 5,000 miles of railway were open for traffic ; during the next 12 years the mileage was doubled, and at present the total length is nearly twice what it was in 1860. The industrial and commercial districts of England and Wales and Southern Scotland are intersected by a close network of railways, and railway lines penetrate even to the remotest corners of the most thinly-populated portions of the country.

There is 1 **mile of railway to every 4 square miles of area in England and Wales. In Scotland there is 1 mile of railway to every 9 square miles of area, and in Ireland 1 to every 11 square miles.**

The ratio for the whole of the United Kingdom is as 1 to 6 (20,000 miles of railway to 120,000 square miles of area).

Upwards **of 864 million passengers, besides enormous quantities of goods and minerals, are carried** every year on British railways, which employ directly about 385,000 men.

The total capital invested in British railways is over 924 millions sterling. The passenger receipts amounted, in 1892, to 35½ millions sterling, and the goods traffic receipts to nearly 43 millions sterling. Of the total receipts—82 millions sterling—England took nearly 70 millions, Scotland 9 millions, and Ireland 3 millions. The average working expenditure in the same was 56%, or a little more than half the gross receipts.

The Rolling Stock of the chief railway companies in the kingdom includes nearly 17,000 locomotive engines, 53,000 carriages for passengers, and con-

F

siderably more than half a million waggons for the conveyance of goods, minerals, and animals. The total *train mileage* is over 310 millions of miles, the working expenses amounting to 43 millions sterling, and the *net receipts* to 35 millions sterling. Nearly all the great English and Scotch railway companies have *running powers* over closely connected lines, so that there are *through trains* between nearly all the great centres of population; and thus one can travel by rail without change of carriage from London to Edinburgh, Perth, or Inverness, &c.

ENGLISH RAILWAYS: Nearly all the main lines radiate from London, which is thus connected by rail with every part of Great Britain.

The Great English Railways radiating from London are the **London and North-Western** (L. & N. W. R.); the **Great Western** (G. W. R.); the **Great Northern** (G. N. R.); the **Midland** (M. R.); the **Great Eastern** (G. E. R.); the **London and South-Western** (L. & S. W. R.); the **South-Eastern** (S. E. R.); the **London, Brighton, and South Coast** (L. B. & S. C. R.); and the **London, Chatham, and Dover** (L. C. & D. R.).

The **North-Eastern Railway** (N. E. R.), the **Lancashire and Yorkshire** (L. & Y. R.), the **Manchester, Sheffield, and Lincolnshire** (M. S. & L. R.), the **Cheshire Lines** (C. L. C.), and the **Cambrian Railway** (C. R.), in Wales, have no lines of their own to London.

The London and North-Western Railway—London terminus, Euston—gives direct communication between London and the West Midlands, the North-West of England, Scotland (West Coast Route), all Wales, and Ireland (by boat).

The London and North-Western main line runs from *Euston Station* through Northampton, Rugby, Stafford, Crewe, Wigan, Preston, Lancaster to *Carlisle*, where it connects with the Caledonian Railway. From the main line, important branches are thrown out in all directions, the principal running (1) from *Rugby* through Coventry, Birmingham, and Wolverhampton to *Stafford*; (2) from *Crewe* to Manchester and *Liverpool*; (3) from *Crewe* through Chester and Bangor to *Holyhead* (for Dublin); (4) *Crewe* to *South Wales*; (5) from *Liverpool* to *Manchester, Huddersfield* and *Leeds*.

The L. & N. W. express from London between Rugby and Crewe attains a speed of 53¼ miles an hour, and the entire distance from London to Carlisle, 299 miles, is accomplished in about 6½ hours.

The Great Western Railway—London terminus, Paddington—radiates from London, and spreads out over the Western and South Western counties of England, and the great iron and coal districts of the Midlands and South Wales.

The Great Western main line runs from *Paddington* through Reading, Didcot, Swindon, Bath, Bristol, Exeter, and Plymouth to *Penzance*, and throws off important branches:—(1) from *Didcot* through Oxford, Warwick, Birmingham, Wolverhampton, Shrewsbury and Chester to *Birkenhead*; (2) from *Swindon* through Gloucester, Cardiff, Swansea, and Carmarthen to *Milford*; (3) from *Bristol*, by the Severn Tunnel, westwards through Cardiff to *Milford*, and northwards through Hereford to *Liverpool*; (4) from *Oxford* through Worcester and Dudley to *Wolverhampton*; (5) from *Chippenham* through Yeovil and Dorchester to *Weymouth*, whence steamers run regularly to the *Channel Islands*.

The "Flying Dutchman" travels between Paddington and Swindon at an average rate of 53¼ miles an hour, and covers the entire distance between London and Plymouth, 245 miles, in 5½ hours.

The Midland Railway—London terminus, St. **Pancras**—connects London with the Midland Counties, West Yorkshire and South Lancashire, its extreme termini being Carlisle, Liverpool, and Bristol.

The Midland main line runs from *St. Pancras* through Bedford, Leicester, Derby, Sheffield, Leeds and Settle to *Carlisle*, where it joins the North British line (to Edinburgh) and the Glasgow and South-Western (to Glasgow). The chief branch line runs from *Derby*, the centre of the Midland System, through Birmingham, Worcester, Cheltenham and Gloucester to *Bristol*, and there is also an important branch from Ambergate Junction through the Peak District to Manchester and Liverpool. The Midland express runs 50 miles—St. Pancras to Bedford—in 52½ minutes, and covers the whole distance from London to Carlisle, 308 miles, in about 7 hours.

The Great Northern Railway—London terminus, King's Cross— runs directly north to York, where it joins the North-Eastern line, which at Berwick connects with the East Coast branch of the North British.

The *Great Northern*, *North-Eastern*, and *North British* are the three links in the East Coast Route between *London* and *Edinburgh*, which is daily traversed by the fastest train in the world, the "Flying Scotsman" covering the entire distance, 395 miles, in 8½ hours, and between Grantham and Retford running at a speed of 55¼ miles an hour, or not far short of a mile a minute.

The North-Eastern Railway joins the Great Northern at York, and the Scotch East Coast line at Berwick.

The main line of the North Eastern runs from *Normanton Junction* through York (where it connects with the Great Northern), Stockton, Durham, Newcastle, and *Berwick*, where it joins the North British line (for Edinburgh, Dundee, and Aberdeen). The longest run, without stoppage, on any British railway, is made by the North-Eastern express between Newcastle and Edinburgh, the distance, 124 miles, being covered in less than three hours.

The Great Eastern Railway—London terminus, **Liverpool Street** —carries enormous quantities of fish and agricultural produce, and has a large through traffic to and from the Continent, *viâ* Harwich.

The Great Eastern has two main lines, one between *London* and *Harwich*, whence there is regular steam communication with the Continent, through Rotterdam ; and the other between *London* and *Yarmouth*, through Cambridge, Ely, and Norwich, or through Ipswich.

The South-Eastern Railway—London terminus, **Charing Cross** —has the largest share of the through traffic with the Continent.

The South Eastern main line runs from *London*, through Tunbridge, Ashford, and Folkestone, to *Dover*—the distance, 76 miles, being covered by express trains in about 2 hours. There are branches from *Tunbridge* to *Hastings*, and from *Ashford* through Canterbury to the sea-side resorts of *Ramsgate* and *Margate*.

The London and South-Western Railway—London terminus, **Waterloo**—gives direct communication between London and the South and West of England, and, by steamer, with the Channel Islands.

The London and South-Western main line runs from *Waterloo*, through Basingstoke, Salisbury, and Exeter to *Devonport*, sending branches to the ports of *Southampton*, *Portsmouth*, and *Weymouth*, whence steamers sail daily to the *Channel Islands*.

The London, Brighton, and South Coast Railway—London terminus, **London Bridge**—has an immense passenger traffic between the metropolis and the pleasure resorts on the south coast, and also a large continental traffic through Newhaven.

The London, Brighton, and South Coast main line runs from *London* through Croydon to *Brighton*—the 5 p.m. train doing the 50 miles in 1 hour and 5 minutes. From Brighton the Company's lines run along the coast—eastwards to Hastings, and westwards to Portsmouth, and thence by steamer to the Isle of Wight.

The London, Chatham, and Dover Railway—London terminus, **Victoria**—has an immense through traffic with the Continent.

The London, Chatham, and Dover main line, *London* to *Dover*, is 78 miles in length, and is traversed by the 3-25 express from Herne Hill to Dover in 1 hour 35 minutes. From *Faversham* a branch line runs to *Ramsgate* and *Margate*.

The Lancashire and Yorkshire, and the Manchester, Sheffield, and Lincolnshire Railways, are the longest of the lines which have no direct communication with London. Connecting, as they do, the greatest centres of the chief textile industries, the passenger and goods traffic is enormous.

The Lancashire and Yorkshire main line runs from *Liverpool* through Wigan, Bolton, Bury, Todmorden, and Wakefield, to *Normanton ;* and also from *Manchester* through Rochdale, Halifax, and Bradford, to *Leeds*. This Company has only 523 miles at work, but the annual revenue exceeds £8,500 per mile (as against about £6,200 per mile on the L. & N. W. R., and £3,600 on the G. W. R.).

The Manchester, Sheffield, and Lincolnshire Railway runs from *Liverpool* through Manchester and Sheffield to the port of *Great Grimsby*, in Lincolnshire, and is thus, like the L. & Y. R., a "cross country" or west-to-east line. The mileage worked is only 344, but the traffic is very large, the total receipts being nearly 2¼ millions sterling a year.

LONDON RAILWAYS : In addition to these great trunk lines, the **Metropolitan**, the **District**, and the **North London Railways** carry an enormous number of passengers to and from the suburbs and the city.

The Metropolitan Railway, 67 miles in work, earns a revenue of over £11,000 per mile ; the District Railway, with 19 miles open, earns nearly £21,000 per mile ; while the North London Railway, only 12 miles in length, earns over £40,000 per mile every year, at an expenditure of considerably less than one-half the amount earned.

SCOTCH RAILWAYS : The Scottish railway system connects with that of England at *Carlisle* and *Berwick*. There are over 3,000 miles of railway in work, earning over 9 millions sterling a year.

At Carlisle, on the **West Coast Route**, the London and North-Western main line joins the Caledonian Railway, and here also the Midland Railway of England connects with the North British system and with the Glasgow and South-Western.

At Berwick, on the **East Coast Route**, the North-Eastern Railway connects with the East Coast branch of the North British.

A line of minor importance runs south from Riccarton Junction, crosses the Cheviots near Peel Fell, and follows the North Tyne valley to Hexham, and thence to Newcastle.

The Caledonian Railway traverses the rich, thickly-populated region between the Forth and the Clyde, and thus has an immense mineral and goods traffic, and a very large passenger traffic.

The main line of the **Caledonian Railway** runs from *Carlisle* to *Edinburgh*, and thence through Stirling and Perth to *Aberdeen*. The C. R. line from Carlisle to Glasgow branches off at Carstairs. Other branches are from *Edinburgh* to *Glasgow ; Perth* to *Dundee;* and *Dunblane*, through Callander (for the Trossachs) to *Oban*, on the west coast. The Caledonian connects at Carlisle with the London and North-Western, and through trains are run from London to Glasgow, Edinburgh, and Aberdeen, &c., by this route. At *Lockerbie*, between Carlisle and Carstairs, a branch goes westwards through Dumfries, Castle Douglas, and Newton Stewart to *Stranraer* and *Portpatrick.*

The Glasgow and South-Western Railway connects at Carlisle with the Midland Railway of England.

The Glasgow and South-Western runs from *Carlisle*, *viâ* Dumfries, Kilmarnock and Paisley to *Glasgow*, with a branch from Glasgow to *Ayr* and *Girvan.*

The North British Railway competes with the Caledonian for the great traffic in the Lowlands, and connects with the English Midland Railway at Carlisle, and at Berwick with the North-Eastern and Great Northern Railways of England.

The North British main lines radiate from *Edinburgh* (1) by the East Coast Route through Dunbar to *Berwick*, with a branch line from *Reston* through Duns and Greenlaw to *St. Boswells*, on the Waverley Route ; (2) by the Waverley Route through Melrose and Hawick to *Carlisle*, with a branch from *Galashiels* through Innerleithen and Peebles to *Edinburgh;* (3) through Falkirk to *Glasgow*, with an alternative route through Airdrie to *Glasgow;* (4) over the *Forth Bridge* and then across the *Tay Bridge* to *Dundee*, and thence to *Aberdeen.*

The Highland, the Great North of Scotland, and the West Highland Railways maintain communication with the north and north-west of Scotland.

The Highland Railway runs from *Perth*, through Dunkeld, Kingussie, Forres, Inverness, and Dingwall to *Wick* and *Thurso;* and branches from *Forres* to *Aberdeen*, and from *Dingwall* to *Strome Ferry* (for Skye). The Great North of Scotland lines run from *Aberdeen* to *Elgin, Fraserburgh, Peterhead*, and *Banff.* The new West Highland Railway runs from *Craigendoran*, on the Clyde, to *Fort William*, at the foot of Ben Nevis.

IRISH RAILWAYS : The Irish railways radiate from Dublin and Belfast. They have a total length of nearly 3,000 miles, and earn over 3 millions sterling a year.

The chief lines in Ireland are (1) the **Great Southern and Western**, which runs from *Dublin*, through Kildare and Maryborough to *Cork* and *Queenstown*, with a branch to *Limerick;* (2) The **Midland Great Western**, which runs from *Dublin*, through Mullingar and Athlone to *Galway;* (3) the **Great Northern**, running from *Dublin* through Drogheda and Dundalk to *Belfast;* (4) the **Belfast and Northern Counties**, from *Belfast*, through Antrim to *Londonderry;* (5) the **Dublin, Wicklow, and Wexford**, from *Dublin*, through Bray, Wicklow, and Arklow to *Wexford;* (6) the **Waterford and Limerick**, joining these ports, with a branch to *Tuam*. The American mails are landed at Queenstown, and sent by the Great Southern and Western express to Kingston, and thence by the London and North-Western Company's steamers to Holyhead.

∴ The foregoing résumé of the railway lines of Great Britain and Ireland is necessarily brief, and the series of names may not be interesting reading, but the student of geography, and especially of commercial geography, should have a clear conception of the geographical position of one town in relation to another, and of the means of transit to and from all parts of the country. The details given of the various lines should not be memorized, but the lines should be traced on the map, and the positions of the termini and chief centres carefully noticed.

INTERNAL WATERWAYS: Before railways were introduced, canals, linking together the chief navigable rivers, formed a most important means of transit, and though now superseded by railways for passenger and the greater part of the goods and mineral traffic, the canals are still largely used for the conveyance of *heavy goods and coals.*

"A country," writes Dr. Yeats, "favoured with internal waterways connecting the leading ports with the manufacturing centres, must be in a better position than one whose means of conveyance are only railways and roads, for every facility for transit is an additional incentive to interchange, and ease of communication is reflected in the imports and exports of a country."

ENGLISH CANALS : There are 3,800 miles of canals in England uniting the numerous navigable rivers, the total length of internal waterways being nearly 5,000 miles, or more than one-third the total length of the railways of the country.

Canals not only connect all the great manufacturing and mining districts of England with each other and with the seaports, and each and all with the great centre—London. It may be said that all efforts made " by the widening or deepening of canals to carry raw materials direct to the manufactories, and manufactured goods direct to the markets," are a direct boon to the home-worker and consumer, and also enable us to compete more successfully with other nations in foreign markets. The most stupendous work of this kind in England is the **Manchester Ship Canal**, which has been constructed, at an enormous cost, from the Cheshire side of the Mersey at Eastham to Manchester. This great work, which has cost over 15 millions sterling, was commenced in November, 1887, and was opened in January, 1894 ; and sea-going vessels up to 7,000 tons burden can now load and unload at Manchester, 37 miles from the sea.

Other important canals are (1) the **Bridgwater Canal**, connecting Manchester and the Mersey, the first large canal made in England, having been completed in 1760 ; (2) The **Grand Trunk Canal**, which joins the Mersey and the Trent ; (3) The **Grand Junction Canal**, which runs from the Thames at Brentford to the Trent ; (4) The **Leeds and Liverpool Canal**, which joins the Yorkshire Ouse and the Mersey ; (5) the **Kennet and Avon**, which connects Bristol with the Thames ; (6) The **Oxford Canal**, joining the Thames and the Trent ; (7) The **Trent and Mersey Canal** ; (8) The **Shropshire Union Canal**, and many other useful waterways.

SCOTCH CANALS are, from the nature of the country, neither numerous nor long, the principal being the Forth and Clyde Canal, the Caledonian Canal, the Crinan Canal, and the Union Canal.

The two principal canals are the **Forth and Clyde Canal**, connecting the Clyde near Renfrew with the Forth near *Grangemouth*, and the **Caledonian Canal**, through Glenmore, the three lochs in which are joined by about twenty-three miles of cuttings, thus affording a passage from the Atlantic to the North Sea without rounding the northern coast of Scotland. The **Crinan Canal**, across the peninsula of Cantire, enables vessels to pass from Loch Fyne to the Atlantic Ocean, without passing round the Mull of Cantire. The **Union Canal** connects Edinburgh and Glasgow, and is now chiefly used for the conveyance of minerals.

IRISH CANALS have a total length of about 300 miles, or rather more than those of Scotland, but scarcely one-tenth those of England.

The **Royal Canal** and the **Grand Canal**, constructed by the Government at a cost of 3 millions sterling, to connect Dublin with the Shannon, the former *via* Mullingar and the latter *via* Tullamore, are splendid examples of canalization. There are also the **Belfast**, **Ulster**, and **Newry Canals**, and other smaller waterways.

POSTS AND TELEGRAPHS : The British postal and telegraphic services are the most complete and efficient in the world. There are also Telephone Exchanges in nearly all the great centres of population.

The **Postal and Telegraphic Services** are Government monopolies, but the **Telephonic Service**, which is rapidly extending, is mainly worked by private companies, as also are the numerous **submarine cables** which connect the United Kingdom with the Continent and America.

There are nearly **20,000** Post Offices, and 25,000 road and pillar letter-boxes in the United Kingdom, and the Postal Service, which has its headquarters at the General Post Office (St Martin's le Grand), in London, employs 72,000 officers on the permanent staff, besides 59,000 persons who do not hold permanent positions.

The enormous number of **1,790¼** millions of letters—an average of 47 letters for every inhabitant—244 million post-cards, 535 million book-packets, and 162 million newspapers, was delivered in the United Kingdom in 1892-3, and during the same period the Parcels Post conveyed not less than 52 million parcels.

The Post Office also does a large banking business. In 1892, the **Post Office Savings Banks**—capital, 75¼ millions sterling—received nearly 25 millions sterling, and paid over 20 millions sterling. In the following year, over

10¼ millions of Money Orders, to the amount of 28½ millions sterling, and 56½ million Postal Orders, to the amount of 21 millions sterling, were issued by the Post Office. The Post Office annually yields a net revenue to the Government of between 2 and 3 millions sterling.

The **Telegraph Department** has over 34,000 miles of line, and 209,000 miles of wire, and in 1893, nearly 70 million messages were sent (59 millions in England and Wales, 7 millions in Scotland, and 3¾ millions in Ireland).

The "**National Telephone Company**," the largest of its kind in the Kingdom, 'give and take' over 2 million messages a week, or more than 100 million messages a year. From any of the numerous Telephone Exchanges or Call-Offices in London, conversation can be carried on with subscribers in Birmingham, Manchester, Liverpool, &c. London and Paris were united by telephone in March, 1891, and there is now telephonic communication through Paris between London and Marseilles and Brussels.

Submarine Cables, across the Channel and the North Sea, keep England in constant 'touch' with the Continent ; and instant communication with the New World is maintained by the cables laid along the bed of the Atlantic, between Valentia and Newfoundland. Other cables and land-lines complete the circuit of the globe, and connect the British possessions with the Mother Country, and every part of the Empire with the rest of the world.

COMMERCE : The commerce of the United Kingdom is by far the most gigantic in the world, the total **Annual Trade** with foreign countries and British Possessions now reaching the enormous amount of **682½ millions sterling**, or one-fifth of the value of the entire commerce of the world.

Until recently, the external trade of Great Britain exceeded that of France and Germany taken together, and was three times greater than that of the United States—the fourth in rank among the great commercial nations of the world. But the trade of each of these three countries, and especially that of Germany, has within the last few years advanced by leaps and bounds ; and although British trade has also advanced, its progress has not been at anything like the same rate as that of our Continental and American rivals. If, therefore, we compare our present trade with that of Germany, France, and the United States, we must admit a relative decline.

According to the latest returns, the total trade or general commerce of Germany amounted to 397 millions sterling, that of France to 320 millions sterling, and that of the United States to 340 millions sterling. Our trade is therefore still larger than that of France and the United States taken together, and nearly equal to that of Germany and France taken together.

Of the **foreign trade** of the United Kingdom **90.5 per cent.** falls to the share of *England*, **8.1 per cent.** to *Scotland*, and **1.4 per cent.** to *Ireland*.

The actual share of each of the three kingdoms in the foreign trade of the country in 1892 was—England, £645¾ millions, or 90½ per cent. ; Scotland, £58 millions, or 8·1 per cent. ; and Ireland, £10¼ millions, or 1½ per cent.

Our foreign trade is carried on chiefly with the *United States, India, France, Australasia, Germany, Holland, Russia, Belgium, British North America, South Africa, Spain, the Argentine Republic,* and *China.* Our annual trade with these countries ranges between 9 millions sterling with China, and 134 millions with the United States.

Trade of the United Kingdom with the chief British Possessions and Foreign Countries in 1892 (in million pounds sterling).

TRADE WITH BRITISH POSSESSIONS.

	Imports.	Exports.	Total.
India	30½	28	58½
Australasia	30½	19¼	49¾
British North America .	14½	7½	22
South Africa . . .	5½	8	13½
Straits Settlements . .	4¾	2	6¾
British West Indies . .	2	2	4
Hong Kong . . .	¾	1¾	2½
Ceylon . . .	4	1	5
All other Possessions . .	5¼	5	10¼
Total . .	97¾	74½	172¼

TRADE WITH FOREIGN COUNTRIES.

	Imports.	Exports.	Total.
United States . . .	108	26½	134½
France . . .	43½	14½	58
Germany	25¾	17½	43¼
Holland . . .	28¾	8¾	37½
Belgium . . .	17	7	24
Russia . . .	15	5¼	20¼
Spain	11	4½	15½
Argentine Republic .	4½	5½	10
Sweden . . .	8¾	2¾	11
Egypt . .	10½	3	13½
Brazil	3½	8	11½
Turkey . . .	5½	6	11½
China . .	3⅝	5¾	9¼
Italy . . .	3¼	5½	8¾
Denmark	8	2½	10½
All other Countries . .	30	29½	59½
Total . .	326	152½	478½
Grand Total . .	423¾	227	650¾

∴ From these numbers, which, however, vary from year to year, sometimes very considerably, we see that about *one-fifth* of our trade is carried on with the United States, and considerably more than *one-fourth* with the British Possessions, while a trade of over 100 millions sterling is done every year with France and Germany.

IMPORTS AND EXPORTS : In 1893, the imports amounted to 405 millions sterling, or over £10 per head of the population, and the exports to 277½ millions sterling, or £7 for every inhabitant; the total trade thus reaching the enormous sum of 682½ millions sterling, or £17 per head of the population.

But vast as the trade is which these values represent, they do not show the whole trade, as the imports and exports of gold and silver bullion and specie, and the value of goods imported for transhipment, are not included. The gold and silver imports amounted in 1893 to £36¼ millions, and the exports to £33 millions, while the goods transhipped were valued at £10 millions. Extensive use is now also made of the Parcels Post—about 1½ million pounds' worth of goods enter and leave the country in this way every year.

"**Balance of Trade** :" British imports invariably exceed the exports by about 100 millions sterling a year.

In 1893, the "Balance of trade" was no less than £128 millions "in favour of the foreigner," that is, apparently so ; but how is it possible for a country like England to continue to buy, year after year, so much more than it sells and yet remain solvent, and in fact actually increase in wealth and comfort? John Stuart Mill remarks that the "vulgar theory deems the advantage of commerce to reside in the exports, as if not what a country gets, but what it parts with by its foreign trade, was supposed to be the gain." And as imports are practically paid for by exports, and *vice versa*, that is, every hundred pounds' worth of raw cotton is really paid for by one hundred pounds' worth of finished cotton goods, and as our imports and exports of bullion and specie on the whole balance each other, the enormous excess of imports over exports must represent goods sent *on credit* in payment of debt. One hundred million pounds' worth of goods would certainly not be sent us on credit, year after year ; they must therefore represent the interest on money lent to foreign countries, or payments for other services rendered. Almost all foreign States have "from time to time borrowed capital from England, to enable them to open up mines, lay railways, reclaim wastes, aid agriculture, and for many other purposes. England has lent them that interchangeable commodity, money, on the condition that they pay interest. When the time for paying the interest comes, *money* may not be procurable, or the creditor prefers payment in 'kind,' and the country liquidates its debt by sending produce. This produce is included under the head of 'imports,' and goes further to swell the difference between exports and imports, as, of course, there have been no goods sent out against these receipts, and this is simply paying an interest on a loan." [1]

British **Imports** consist chiefly of *articles of food* and *raw materials* for our manufactures, while British **exports** are principally *manufactured goods, coal, metals* and *chemicals*.

What the staple articles of British Trade and their actual and relative value are, may be seen from the following summary of the imports and exports of the United Kingdom for the year 1893.

---

1. The various causes of the excess of imports over exports are fully explained in Chap. IV. of | the "Golden Gates of Trade," by Dr. Yeats (London : George Philip & Son)

Summary of Imports and Exports of the United Kingdom for 1893.

IMPORTS.	£	EXPORTS.	£
1. Animals, living (for food)..	6,351,000	1. Animals, living	629,000
2. Articles of food and drink, duty free	144,456,000	2. Articles of food and drink..	10,603,000
„ Articles of food and drink, dutiable	24,987,000	3. Raw materials	17,168,000
„ Tobacco, dutiable	3,556,000	4. Articles manufactured, or partly so :—	
3. Metals	20,609,000	(a) Yarns and textile fabrics	96,608,000
4. Chemicals, dye stuffs, &c..	6,353,000	(b) Metals and metal-wares	30,866,000
5. Oils	7,400,000	(c) Machinery and mill work	13,970,000
6. Raw materials for textile manufactures	67,976,000	(d) Apparel, &c.	9,564,000
7. Raw materials for other in-dustries and manufactures	40,976,000	(e) Chemicals and medicinal goods	8,695,000
8. Manufactured articles ..	65,906,000	(f) Other articles wholly or partly manufactured ..	29,347,000
9. Miscellaneous articles ..	15,834,000	(g) Articles sent by Parcels Post	1,042,000
„ by Parcels Post	619,000		
Total Imports, £405,067,000		Total Exports (British Produce) £218,496,000	

To the exports of British produce must be added the exports of Foreign and Colonial produce, of the value of nearly £60,000,000, and the value of foreign merchandise transhipped at British ports, £10,000,000.

CUSTOMS DUTIES: "The United Kingdom is a free trading country, the only imports on which customs duties are levied being *chicory, cocoa, coffee, dried fruits, plate, spirits, tea, tobacco,* and *wine*—spirits, tobacco, tea, and wine yielding the bulk of the entire levies. In 1892, duty was levied on goods of the value of £29,898,344, out of a total of £423,793,882 imports, or about 7 per cent. of the total imports."—*The Statesman's Year Book,* 1894.

FOOD IMPORTS: The principal articles of food and drink imported into the United Kingdom (with their value, in 1893, in millions of pounds sterling) are—*grain* and *flour* (51¼), raw and refined *sugar* (22), *dead meat* (22¼), *animals* (6¼), *butter* and *margarine* (16¼), *tea* (10¼), *fruits* (6), *wine* (5¼), *cheese* (5), *coffee* (4), and *eggs* (3¾).

The extent of our dependence on foreign and colonial supplies of *food-stuffs* may be gathered from the fact that, on an average, every person in the United Kingdom requires, in addition to the food produced at home, no less than 252 lbs. of wheat and flour, 9 lbs. of rice, 14 lbs. of bacon and hams, 6 lbs. of butter, 6¼ lbs. of cheese, 5½ lbs. of tea, 78 lbs. of sugar, and 35 eggs.

Grain and Flour: In addition to over 50 million bushels of wheat grown in the country, about 124 million bushels of wheat and flour are imported every year from other countries—wheaten bread being the staple food of the British people.

The seven great sources of wheat (with the quantities in millions of quarters, imported in 1893) are the *United States* (129), *Russia* (40¼), *Argentina* (31¼), *India* (24¾), *Canada* (12½) *Australasia* (10¼), and *Chili* (10¼). Of 375 million quarters of flour imported in 1893, nearly 72 million quarters came from the United States, where the *milling industry* is highly developed.

Sugar : About 11 million lbs. of refined sugar, and 16 million lbs. of raw sugar, valued at 22 millions sterling, are annually imported.

Raw sugar is sent chiefly from the *East* and *West Indies, Brazil, British Guiana*, and *India*, but the cultivation of the sugar-cane has been largely checked in recent years, owing to the enormous development of the beet-root sugar industry on the continent. Germany alone now sends us refined sugar to the value of over 9½ millions sterling a year, and France about 1½ million pounds' worth. Belgium sends ¾ million pounds' worth, and Holland about 2 million pounds' worth of beet-root sugar. The system of *bounties*, by which the continental sugar industry has been fostered, has practically ruined the sugar-refining trade in this country.

The Fresh and Tinned Meat trade has largely increased within the last few years, and the meat imports now amount to over 22 millions sterling a year.

Enormous quantities of *fresh and tinned beef* come from the United States : *fresh and tinned mutton* from New Zealand, New South Wales, and Argentina ; *rabbits* are sent from Belgium, and *tinned rabbits* from New Zealand, &c. In fact, almost every description of animal food is poured into this country from all the great meat-producing countries all over the world. From the United States alone we received, in 1892, over 8 million pounds' worth of bacon and hams, besides nearly 4¼ million pounds' worth of fresh beef.

Live Animals are sent in immense numbers from the United States and Canada, and very largely from Holland, Denmark, and Germany.

Over 6¾ million pounds' worth of oxen, sheep, and lambs were imported in 1893. In the previous year, the live cattle sent from the United States were valued at 7½ millions sterling, and in the same year Canada sent 1½ million pounds' worth of oxen. Live animals are almost all landed at London, Liverpool, Newcastle, Hull, and other East Coast ports.

Butter and margarine, of the value of 16½ millions sterling, **cheese** (5 millions sterling), and **eggs** (3¾ millions sterling), were imported in 1893.

Butter and margarine : Over 224 million lbs. of *butter*, and 114 million lbs. of margarine, are imported from the Continent into the English market, in addition to the large quantities of butter sent from Ireland and the Channel Islands.

Cheese : About 225 million lbs. of cheese were sent to this country in 1893, chiefly from the United States, Canada, and Holland.

Eggs : Over eleven hundred million foreign eggs were also consumed, in addition to an enormous home production.

Tea, from *India, China*, and *Ceylon ;* **coffee**, from *Brazil, Central America, India*, and *Ceylon ;* and **cocoa**, from the *West Indies*, are the principal non-alcoholic beverages in the United Kingdom.

Tea, which formerly was almost exclusively supplied by China, is now most largely supplied to the British market by India and Ceylon. Of the 250 million lbs. imported in 1893, no less than 115 million lbs. came from India, 72½ million lbs. from Ceylon, and 50 million lbs. from China.   The British import of tea from China is decreasing at an extraordinary rate ; the quantity received in 1892 being only one-third the quantity imported in 1878.   We also receive about 1½ million lbs. through Holland (from the Dutch East Indies), and about 4¼ million lbs. from other countries.

Besides the 5½ lbs. of tea per head of the population received in 1893, we also imported 4 million pounds' worth of **coffee**, and about a million pounds' worth of **cocoa**, the latter almost exclusively from the island of Trinidad.

Fruits : Over 6 million pounds' worth of fruits of all kinds were imported in 1893 from the Mediterranean countries, the United States, and the West Indies.

Millions of **oranges, lemons, grapes, &c.,** are sent by fast steamers from Spain and Portugal, Italy, and France, while Greece sends us annually about 1½ million pounds' worth of *currants*, and Asia Minor supplies about three-quarters of a million pounds' worth of *raisins* and *figs*.   Large quantities of *tinned fruits* come from the United States, and there is every prospect of the rapid growth of a large fruit trade with our Australasian colonies.

Wine is also a large import—about 15 million gallons, valued at about 5 millions sterling, being received every year.

The wine consumed in this country, or re-exported, is derived chiefly from France (claret, champagne, &c.), Portugal (port), Spain (sherry), Italy (Marsala, &c.), Austria-Hungary (tokay, &c.), Germany (hock), Australia (tintara, &c.), and the Cape Colony (Constantia, &c.)

Raw materials for our manufactures, metals and minerals, oils, chemicals, dye-stuffs, and tanning materials, with the articles of food and drink already mentioned, form the bulk of the British import trade.   The imports of manufactured articles, or finished goods, only amount to about one-seventh of the whole.

The raw materials for our textile industries, which include raw **cotton, wool, flax, hemp,** and **jute,** are imported to the value of between 85 and 90 millions sterling a year.   Raw cotton alone amounts to fully one-half, and wool to about one-third (in value) of the total import of textile materials.

Raw Cotton : Nearly 2,000 million lbs. of raw cotton are now imported every year.   Three-fourths of it are sent from the *United States*, the rest comes from *Egypt, British India, Turkey,* and other countries.

In 1893, the value of the raw cotton imported and retained for home consumption, amounted to 30½ millions sterling.   Between 200 and 270 million lbs. of raw cotton are re-exported to other countries every year.   Nearly all the raw cotton shipped from New Orleans, Bombay, Alexandria, and other cotton ports, is landed at Liverpool, the greatest cotton emporium in the world, for transport by rail to Manchester, and the surrounding cotton-manufacturing towns. Now that the **Manchester Ship Canal** is open, much of the cotton hitherto unloaded at the Liverpool docks, and forwarded by rail, will be taken direct to Manchester, and thence distributed to the factories in the neighbourhood. This great Canal should thus affect very favourably the staple industry of South Lancashire, and also foster and develop new industries in the district through which it passes—it being practically an extension, 36 miles in length, of the Liverpool docks into the heart of the most active and enterprising industrial region in the world.

WOOL : The annual import of wool now amounts to nearly **700** million lbs., and considerably more than half the total quantity comes from **Australia and New Zealand.**

The **total quantity** of wool—*sheep, lamb*, and *alpaca*—imported into the United Kingdom in 1893 was **692** million lbs., one half of which was exported to other countries, and the other half retained for home consumption. Of the total import, over **431** million lbs. came from AUSTRALASIA, which now surpasses all other wool-producing countries in regard to both quantity and quality. Nearly all the Australasian and South African wool is consigned to LONDON, *the greatest wool-market in the world;* but as *wool-markets* have been established at the great wool ports of Australasia—SYDNEY, MELBOURNE, ADELAIDE, BRISBANE, and WELLINGTON—much of the wool is now sent direct to the Continental ports of ANTWERP, HAMBURG, and MARSEILLES, and to America, chiefly to NEW YORK. The value of the wool imported in 1893 was nearly 24½ millions sterling.

HIDES and **SKINS, and LEATHER,** dressed and undressed, are imported to the value of 13¾ millions sterling a year.

The trade in **raw** hides and skins, and in dressed and undressed leather, is enormous, the British leather industry ranking next in importance to the great textile and iron industries. The raw material is obtained principally from *British India*, the *Continent* (France, Belgium, &c.), *South Africa*, the *United States*, and *South America*. The import of leather alone amounted, in 1893, to over 6½ millions sterling.

FLAX, HEMP, and **JUTE** are imported to the value of about 8 millions sterling a year.

Flax is largely grown in Ireland, but more is imported from Russia and other European countries. Hemp is supplied by Russia and Italy, and Jute is obtained from British India.

Among a host of other materials for our manufactures, obtained from other countries, are **metals,** of the annual value of **20** millions sterling, **chemicals, dye-stuffs** and **tanning materials,** and **oils** and **seeds,** to the value of **21** millions a year.

The chief metal imports, with their value in 1893 (in millions sterling), were copper ore (3½), and copper partly wrought (2); iron ore (2½), iron bars and manufactures (3½); lead (1½); silver ore (3); tin (3); zinc and zinc goods (1½). The chemicals, dye-stuffs, and tanning materials imported in 1892 were valued at 6½ millions sterling, the oils at 7½ millions, and the seeds at 7 millions sterling.

MANUFACTURED GOODS are annually imported to the value of over **60** millions sterling, about **one-seventh** of the value of the total imports.

Over 11 million pounds' worth of silk manufactures are imported every year, chiefly from France, the greatest silk-manufacturing country in the world ; but large quantities of silk goods, of German manufacture, also reach this country through Holland and Belgium. Fancy silks are obtained from China and Japan, India, and Persia.

Woollen goods and yarns are also imported to about the value of nearly 10 millions sterling—chiefly from Germany through Belgian and Dutch ports. A great deal of French and Belgian cloth and linen is placed on the British market, and from the same countries we also receive many million pairs of boots, shoes, and gloves every year.

Many of the miscellaneous imports are of considerable importance in the arts and manufactures.

Such are alfa or esparto grass, so extensively used in papermaking, imported from Algeria, Spain, &c. ; petroleum, from the United States and Russia ; ivory, from Africa and India ; ostrich feathers, from South Africa ; &c. &c.

BRITISH EXPORTS consist principally of **manufactured goods** and **minerals**, which amount to nearly *four-fifths* of the whole ; the remaining *one-fifth* being **re-exports of foreign and colonial produce**.

Four-fifths of the exports from the United Kingdom are thus finished products of British manufacture, and minerals obtained from British mines, while one-fifth consists of raw materials and finished products of foreign or colonial origin, brought, along with goods and produce for our own use, into our ports, and thence distributed to other countries. This item in our Trade Returns does not include the value (over 10 millions sterling a year) of the foreign merchandise merely transhipped at British ports.

The total value of **British exports** in 1893 amounted to 277½ **millions** sterling—the exports of *British produce* amounting to 218½ millions, and the re-exports of *foreign and colonial produce* to 59 millions—an export trade amounting, on an average, to £7 6s. per head of the total population.

The principal articles of export (Home Produce) in 1893 (with their value in millions sterling) were :—Cotton goods and yarn (63¾) ; woollen and worsted goods and yarn (21) ; linen and jute manufactures and yarn (8¼) ; apparel and haberdashery (5¾) ; iron and steel (20½) ; hardwares and cutlery (2) ; copper (3) ; machinery (14) ; coals, cinders, fuel, &c. (14½) ; chemicals (8½). From the summary of British exports already given, it will be seen that we exported articles of food and drink (Home Produce) to the amount of 10½ millions sterling, and that British animals, chiefly horses, to the value of £630,000, were sent abroad in the same year.

The chief markets for **British manufactures** and other products are *India, Australasia, Canada*, and *South Africa*, within the empire ; and the *United States, Germany, France*, the *Argentine Republic, Holland, Italy, Belgium, Brazil, Turkey*, and *China*, among foreign countries.

About one-third of the export trade of Great Britain and Ireland is done with India and the British Colonies and Dependencies, and *two-thirds* with foreign countries, of which the United States is by far our best customer. India and the United States each take over 30 million pounds' worth of British goods every year, and Australasia and Canada together buy an equal amount. Germany takes 19 millions' worth, and France about 15 millions' worth of British goods annually.

Cotton Manufactures : 66 million pounds' worth of *cotton goods* and *yarn* were exported from the United Kingdom in 1893.

The cotton goods manufactured in Lancashire and Lanarkshire in such vast quantities are sent all over the world ; fully four-fifths of the piece-goods, and about one-third of the yarn exported from the United Kingdom, pass through LIVERPOOL, which is thus not only the chief port of entry for the raw material, but also the principal outlet for the finished product.

India takes about 15½ million pounds' worth of British cotton manufactures every year ; **China** (including Hong Kong and Macao), 4 ; **Turkey** (in Europe and Asia), 4½ ; **Germany,** 2½ ; **Holland,** 2¼ ; **Brazil,** 3¼ ; the **United States,** 2½ ; and **Belgium,** 1.

Woollen Manufactures : The annual export of *woollen goods and yarn* amounts to about 21 millions sterling—about two-thirds of the value of the cotton manufactures sent abroad every year.

The value of the woollen goods exported in 1893, principally through Liverpool, London, and Hull, was 16½ millions sterling, and of the woollen and worsted yarn, 4½ millions sterling. The countries to which they are chiefly sent are the **United States, France, Germany, Holland,** and **Belgium.**

Linen and Jute Manufactures and yarn, to the value of 8 millions sterling, are sent abroad every year, principally to the United States.

In 1892, the United States imported over 2¾ million pounds' worth of British linen goods—the rest being purchased chiefly by France, Germany, Australia, Canada, Japan, and Belgium. More than half of the linen goods exported passes through Liverpool.

Apparel and **Haberdashery** are exported, chiefly through London, to the value of 5¾ millions sterling a year.

The British Colonies and Dependencies, especially Australia and South Africa, are the best customers for all kinds of clothing and haberdashery ; among foreign countries, the United States and France buy more British-made clothing than any other.

Metals : The exports of **metals** and **metal goods** rank next in value to those of cotton goods and yarn, more than 60 million pounds' worth of *iron* and *steel goods, hardwares* and *cutlery, copper* and *machinery,* being sold every year in the chief foreign and colonial markets.

The iron and steel **exports,** which amounted to 20½ millions sterling in 1893, included no less than 5 million pounds' worth of *tin plates.*

In the same year the exports of *hardwares* and *cutlery* amounted to 2 millions sterling, and of *copper* to 3 millions, and machinery to no less than 14 millions sterling. The heavy exports of British machinery and steam engines to India, France, Germany, the United States, and other countries, are really the chief cause of the enormous advance made in these countries within recent years in the manufacture of textile fabrics and other goods previously imported from this country.

Coal : 14½ million pounds' worth of coal was exported in 1893, principally through *Cardiff*—the world's greatest coal port—and the *Tyne Ports*.

In three years, 1890 to 1893, the coal export decreased from 19 millions to 14½ millions sterling. France takes about 5 million tons of British coal every year ; Italy and Germany, each about 3¾ million tons ; and Russia, Sweden, Denmark, Spain, and Egypt, about a million and a half tons each. Vessels to the Baltic ports from Newcastle, and to the Spanish ports from Cardiff, take out coal and bring back timber from Russia and Scandinavia, and iron ore from Spain.

Chemicals, dye-stuffs, and tanning materials are annually exported to the value of between 8 and 9 millions sterling.

British chemicals are sent principally to the United States and the Continent.

Other important items in the **Export Returns** (with their annual value in millions sterling) are :—leather and leather goods (3¼), silk manufactures (2½), books, paper and stationery (3½), india-rubber goods (1¼), earthenware (2), fish (1½), tea (1½), coffee (2¼), beer and ale (½).

Leather and leather goods are sent chiefly to France, the United States, Germany, Holland, and Belgium. Silk manufactures, silk waste and raw silk, are sent to the United States, France, Australasia and India. Books, paper and stationery are sent principally to the United States and the British Colonies in Australasia, &c. India-rubber goods and the raw material are exported chiefly to the Continent and the United States. British-made Earthenware goes chiefly to the United States. Large quantities of Fish are sent to the Catholic countries of Europe. Tea, Coffee and Cocoa are re-exported in large quantities to Germany, Holland, France, &c., and British Beer and Ale are sent all over the world, but most largely to our own Colonies and Dependencies.

The **British Mercantile Marine** exceeds the merchant navies of all other nations taken together, and most of the sea-borne goods and produce, not only of our own country, but also of foreign countries, is carried in British ships, which are thus the "ocean carriers" of the world.

Britain possesses more than half the merchant service of the world. British "liners" constantly traverse the great Ocean Highways almost with the regularity of express trains, while British cargo steamers and sailing vessels load and unload in almost every port in the world. "Wherever a cargo is expected, you will generally find a British vessel lying to bring it ; wherever a cargo is waiting, there is a British ship ready to load it." Protected by the *most powerful navy* the world has ever seen, our immense fleet of merchant vessels, largely owned by wealthy and enterprising commercial companies, has

practically no rival to contest with it the supremacy of the sea.   British merchant vessels are just so many links in the golden chain of trade which binds
the world together ; they are, as Mr. Gladstone finely says, "like the shuttle on
the loom, weaving the web of concord between the nations of the earth."

The **Mercantile Navy** of Britain consists of nearly **22,000** steam
and sailing vessels, of over $8\frac{1}{2}$ million tons (net tonnage), manned
by about **250,000** men.

In 1893, 13,329 sailing vessels, of 3,038,260 tons, and 8,088 steamers, of
5,740,243 tons, were *registered* as belonging to the United Kingdom.   Adding
those belonging to the colonies, the total number of vessels belonging to the
British Empire in the same year was 36,000, of over 12,000,000 tons.

A recent Return of the *shipping owned in each country* of the world (excluding
vessels of less than 100 tons) shows that 11,906 steamers and sailing vessels,
of 12 million tons, were British ; 3,357, of 2,000,000 tons, American ; 1,345,
of 1,082,674 tons, French; 1,864, of 1,678,446 tons, German ; 3,394, of 1,665,477
tons, Norwegian ; out of a total for all countries of 32,326 vessels, of 22,939,958
tons.

The **total number of vessels** in the home and foreign trade that
entered the ports of the United Kingdom in 1892 was **376,000**, of
87 million tons; while **343,000** vessels, of $81\frac{3}{4}$ million tons, *cleared*
in the same year.   The **total tonnage**, in cargo and ballast, which
entered and left British ports in one year, thus reached the enormous amount of $168\frac{3}{4}$ **million tons**, an average 'movement' of
nearly $4\frac{1}{2}$ tons per head of the entire population.

Coasting Trade : The total number of vessels that entered and
cleared *coastwise* in 1892 was **596,000**, carrying **93** million tons.

Of these, 314,000 vessels, of 49½ million tons, *entered*, and 281,000 vessels,
of 43½ million tons, *cleared* in the coasting or home trade.

Foreign Trade : The total number of vessels that entered and
cleared in the *foreign trade* in 1892 was **124,000**, carrying about 76
million tons.

Of the 76 million tons entered and cleared in the foreign trade
at British ports, $54\frac{1}{4}$ million tons were in *British*, and $21\frac{1}{2}$ million
tons in *foreign* vessels.

Of the foreign tonnage, **Norway** stands first, with 5¼ million tons; **Germany**,
second, with 4 million tons ; then **Holland, Denmark, France,** and **Sweden,**
each with nearly 2 million tons ; all other countries, except Spain (1¼ million
tons), less than a million each ; the United States being the *eleventh* on the
list, with only 222,000 tons entered and cleared, or less than *one-fourth* that of
Belgium, which shows that the enormous import and export trade between
the United Kingdom and the United States is carried almost entirely in
British vessels.

More than half the foreign trade of the United Kingdom is carried on through the four great ports of London, Liverpool, Cardiff, and Newcastle.

The next largest ports *in order of tonnage* are Hull, Glasgow, Newport, North and South Shields, Sunderland, Southampton, Middlesborough, Swansea, Dover, Leith, Grimsby, and Harwich, each of which have a 'movement' of over a million tons a year. Less than a million tons entered and cleared at Hartlepool, Bristol, Dublin, and Belfast.

Total Tonnage (in millions of tons,[1] excluding coastwise) entered and cleared at British Ports in 1892:—

London	14	Newport	1¾	Middlesborough	1¼
Liverpool	11	Southampton	1¾	Bristol	1
Cardiff	9¾	Grimsby	1½	Hartlepool	¾
Newcastle	4¼	Leith	1½	Belfast	½
Hull	3¾	Sunderland	1½	Greenock	½
N. & S. Shields	3¼	Swansea	1½	Dundee	¼
Glasgow	3	Grangemouth	1¼		

"**The Customs Returns** show that there is an ever-increasing tendency of concentration of trade within a few great centres of commerce."[2]

As customs duties are only levied on a few articles, such as *tea, tobacco, wine* and *spirits*, &c., the receipts also show the extent of the import trade in these commodities at the various ports.

One-half the customs receipts of the United Kingdom is collected at the port of London, *one-fourth* at Liverpool and other English ports, *one-twelfth* at Scotch ports, and *one-tenth* at Irish ports.

The actual amounts in 1890 were—**London** 10 millions sterling, **Liverpool** 2½ millions, **other English ports** 2¼ millions, **Scotland** 1¾ millions, and **Ireland** 2 millions. Only one port in England, besides London and Liverpool, Bristol (the eighteenth on the list of British ports in order of foreign trade tonnage), has customs receipts of over half a million a year.

Value (in millions sterling) of the Foreign Trade (imports and exports of British, foreign and colonial produce) at the chief British seaports in 1889:—

London	232½	Newcastle	12¼	Belfast	3
Liverpool	226¼	Goole	10¾	Hartlepool	3
Hull	48½	Bristol	9¼	Newport	2¾
Glasgow	28	Dover	9	Dublin	2½
Harwich	19¾	Cardiff	9	Plymouth	1½
Southampton	18½	Swansea	6¼	Sunderland	1½
Folkestone	14½	Dundee	5½	Cork	1
Newhaven	14¼	Middlesborough	4½	Aberdeen	1
Leith	13¾	Greenock	4	Portsmouth	½
Grimsby	13¼	Grangemouth	3¼		

1. Nearest approximate round number. | 2. *Statesman's Year Book*, 1891, p. 84.

London and Liverpool are thus the chief inlets and outlets of the gigantic foreign trade of the United Kingdom.

The value, in 1890, of the total trade of Liverpool was 227 millions sterling, while that of London amounted to 232½ millions ; but the value of the *imports* of London exceeded those of Liverpool by 33½ millions, while the value of the *exports* of Liverpool exceeded those of London by nearly 28 millions.

The imports of London amounted, in 1890, to 144¾ millions, and those of Liverpool to nearly 111¼ millions, or 33½ millions less than the imports of London.

The exports of London in the same year were valued at 87¾ millions, and those of Liverpool at 115¾, or about 28 millions more than the exports of London.

If we compare the number of vessels and tonnage registered at these two great ports, we find that, in 1889 (Dec. 31st), London had 2,577 sailing and steam vessels of 1,327,726 tons, as against 2,313 vessels of 1,881,862 tons belonging to Liverpool. That is to say, at that date Liverpool had 264 fewer vessels, but half a million more tonnage than London ; but further, while Liverpool has 26 steamers of 3,000 tons and upwards, London has only 4 such vessels.

In respect of **tonnage** registered as belonging to the port, Liverpool therefore stands first, not only among British ports, but absolutely the first in the world.

These two great ports may also be compared with regard to the nature and origin of their exports. Thus, Liverpool exports twice as much (in value) *British produce* as London ; but, on the other hand, London re-exports nearly three times (in value) as much *foreign and colonial produce* as Liverpool.

Further, the customs duties collected at the Port of London being about 10 millions sterling a year, while those received at Liverpool amount to 2½ millions, it follows that London must import dutiable articles—*tea, coffee, cocoa, wine, tobacco,* &c.—to four times the value of those landed at Liverpool.

The difference also extends to other than dutiable articles ; thus, London imports seven times more wool than Liverpool ; but, on the other hand, while enormous supplies of raw cotton pour into Liverpool, it hardly enters into London imports.

LONDON, with its vast population of 5½ millions, is the greatest commercial and business city in the world, and its markets and exchanges regulate the production and control the price of almost every commodity. The chief centre of credit as well as commerce, the metropolis is virtually "the banking-house of the commercial world." As a money market, London is absolutely unrivalled and always has at immediate command more capital than could be furnished by all the other money markets of the world.

The **Port of London** extends from *London Bridge* to *Tilbury,* on the north side, and to *Gravesend,* on the south side of the Thames, a distance of 30 miles, and the gigantic docks which line both banks of the Thames, between *London Bridge* and *Blackwall,* are crowded with steamships and sailing vessels from all parts of the world. Along the quays are miles of sheds and warehouses, filled with the products of every clime, and thence distributed *coastwise*, or *by rail* and *canal*, to every part of the British Isles, or *re-shipped to other countries* along with the 50 million pounds' worth of finished goods annually obtained from her own workshops and factories—for London is a great manufacturing city, as well as the chief commercial centre of the kingdom—or sent to the Thames by rail, canal, and coasting vessels for shipment abroad.

The **Coasting Trade of London** is larger than that of any other port in the world. Regular lines of steamers run between London and all the principal, and several of the smaller, "home" ports. A vast number of sailing vessels as well as steamers are constantly employed as ' feeders ' and ' distributors ' of the immense trade with foreign countries and British possessions which centres in the metropolis. (The foreign trade of Queenborough, in the Isle of Sheppey, is also included in the returns for the port of London).

The **Foreign Trade of London** is more extensive than that of any other port in the kingdom, and, indeed, in the world. Liverpool and New York are the only ports which approach London in the magnitude and value of their foreign trade. The export trade of Liverpool, however, is much larger (£28 millions more in 1890) than the export trade of London, but the foreign and colonial produce and goods which are brought into the Thames are £33½ millions greater in value than those which enter the Mersey.

Of the numerous STEAMSHIP COMPANIES which carry on the greater part of the regular Foreign Trade of London, the most important are the *Peninsular and Oriental Steam Navigation Co.* (P. & O.), the *Orient*, and other Lines to the **Mediterranean Ports, India, China, Japan, Australia,** and **New Zealand,** *via* the Suez Canal ; the *British India Steam Navigation Co.* to **India** and **East Africa** (Zanzibar) ; the *Castle Line*, *via* Madeira and St. Helena, to **South Africa** (Cape Town, Mossel Bay, Port Elizabeth, East London and Port Natal), and **East Africa** (Delagoa Bay, Mozambique, and Zanzibar, calling also at Mauritius and Madagascar).

Some of the *Pacific Steam Navigation Co.'s* boats and other vessels sail from London to Brazil, the Argentine Republic, and the West Coast of South America ; and vessels belonging to other Liverpool companies trade between London and the United States and Canada ; while *Shaw, Savile and Co.*, and the *New Zealand Shipping Co.* do a large direct trade between London and New Zealand.

The *General Steam Navigation Co.'s* steamers sail regularly to the North Sea ports (Ostend, Antwerp, Rotterdam, Amsterdam, and Hamburg), and also to Boulogne and Bordeaux, Oporto, and the chief Spanish and Mediterranean Ports.

LIVERPOOL rivals London in the magnitude and value of its foreign commerce. It is the greatest cotton port in the world, the headquarters of the great "Atlantic liners," and the principal centre of trade with America and the West Coast of Africa.

The **Coasting Trade of Liverpool** is very large. Passenger and cargo steamers run regularly between Liverpool and *Glasgow, Belfast, Dublin, Waterford, Cork, Bristol, Plymouth, Southampton,* and *London ;* while smaller steamers and sailing vessels bring goods or produce from almost every port in the British Isles. Boats run daily to Dublin and Belfast, and on alternate days to Cork and Waterford ; they bring back large supplies of *provisions* and *dairy produce.*

The **Foreign Trade of Liverpool** has so increased with the development of the cotton and woollen manufactures of *Lancashire* and *Yorkshire*, of which the Mersey is the natural outlet, that it bids fair to eclipse in absolute amount and actual value that of London, at present the greatest seaport in the world. The *difference* (in value) in the total trade of these two great ports was, as we have already stated, only 5½ millions sterling in favour of London, but while the import trade of Liverpool was 33½ millions less, its export trade was 28 millions more than that of London. As a centre of distribution of *Australian, East Indian*, and *Continental* commodities, London stands first; but as an outlet to British manufactures and 'home produce' generally, Liverpool is far more important than any other port in the kingdom.

The **mail and passenger traffic** with the **United States** is carried on by the magnificent fleets of the *Cunard*, the *Inman and International*, the *White Star*, the *Anchor*, the *National*, and the *American Lines*. Mails and passengers for **Canada** and the **East**, by the Canadian-Pacific route, are conveyed by the vessels of the *Allan* and the *Dominion Lines*, and for **South America** by the *Pacific Steam Navigation Co.*, the *Holt Line*, &c. The *African Steamship Co.*, and the *British and African Steam Navigation Co.*, despatch mail steamers regularly to the **West African** ports, and several Lines—the *Moss, Papayanni*, and *Cunard*—trade with the **Mediterranean** ports. The *West India and Pacific Steamship Co.'s* mail steamers do most of the regular trade with the **West Indies**; and the *Hall*, the *Clan*, the *Anchor* and other lines, do a large direct trade—*via* the Suez Canal—with the **East Indies** and **China**. Regular lines of steamers also run between Liverpool and **New Orleans** and other cotton ports in the Southern States.

The Liverpool steamers take out, besides passengers and mails, enormous quantities of *British manufactures*, especially cotton and woollen goods, and machinery and metal wares, and bring back *raw cotton, grain, meat, provisions*, &c. from the United States and Canada; *sugar, coffee, cocoa, fruits, mahogany*, &c., from the West Indies; *wool, hides* and *skins, sugar* and *coffee, fresh* and *preserved meat, guano* and *nitrate*, from South America; *palm oil, ivory, oil nuts*, &c. from West Africa; and *fruits* and *wine, grain, cotton*, &c. from the Mediterranean ports.

In the "**Liverpool Channel**," as the estuary of the Mersey is called, all the fleets of the world could ride safely at anchor, but the requirements of trade necessitated the construction of a series of magnificent docks, which extend for 6 miles along the Liverpool side of the river. The port also includes the Birkenhead docks on the Cheshire side. The Liverpool docks have a water area of 370 acres, and those of Birkenhead 170 acres; the quay walls on both sides of the river are about 40 miles in length. The "bar" at the mouth of the river is a serious obstacle, as it prevents the ingress and egress of the larger vessels except at high water; but it is now being dredged, and the channel may ultimately be sufficiently deepened to admit the largest vessels at all hours. The **Manchester Ship Canal** enters the Mersey at Eastham, on the Cheshire side of the river.

Hull and Glasgow rank next to London and Liverpool as regards the value of their *import* and *export* trade, and both have also a very large *coasting* trade.

HULL is the *fifth* in the list of British ports in order of tonnage entered and cleared, but the *third* as regards the value of its foreign trade, which is carried on (largely by the *Wilson Line* of steamers) with the **Norwegian** and **Baltic** ports, the great German ports of **Hamburg** and **Bremen**, and also with **Rotterdam, Amsterdam, Antwerp, Bordeaux, Lisbon,** and the **Mediterranean, East Indian** and other ports. The imports and exports of Hull each amount in value to about 20 millions sterling a year. Its chief rival in the trade with the Baltic and Northern Europe, and in the North Sea fishing trade, is **Grimsby,** which, being the East Coast terminus of an important "cross country" railway—the *Manchester, Sheffield and Lincolnshire Railway*—has exceptional facilities for the conveyance and distribution of fish and other products landed at the mouth of the Humber, instead of being taken up the river to Hull.

GLASGOW, the second city in the kingdom in point of population, is the *fourth* among British seaports in order of the value of its trade. Being itself a great manufacturing town, and the chief centre of the rich coal and iron district of Southern Scotland, surrounded by a cluster of towns and villages with numerous and varied industries, and situated on a tidal river deepened sufficiently to admit the largest vessels, Glasgow possesses unequalled advantages for carrying on an extensive foreign commerce, and a very large home trade. The chief commercial, as well as the greatest manufacturing city in Scotland, Glasgow communicates by rail with all parts of Great Britain, and by sea with Liverpool, Belfast, Dublin, and other Irish and English ports. During the season, pleasure steamers run between the Clyde and the Highlands and Western Islands.

The foreign trade of Glasgow is large enough to support several regular lines of steamers, such as the *Allan* and *Donaldson Lines* to **Canada**; the *Anchor* and *State Lines* to the **United States**; the *Anchor, MacIver,* and *Clan Lines* to **India**; the *British Line* to **Cuba**; the *Gulf Line* to **South America**; the *Clan Line* to the **West Coast of Africa**, &c. Glasgow and the minor Clyde ports— Govan, Renfrew, Dumbarton, Port Glasgow, and Greenock—have also an enormous shipbuilding trade, and in some years the Clyde yards have turned out more shipping than the rest of the world together.

Cardiff and Newport, the Tyne Ports and Sunderland, are the greatest *coal ports* in the world.

The exports from both Cardiff and the Tyne are about five or six times the amount (tonnage) of the imports; that is, vessels unload foreign and colonial produce at other ports, and proceed to Cardiff or the Tyne in ballast for cargoes of coal, or coal and metal goods.

CARDIFF is now the first coal-exporting port in the world. In 1893 it shipped over 10 million tons to foreign and colonial ports, or 6½ million tons more than the total quantity shipped to foreign ports at Newcastle.

Cardiff is the outlet for the *metal* as well as *mineral* products of the Taff valley, as **Swansea** is for the Tawe basin, and **Newport** for the Usk valley. Swansea sent out 1¼ million tons of coal to home and foreign ports in 1892, and Newport nearly 2 million tons, or, together, considerably less than one-fifth the coal export of Cardiff; but Newport now exports more *iron* and *tin-plates* than Cardiff, and Swansea also sends out much larger quantities of *copper* and *iron*.

NEWCASTLE ranks next to Cardiff as a coal port—total export to foreign countries, in 1892, 4 million tons—and is also the chief outlet for the *engineer-*

ing and *chemical* industries of the district. **Gateshead**, on the opposite side of the Tyne, is practically a part of Newcastle; and lower down the river, on the Durham side, stands **Jarrow**, with its immense *chemical works* and *shipbuilding yards;* and at the mouth of the river are the coal ports of **North and South Shields**; while on the coast, 8 miles to the south, is the great shipbuilding and coal port of **Sunderland**.

Both Sunderland and the Tyne Ports trade largely with the Baltic and Norwegian ports, and also with the German, Dutch, and Belgian ports, on the opposite side of the North Sea. The vessels take out *coal* and *British manufactures,* and bring back *corn, cattle, butter, eggs, iron-ore,* &c. They have also a very large trade in coal, coastwise, with London.

The **Channel Ports** of England are of the first *naval,* and of very considerable *mercantile,* importance. **Portsmouth** and **Devonport** are the chief *naval arsenals* of the kingdom; **Southampton** and **Plymouth** are important *mail-packet stations*; while **Dover**, **Folkestone**, and **Newhaven** are practically *outports of London,* most of the enormous passenger and goods traffic between London and the Continent passing through them.

DOVER is a strongly fortified harbour of refuge on the narrowest part of the Channel which divides England from the Continent, for which it is also the chief *mail-packet station.* Swift steamers run daily between Dover and **Calais**, on the opposite coast of France, and **Ostend**, on the Belgian coast. Most of the through **traffic between London and the Continent** passes through this port, but **FOLKESTONE**, which maintains constant communication with **Boulogne**, and **NEWHAVEN** with **Dieppe**, have become formidable competitors for both passenger and goods traffic to and from the Continent. Recent returns, in fact, show that in respect to the value of their imports and exports, Newhaven is the *eighth*, Folkestone the *tenth*, and Dover the *twelfth*, on the list of British ports. Newhaven and Folkestone together have nearly as large a trade (in value) as Glasgow, while **HARWICH**, another great outport of London on the Essex coast, has an import and export trade of much greater value than any of the Kentish outports, amounting in 1889 to 3 millions sterling more than that of Southampton, and 5 millions more than that of the Tyne Ports.

PORTSMOUTH is the chief naval arsenal of the United Kingdom, and its trade as a seaport is practically limited to the requirements of the *arsenal* and *shipbuilding* yards at **Portsea**, and the *victualling yard* at **Gosport**. The harbour, in which the entire British Navy could shelter and refit, is strongly fortified. Almost exactly opposite Portsmouth, on the other side of the English Channel, is **Cherbourg**, one of the strongest fortresses in the world, and the chief naval port of France. But the trade of Portsmouth, as a seaport, is insignificant compared with that of the great mercantile port at the head of the adjoining inlet of Southampton Water.

SOUTHAMPTON ranks *sixth* in order of the value of its trade among British ports, but *thirteenth* in order of tonnage. This port is the chief mail-packet station on the south coast of England, and regular lines of steamers sail hence to all parts of the world. The *Union Steamship Co.'s* boats run regularly to all the South African ports, *via* Lisbon or Madeira; the *North German Lloyd* steamers from Bremen to America, the East Indies, and Australia, call here; while the Atlantic steamers of the *Royal Mail Steam*

Packet Co. carry mails, passengers, and cargo to the **West Indies and Central America** (**Colon** for **Panama**), and their Pacific boats run north along the coast from Panama to **San Francisco** and **Victoria** or **Vancouver**, and south to **Guayaquil**, **Callao** (the port of Lima), and other ports.

PLYMOUTH, although it has scarcely a tenth part (in value) of the trade of Southampton, is yet an important mercantile and naval station. Its triple harbour is protected by a magnificent artificial breakwater, 5,000 feet long, and defended by strong fortifications. **Devonport** contains the *naval dockyard :* the *victualling yard* is at **Stonehouse**.

BRISTOL, formerly second only to London, has still a very large import trade, but its export trade is relatively small, there being no great manufacturing or mining towns near it, as is the case at Liverpool, Hull, Glasgow, and Cardiff. Since the construction of the docks at **Avonmouth**, on the north side of the mouth of the river, and at **Portishead**, on the south side, both communicating by rail with the city itself 8 miles up the river, the trade of Bristol with the **United States** (the *Great Western Line* of steamers runs regularly to New York) has greatly increased. It also trades extensively with the North Sea ports (Amsterdam and Rotterdam), with Cardiff and other Welsh ports, Liverpool and London, Dublin, Belfast, Waterford, and Cork. The Irish trade of Bristol, especially in provisions, dairy produce, &c., is very large.

The great **Fishing Ports** of England are on the East Coast— **Yarmouth** being the chief seat of the English *herring fishery.* **Grimsby, Hull, Lowestoft, Whitby, Scarborough,** and **Bridlington** are all actively engaged in the extensive fisheries carried on in the North Sea.

Most of the fish caught is landed at **London, Grimsby, Hull, Lowestoft,** and **Yarmouth**—the great London fish-market (Billingsgate) being largely supplied by swift steamers which run regularly between the fishing fleet and the Thames ; but much larger quantities of all kinds of fish are despatched by rail from Yarmouth, Lowestoft, Grimsby, Hull, &c., both to London and the inland towns and districts. The *herring fisheries* of Scotland, which centre at **Wick** in Caithness, **Peterhead** and **Fraserburgh** in Aberdeenshire, and **Lerwick** in Shetland, exceed in value those of England.

SCOTCH PORTS : Besides Glasgow and the Clyde ports, **Leith, Dundee, Grangemouth,** and **Aberdeen** carry on an extensive *coasting trade* and a very large *foreign trade* with the Baltic and Western Europe, and with the British Colonies and India.

From **LEITH**, the port of Edinburgh, regular lines of steamers run to **Copenhagen, Hamburg, Bremerhaven** (for Bremen), **Amsterdam, Rotterdam,** and **Antwerp**; and from the adjoining port of **GRANTON**, steamers sail regularly during the summer to **Norway** and **Iceland**. Both ports do a large coasting trade with **London** and other ports.

GRANGEMOUTH and **Bo'ness**, both situated on the south side of the Firth of Forth, the former at the mouth of the Forth and Clyde Canal, and the latter (five miles to the east of Grangemouth) being a terminal port of the North British Railway, are the eastern outlets of the products of the densely-peopled manufacturing region of the Forth and Clyde, which they export to the **Baltic ports and the British Possessions.**

DUNDEE and **ABERDEEN** are largely occupied, the latter almost entirely, in the coasting trade. The foreign trade of Dundee is about 5 millions sterling a year, that of Aberdeen scarcely amounts to 1 million. Dundee imports enormous quantities of *jute* from **British India** for its great jute works, the largest in the kingdom, employing altogether over 40,000 operatives.

IRISH PORTS: The direct foreign trade of **Irish ports** is small; no less than 98 per cent. of the total trade of Ireland is done with Great Britain.

Dublin, Belfast, Cork, Waterford, Limerick, and **Londonderry** do nearly all the trade with Great Britain—the goods traffic is mainly through **Liverpool, Bristol,** and **Glasgow,** and the passenger traffic through **New Milford, Holyhead, Fleetwood,** and **Stranraer.**

∵ There are numerous other ports in the British Isles, with comparatively small tonnage, which are yet important as the outlets or feeders of the staple industries of many districts. Thus the port of **Beaumaris** (which in the Customs Returns includes *Bangor*, the port of the Penrhyn slate quarries, *Port Dinorwic*, the port of the Llanberis quarries, and Carnarvon, that of Nantlle, &c.) probably exports more *slates* than any other port in the world. The port of **Poole,** again, on the south coast of England, sends very large quantities of the best *clay* from the adjacent Purbeck Hills to the Mersey, and thence forwarded to Staffordshire and other places for the manufacture of superior *china.* Several Scotch ports, besides Aberdeen, have very large exports of *granite;* Portland, in the south of England, supplies *building stone.*

CONSTITUTION and **GOVERNMENT:** The United Kingdom of Great Britain and Ireland constitutes, in form of government, a **Hereditary and Limited Monarchy.** The *executive power* is nominally in the hands of the Sovereign; the *legislative power* is divided between the Sovereign, the House of Peers, and the House of Commons—the last being a Representative Assembly, elected by qualified classes of the people at large. The **House of Commons** alone has the right to regulate the taxes and expenditure of the kingdom, and the Ministers of the Crown are responsible to it for their public proceedings. The people of the British Islands thus enjoy the blessings of a **free Constitution.** The expression of opinion is free to all classes.

"The present form of Parliament, as divided into two Houses of Legislature, the Lords and the Commons, dates from the middle of the fifteenth century."

The **House of Lords** consists of all the *Peers of England,* 16 Scottish Representative Peers, elected for each Parliament, and 28 Irish Representative Peers, elected for life, besides 24 Bishops and 2 Archbishops, numbering **537** in all, in 1890.

The **House of Commons** consists of **670** members—**465** for *England,* **30** for *Wales,* **72** for *Scotland* and **103** for *Ireland.* Members of Parliament are elected for each Parliament by the electors; in 1890, there were 6 million registered electors, or one to about every six of the population. The House of Commons *must* be dissolved every seven years, and a new one elected, and it *may* be dissolved by the Sovereign at any time; but this right is now seldom exercised by the Sovereign, except with the advice and consent of the Cabinet.

The **Executive Power**, nominally in the hands of the Sovereign, is practically vested in the **Cabinet**, which at present consists of the following **Ministers** :— The Prime Minister, who is also the First Lord of the Treasury and Lord President of the Council, the Secretary of State for Foreign Affairs, the Lord High Chancellor, the Chancellor of the Exchequer, the Secretary of State for the Home Department, the Secretary of State for Scotland, the Secretary of State for War, the Secretary of State for the Colonies, the Secretary of State for India, the First Lord of the Admiralty, the Lord Chancellor of Ireland, the Chief Secretary to the Lord Lieutenant of Ireland, the Chancellor of the Duchy of Lancaster, the President of the Board of Trade, the Lord Privy Seal, the President of the Local Government Board, the President of the Board of Agriculture, the Vice-President of the Committee of Council on Education, the Postmaster-General, and the First Commissioner of Works. The heads of other Government Departments are not in the Cabinet.

"Our Government, like other Governments, has a great deal to do besides making laws. It has to administer the affairs of an enormous empire. It has to conduct the foreign relations of an empire not less powerful than any empire in the world. It has to conduct the relations of that empire with every foreign State, civilised or uncivilised, all over the world. It has to administer the affairs of a great Indian Empire—a task the like of which does not devolve upon any other Government in the world. It has to regulate the relations of the mother country with vast self-governing colonies, to provide to a certain extent for the government of those great communities, to watch over their interests, and to regulate their relations with each other, and with ourselves. Our Government has vast naval and military departments to administer, not to speak of such unconsidered trifles as the Post-Office and Telegraph Department, which are in themselves departments as great as any of those railway or industrial companies with whose affairs we are acquainted. Above all, it has to superintend the collection and the expenditure of an enormous annual revenue." [1]

FINANCE : The annual **Revenue**, which is chiefly derived from the *Customs, Excise, Stamps* and *Taxes*, and the *Post Office* and *Telegraphs*, and **Expenditure**, mainly on account of the *Public Debt*, the *Army and Navy* and the *Civil Services*, each amounts to about 90 millions sterling ; while the **National Debt** amounts to 671 millions.

In 1892-3, the **National Income** amounted to 90¼ millions sterling, to which the Customs contributed 19¾ millions, Excise 25¾ millions, Stamps 13¾ millions, Income Tax 13½ millions, Post Office and Telegraphs 12½ millions.

The principal branches of **Expenditure** in 1892-3 were :—Interest and charges on the National Debt, 25 millions ; Army and Navy, 31¾ millions ; and Civil Services—Public Works, Law and Justice, Education, Science and Art, &c.—17¾ millions.

FORCES : The British **Army** is small compared with the huge armies of continental powers, but the **Navy** is the largest and most powerful in the world.

The **Regular Army** consists of about 220,000 soldiers (of whom 75,000 are on service in India, about 30,000 in the Colonies, and 3,000 in Egypt), with an

---

1. Extract from Lord Hartington's speech, August 12, 1890.

Army Reserve of 80,000 men.   The Auxiliary Forces include 140,000 Militia, 12,000 Yeomanry, and 260,000 Volunteers.   The British forces available in case of war thus number considerably over **700,000** men.

The **Navy** is the chief defence of the home country and its colonies and dependencies in all parts of the world, and so long as it is unquestionably the most powerful in the world, it assures to us the **supremacy of the sea**.   The Navy has been greatly strengthened by the addition of new ironclads, cruisers, torpedo boats, &c., and now includes 45 battleships, 18 port defence ships, 35 first class cruisers, 63 second class and 189 third class cruisers, and 136 torpedo craft, including 42 torpedo-boat "destroyers."   The fleet is at present manned by about 76,000 seamen and marines.

RELIGION and **EDUCATION**: There is perfect *religious equality* and absolute *freedom of worship* in our country. *Elementary education* is compulsory in both Great Britain and Ireland, and was made free in Scotland in 1889, and in England and Wales in 1891. *Higher education* is amply provided for by public and private schools, colleges, and universities.

The **Established Church of England** is *Protestant Episcopal*, and is under the government of 2 archbishops and 32 bishops.   "The sovereign is by law the supreme governor of the church, and in the theory of English law, every Englishman is a member of the Church of England"; but the number actually belonging to it does not probably exceed 13½ millions, or about one-half of the population of England and Wales; of the rest, about 10 millions belong to various Protestant Sects—*Methodists*, *Baptists*, &c., and there are about 1¼ million *Roman Catholics* and 70,000 *Jews* (40,000 of them in London).

The **Established Church of Scotland** is Presbyterian in form, and is under the supreme control of the *General Assembly*, which meets annually at Edinburgh, under the presidency of an elected *Moderator*, the Sovereign being represented by the *Lord High Commissioner*.   Of the dissenting Presbyterian bodies, the largest are the **Free Church** (with, in 1890, over a million members and adherents), and the **United Presbyterian Church**.   About 80,000 of the Scottish people belong to the *Episcopal Church*, while 326,000 are *Roman Catholics*.

There is no **State Church in Ireland**.   The majority of the people belong to the *Roman Catholic Church*, which is ruled by 4 archbishops and 23 bishops. The *Protestant Episcopal Church* of Ireland, formerly the Established or State Church, has about 620,000 members, and there are about half-a-million *Presbyterians*, *Methodists*, &c.

As regards education, Scotland is far in advance of both England and Ireland. **Elementary Education** is *compulsory* in each country, and was made *free* in Scotland in 1889, and in England and Wales in 1891.   In England and Wales, Voluntary Schools connected with the Church of England, the Wesleyan, and Roman Catholic churches, &c., are much more numerous than those directly under School Boards.   In Ireland, elementary education is controlled by the "Commissioners of National Education."   All Public Elementary Schools receive Government Grants.   The total grants to Primary Schools, in 1893, amounted to nearly 8 millions sterling; the average number of children in attendance being about 5 millions.   In addition to the Government Grants, the fees, rates, voluntary subscriptions, &c., amounted to about 3 millions sterling.   The total **expenditure on elementary education** in the British Isles is no less than 11 millions a year.

Higher Education is provided for by the Universities of *Oxford, Cambridge, Durham, Victoria,* and *London* (an examining body only) in England ; *Edinburgh, Glasgow, St. Andrews,* and *Aberdeen* in Scotland ; and *Dublin* in Ireland.   Besides these Universities, there are University Colleges at *London, Liverpool, Manchester, Leeds, Birmingham, Bristol, Newcastle, Nottingham,* and *Sheffield* in England ; at *Aberystwith, Bangor,* and *Cardiff* in Wales; and *Dundee* in Scotland.   In Ireland higher education is given in the Queen's Colleges at *Belfast, Cork,* and *Galway.*

The Public and Grammar Schools, as well as the Private Schools and Colleges of England and Wales, are not under Government control ; in Scotland a large number of higher-class schools are inspected, and in Ireland there are about 1,500 superior schools.   There are numerous Training Colleges for elementary teachers, and a large number of Science and Art Classes in connection with the Science and Art Department at South Kensington.   Medical Schools are attached to most of the large hospitals and some of the universities and schools, and there are several Engineering and Agricultural Colleges, and Naval and Military Schools.

THE BRITISH EMPIRE.—Besides Great Britain and Ireland, the British Empire embraces a vast number of Colonies, Protectorates, and Dependencies, including amongst them territories in every quarter of the globe.

In Europe, the British flag floats over *Gibraltar,* which commands the entrance to the Mediterranean, and *Malta,* a fortified coaling-station and entrepôt for British goods.

The British Empire in India extends over a territory nearly one-half the area and about three-fourths the population of the Continent of Europe.   Our *Indian Empire* embraces twelve Provinces and Districts under direct British rule, and a large number of tributary Native States.

Other British Possessions in Asia are *Ceylon,* a large island in the Indian Ocean to the south-east of India ; *Aden,* on the south coast of Arabia, with the islands of *Perim* and *Kuria Muria ;* the *Straits Settlements* and *Protectorates* in the *Malay Peninsula ; Hong-Kong* and *Kowlun* in China ; and *British North Borneo, Sarawak, Brunei,* and *Labuan* in the East Indian Archipelago.

In Africa, we have the West, South, and East African Colonies and Protectorates on the mainland, together with the islands of *Ascension* and *St. Helena* off the west coast, and *Mauritius, Zanzibar, Pemba, Seychelles, Amirantes,* and *Socotra,* off the east coast.   *British West Africa* includes Gambia, Sierra Leone, the Gold Coast, Lagos, the Niger Coast Protectorate, and the Niger Territories, with its dependencies, Sokoto, Gando, Bornu, and Adamawa.   *British South Africa* embraces the Cape Colony, Natal, Bechuanaland, Basutoland, Zululand, and British Zambesia.   *British East Africa,* a vast region extending from the coast to the Victoria Nyanza and the Upper Nile, which includes Ibea, Uganda, &c., and *British Central Africa,* which includes the inland districts between the Zambesi and Lake Nyasa, complete our possessions on the mainland.   The islands of *Zanzibar* and *Pemba* were declared a British Protectorate in 1890.

The British Empire in America includes the vast *Dominion of Canada, Newfoundland,* the *British West Indies, British Honduras, British Guiana,* and the *Falkland Islands.*

British Australasia is formed of the five great colonies into which Australia is divided, namely, *New South Wales, Victoria, Queensland, South Australia,* and *Western Australia,* together with the islands of *Tasmania, New Zealand, Fiji,* and the south-western part of *New Guinea.*

In the Western Pacific, a number of small islands and island groups are British Possessions or British Protectorates, but are not included in any colony.

Summary of the British Empire—1894.

	Europe.	Asia.	Africa.	America.	Australia.
Area (sq. miles) .	121,600	1,947,000	2,477,000	3,614,000	3,175,000
Population .	38,000,000	290,000,000	40,000,000	6,800,000	4,285,000

∴ The areas and population of British Asia and Africa include the latest estimates of the Protectorates and Spheres of Influence.

The total area of the British Empire is thus 11 million square miles, or more than *one-fifth* of all the land of the globe, while the population numbers over 380 millions, or about one-fourth of the total population of the world.

The British Possessions and Dependencies in Asia alone have an area of nearly 2 million square miles, and a population of over 290 millions ; while the Protectorates and Spheres of Influence, defined within recent years in the Dark Continent, are estimated to add 2¼ million square miles more of land, with perhaps 36 million people, to the already extensive British territories in Africa. Excluding India, the *Colonies* proper have an area of nearly 7¼ million square miles, and a population of 20 millions.

The commercial and political value to the Mother Country of her Colonies and Dependencies, acquired by *conquest* or *treaty, purchase* or *settlement,* is incalculable, and without such boundless *fields for emigration* "under the flag" as Australia and New Zealand, Canada and Southern Africa present ; such *markets for British goods* as we find in India and the Colonies generally ; such facilities for the collection of raw materials from, and the distribution of our manufactures to surrounding countries as are at our command in great *entrepôts* like Singapore and Hong Kong, or *trading stations* as on the West Coast of Africa, *coaling stations* for our merchant steamers as at Aden and other places, and fortified *stations for our men-of-war* on all the great ocean highways—without such possessions and dependencies in all parts of the world, Britain could never have acquired her present predominance either in the commercial or in the political world.   One-fourth of the entire trade of the United Kingdom is with India and the Colonies; and were the rest of the world closed to our commerce, there is no product which we now derive from foreign countries that could not be supplied by one or other of our trans-oceanic possessions, the development of which would vastly increase the requirements of what are, even now, the most valuable markets for British manufactures in the world.   Trade follows the flag, and "colonial trade is safer and steadier than ordinary foreign trade."

The Isle of Man and the Channel Islands, though not Colonies, are virtually self-governing Dependencies of the British Crown.

THE ISLE OF MAN.

The Isle of Man is situated in the Irish Sea, nearly equidistant from England, Scotland, and Ireland. It is about 33 miles in length, 8 to 13 miles in breadth, about 100 miles in circumference, 227 square miles in area, with a population of 54,000, an average of nearly 250 persons to the square mile, or half the density in England. Fully one-half of the people live in the four principal towns—Douglas, Ramsey, Castletown, and Peel.

"The central part of the island is mountainous and beautifully diversified; streams flowing through leafy glens with precipitous sides, form numberless cascades." Snaefell, 2,034 feet in height, is the highest point in the island; 18 other summits rise above 1,000 feet. Between Sulby Glen, north of Snaefell, and the northern coast, is a low plain called the Curragh; another tract of nearly level country extends from the south-eastern base of the mountains to the south coast. To the north, the island ends in a low headland—the Point of Ayre; near Spanish Head, the extreme point on the southern coast, is a small island called the Calf of Man, separated from the larger island by a narrow channel called Calf Sound.

The soil is thin and poor; only about a third of the island is in tillage, the rest is in pasture or is unenclosed mountain-land. The Manx *fisheries* are, however, extensive and valuable, and large quantities of herring, cod, and mackerel are exported. The *mineral products* include lead, zinc, and slate. The lead mines at LAXEY and FOXDALE are very productive.

"The natives of this island belong to a mingled race of *Celts* and *Norwegians*, and the language, in which the Celtic element predominates, is known as Manx; it is still spoken on the island, and all laws are promulgated in that tongue."

The island is governed by a Lieutenant-Governor (who represents, and is appointed by, the Sovereign), a Council of the chief officials, and a popularly-elected House of Keys, who together form the *Tynwald*, probably the most ancient legislative body in the world. Revenue, 1892, £72,000; Expenditure, £58,000.

DOUGLAS (18), on the east coast, is the capital and the largest town in the island; it is connected by rail with Ramsey (4), a pretty town on the north-east coast, the fishing port of Peel on the west coast, and Castletown, the ancient capital, on the south coast. From the tower of Castle Rushen, anciently the residence of the Kings of Man, Snowdon in Wales, the Mourne Mountains in Ireland, and the Cumbrian Mountains in England, may be seen.

There is regular steam communication between Douglas and Liverpool (75 miles distant), Barrow (45), and Fleetwood (54), and a direct service (in summer) also between Ramsey and Liverpool and Whitehaven.

THE CHANNEL ISLANDS.

The Channel Islands—Jersey, Guernsey, Alderney, and Sark—*geographically* belong to France, but *politically* have been depen-

dencies of the British Crown ever since the Norman Conquest.
They have a total area of 75 square miles, and a population of
about 90,000, an average density of no less than 1,200 per square
mile, or twice as much as Belgium, the most densely-peopled
country in Europe.

" The Channel Islands are in no sense Colonies.   They are the only remain-
ing part of the ancient duchy of Normandy, retained by the Queen as the
Representative of its Dukes.   They are governed by their own laws, and the
intervention of the British Parliament in their affairs is extremely rare.   In
Professor Freeman's words (Historical Geography of Europe, chap. xiii.),
' Practically the islands have during all changes remained attached to the
English Crown, but they have never been incorporated with the kingdom.'
Any business connected with them passes through the Home Office, not the
Colonial Office."

JERSEY ("grass isle"), the largest and most important of the Channel
Islands, and also the most southerly, lies about 16 miles from the French coast.
It has a resident population of about 52,000, who carry on an active trade with
England in *fruits* and *vegetables*.   The capital, St. Helier, is a busy seaport as
well as a fashionable watering-place.

GUERNSEY ("green isle") surpasses Jersey in the beauty of its coast-
scenery, and its climate is more equable.   St. Peter's Port, the capital, stands
on a picturesque harbour on the eastern side of the island.

ALDERNEY, the third island in size, lies about 20 miles north of Guernsey,
and is divided from the French coast by a channel 7 miles in width—the *Race
of Alderney*.   Jersey and Guernsey are both fortified, but not so strongly as
Alderney—the strategic position of which is so important (Cherbourg, the
chief naval arsenal of France, being only 25 miles distant, while Portsmouth, the
principal English arsenal, is 90 miles off) that the British Government has
expended vast sums on its defences, which consist of a series of strong forts
and batteries.   Over a million sterling has been expended on Braye Harbour,
which will ultimately form a harbour of refuge for British ships on the French
side of the Channel, and a basis for naval operations in case of war with France.

SARK, which lies about 7 miles east of Guernsey, consists of two parts—
Great and Little Sark—united by a natural causeway about 450 feet in length,
but only 5 to 8 feet in width, with a nearly perpendicular descent of 380 feet on
one side and a steep slope on the other.   The still smaller island of Herm,
between Sark and Guernsey, contains an inexhaustible supply of the finest
granite.

The mild and equable climate of the Channel Islands, and their beautiful coast-
scenery and picturesque rural landscapes, attract great numbers of visitors, and
the passenger and goods traffic between the islands and England is sufficient
to command two direct daily services—from Southampton and Weymouth to
St. Pierre and *St. Helier*, and a weekly service from Plymouth and, during
the summer, from London.

The native language is an old Norman *patois*, and the official language is
still French, but English is generally spoken.   The local Legislatures or States
of Jersey and Guernsey are under the control of the Lieutenant-Governor, who
represents the Queen.   Taxation is light, and trade is extremely active—the
imports and exports from England amounting to about 1½ millions sterling
a year, or nearly £17 per head of the population.

GEORGE PHILIP AND SON, PRINTERS, LONDON AND LIVERPOOL.

PART II.

THE BRITISH COLONIES AND DEPENDENCIES.

CONTENTS.

INTRODUCTION.

BRITISH POSSESSIONS IN AFRICA.

CONTENTS.

BRITISH POSSESSIONS IN OCEANIA.

NOTE.—The figures within parentheses, immediately after the names of cities and towns, are the number of the inhabitants in thousands, thus, Valletta (65), denotes that Valletta has a population of 65,000.

THE GEOGRAPHY

OF

THE BRITISH COLONIES

AND FOREIGN POSSESSIONS.

INTRODUCTION.

THE BRITISH ISLES.—The British Isles consist of Great Britain, Ireland, and numerous smaller adjacent islands, situated in the Atlantic Ocean, off the western side of the European continent.

Great Britain consists of England, Wales, and Scotland, and is the largest island in Europe, being 600 miles in length, and having an area of nearly 89,000 square miles.

Ireland lies to the west of Great Britain, and is divided from it by the Irish Sea. Great Britain is nearly *three times* the size of Ireland, the area of which is 32,500 square miles.

Of the numerous islands and islets adjoining Great Britain and Ireland, the principal are the *Isle of Wight*, on the south ; the *Orkney* and *Shetland Islands*, on the north ; the *Hebrides*, off the west coast of Scotland ; *Anglesey* and the *Isle of Man*, in the Irish Sea ; and *Achil* and *Aran Islands*, off the west coast of Ireland.

England, Wales, Scotland, Ireland, and the adjacent islands, constitute politically the **United Kingdom of Great Britain and Ireland.**

Wales, Scotland, and Ireland were formerly distinct countries from England. The conquest of Ireland commenced in 1170, and virtually ended when Limerick was surrendered in 1691. Wales was conquered in 1282, and formally annexed in 1536. The crowns of England and Scotland were united in 1603 ; in 1707, England and Scotland were united under the same Parliament ; and in 1801 the Parliaments of Great Britain and Ireland were united ; hence the name, " The United Kingdom of Great Britain and Ireland."

Though comparatively small in *area*, and surpassed by many other countries in point of *population*, the United Kingdom of Great Britain and Ireland is the **greatest commercial and manufacturing country in the world.**

The **total area** of the United Kingdom is a little over **120,000 square miles**, or rather more than *one-half* the area of France or Germany, and scarcely *one-seventeenth* part of the Russian territories in Europe.

As regards **number of inhabitants**, Great Britain and Ireland contain a few thousand more people than France, but 10 millions less than Germany, about 24 millions less than the United States, and 57 millions less than Russia, the most populous of European countries.

According to the Census Returns for April, 1891, the population of the *United Kingdom* (inclusive of the islands in the British Seas) amounted to **38 millions**, of which *England* and *Wales* contained **29 millions**, *Scotland* a little over **4 millions**, and *Ireland* nearly 4¾ millions.

The **population of Great Britain has nearly trebled since 1801,** but that of Ireland, which increased by 3 millions in 40 years (1801-41), decreased also by 3 millions during the next 40 years (1841-81), and now the island contains fewer people than it did ninety years ago.

The **density of population** in the United Kingdom is *greatest* in England and Wales, where, according to the latest returns, it amounts to 498 per square mile, and *least* in Scotland—134 per square mile. In Ireland, the density amounts to 145 per square mile. In 1891, England itself had 540 persons to each square mile, as against 206 in Wales, 134 in Scotland, and 145 in Ireland. In the same year, the average density of population in the British Isles, as a whole, was 316 per square mile.

Emigration from the United Kingdom, and especially from Ireland, has been very active during the present century, and between 1815 and 1889 upwards of 10¼ million persons of British origin emigrated, the majority of them settling in the United States of North America. During the last 35 years, nearly 7 million British emigrants left their native land; 4½ millions going to the United States—of these nearly 2 millions were English, 360,000 Scotch, and over 2¼ millions Irish—the rest settling chiefly in Canada, Australasia, and South Africa.

The industrial and commercial supremacy of the United Kingdom is due to the splendid *geographical position* and almost unparalleled *development of coast-line*, a favourable *climate* and an *abundant supply of coal*, the most essential of all minerals, and *iron*, the most useful of all metals, the *energy* and *enterprise* of the people, ample *capital* and efficient *labour*, unrivalled *facilities for carrying on industrial operations* on a vast scale, *colonies* and *dependencies* in all parts of the globe, and a *mercantile marine* larger than the merchant navies of all other countries taken together.

The geographical position of the British Isles is undoubtedly the best in the world, because, although "detached from the great continental masses, they are peculiarly and influentially situated with reference to them," occupying, as they do, a central position among the countries of the world—the **exact centre of the land hemisphere** being within a few miles of Falmouth—and also most favourably placed with regard to the chief industrial and trading countries, with which they easily communicate by what Professor Seeley happily terms "an incomparable road-system," namely, the sea.

The development of coast-line is such that no part of the country is far from tidal waters, and products are thus readily received from, and sent to, all parts of the world—this interchange being all the more easily carried on because of the **abundance of seaports,** and the **unrivalled facilities for internal communication** by roads, railways, and navigable rivers.

Another peculiarity is the **remarkable contraction in breadth** of both Great Britain and Ireland at various points by deep indentations on opposite coasts. Thus, between the tidal waters of the

1. "In the British Isles there are more than 70 seaports with a depth of at least 25 feet at high water, and most of these are situated in the vicinity of the great seats of production. In view of the increasing size of the shipping of the present day, this large number of deep harbours is a matter of peculiar importance."—(Hand-book of Commercial Geography—Chisholm'.

Thames and the Severn, the distance is under 100 miles; while Hull, on the Humber, is 113 miles from Liverpool, on the Mersey. The mouth of the Tyne is only 60 miles from the head of the Solway Firth, and the Firths of Forth and Clyde are divided only by 32 miles of land. In Ireland, there are similar contractions between Dundalk and Sligo and Dublin and Galway.

The numerous navigable rivers, with their broad and deep estuaries, into which the tide penetrates very far, together with an admirable, and so far complete, system of canals, which are much used for the conveyance of minerals and heavy goods, are all favourable to the activity and growth of British commerce.

The entire extent of the coast-line of Great Britain and Ireland, including the larger indentations, is greatly in excess of the direct distance between the extreme points of the land, while the true salt-water coast-line has a far greater extent, and the line of tidal influence is still more enormous in proportion to the area of the country. Further, the British Isles enjoy the advantage of higher tides than most other countries, and this enables vessels of considerable burden to penetrate almost to the heart of the country.

The climate of the British Isles is remarkably mild and equable, as compared with that of continental countries under the same parallels of latitude. The variation of temperature is comparatively slight, and in no other part of the world do the isotherms of 46° and 50° F. reach so far north as in this country, the truly oceanic climate of which is the most favourable for industry and trade in the world. Out-door work and railway traffic are carried on with little interruption all the year round, and the tidal estuaries and navigable rivers are never frozen over even in the severest winters.

Of the energy and enterprise of the British people, industries on a vast scale and a world-wide commerce are the best evidence. The national characteristics of the British people are, according to a German writer,[1] solidity, energy, endurance, enterprise, strict respect for the law, and great industry; while a French *savant*[2] declares that the British race is extraordinarily vigorous, and that in physical strength, practical intelligence, mental soundness and tenacity of purpose, it is the equal of any on the globe. The British, he adds, possess inventive genius, the love of adventure, the innate instinct of trade, a passion for success, and imperturbable courage.

---

1. Carl Zehden.   2. Onésime Reclus.

The wide extension of the English language, which is assuming more and more the character of a universal language, is an important factor in the development of British trade and industry. Ample capital, for carrying on industrial and commercial operations on a vast scale, is readily drawn from the enormous wealth of the country, while an abundant supply of efficient labour is secured by a dense and prolific population.

Between the United Kingdom and the British Colonies and Dependencies in all parts of the world, an interchange of productions naturally arises, and the bulk of the trade of Great Britain is done with them and the kindred English-speaking States of North America.

The British mercantile marine is larger than that of any other country, and, indeed, exceeds those of all other countries taken together. England possesses more than half the merchant service of the world, and her vessels have become the "ocean-carriers of the world."

THE BRITISH EMPIRE.—Besides Great Britain and Ireland, the British Empire embraces a vast number of Colonies, Protectorates, and Dependencies, including amongst them territories in every quarter of the globe.

In Europe, the British flag floats over *Gibraltar*, which commands the entrance to the Mediterranean, and *Malta*, a fortified coaling-station and entrepôt for British goods.

The British Empire in India extends over a territory nearly one-half the area and about three-fourths the population of the Continent of Europe. Our *Indian Empire* embraces 12 Provinces under direct British rule, and a large number of tributary Native States.

Other British Possessions in Asia are *Ceylon*, a large island in the Indian Ocean to the south-east of India : *Aden*, on the south coast of Arabia, with the islands of *Perim* and *Kuria Muria;* the *Straits Settlements* and *Protectorates* in the *Malay Peninsula ; Hong-Kong* and *Kowlun* in China ; and *British North Borneo, Sarawak, Brunei,* and *Labuan*, in the East Indian Archipelago.

In Africa, we have the West, South, and East African Colonies and Protectorates on the mainland, together with the islands of *Ascension* and *St. Helena* off the west coast, and *Mauritius, Zanzi-*

bar, Pemba, Seychelles, Amirantes, and *Socotra,* off the east coast. *British West Africa* includes Gambia, Sierra Leone, the Gold Coast, Lagos, the Niger Coast Protectorate, and the Niger Territories (with the dependencies of Sokoto, Gando, Bornu, and Adamawa). *British South Africa* embraces the Cape Colony, Natal, Bechuanaland, Basutoland, Zululand, and Southern Zambesia. *British East Africa,* a vast region extending from the coast to the Victoria Nyanza and the Upper Nile, and *British Central Africa,* or Northern Zambesia, which includes the inland districts between the Zambesi and Lake Nyassa, complete our possessions on the mainland. The islands of *Zanzibar* and *Pemba* were declared a British Protectorate in 1890.

The **British Empire in America** includes the vast *Dominion of Canada, Newfoundland,* the *British West Indies, British Honduras, British Guiana,* and the *Falkland Islands.*

British Australasia is formed of the five great colonies into which Australia is divided, namely, *New South Wales, Victoria, Queensland, South Australia,* and *Western Australia,* together with the islands of *Tasmania, New Zealand, Fiji,* and the south-western part of *New Guinea.*

In the Western Pacific, a number of small islands and island groups are British Possessions or British Protectorates, but they are not included in any colony.

Summary of the British Empire.

	Europe.	Asia.	Africa.	America.	Australasia.
Area (sq. miles) .	121,600	1,947,000	2,477,000	3,614,000	3,175,000
Population . .	38,000,000	290,000,000	40,000,000	6,800,000	4,285,000

∵ The areas and population of British Asia and Africa include the latest estimates of the Protectorates and Spheres of Influence.

The **total area** of the British Empire is thus **11** million square miles, or more than *one-fifth* of all the land of the globe, while the **population** numbers over **366** millions, or about one-fourth of the total population of the world.

The British Possessions and Dependencies in Asia alone have an area of nearly 2 million square miles, and a population of over 290 millions ; while the Protectorates and Spheres of Influence, defined within recent years in the Dark Continent, are estimated to add 2 million square miles more of land, with perhaps 35 million people, to the already extensive British territories in Africa. Excluding India, the *Colonies* proper have an area of nearly 7¼ million square miles, and a population of 20 millions.

THE BRITISH EMPIRE.

	Area in sq. m.	Population.		Area in sq. m.	Population.
In Europe :—			**In Africa :—contin.**		
The United Kingdom . . .	121,481	37,828,153	Lagos . . .	1,071	100,000
England and Wales	58,186	29,001,018	The Niger Coast Protectorate		
Scotland . .	30,417	4,033,103	The Niger Territories . .	500,000	20,000,000
Ireland . . .	32,583	4,706,162			
The Isle of Man .	227	55,598	Ascension . .	35	360
The Channel Islands . .	75	92,272	St. Helena . .	47	4,116
			Mauritius . .	705	360,000
Gibraltar . .	2	25,755	Zanzibar and Pemba	985	377,986
Malta . . .	117	165,662	Socotra . . .	1,382	10,000
In Asia :—			**In America :—**		
India . . .	1,587,104	286,696,960	The Dominion of Canada . .	3,456,383	4,829,411
British India .	944,108	220,530,000	Newfoundland .	42,200	197,335
Feudatory States .	642,996	66,107,860	The British West Indies :—		
Ceylon . . .	25,364	3,008,239			
Cyprus . . .	3,584	209,291	Jamaica . .	4,424	639,491
Aden and Perim .	70	41,910	The Bahamas .	5,800	48,000
The Straits Settlements . .	1,472	506,577	The Leeward Islands . .	701	129,760
Brit. North Borneo	30,000	150,000	The Windward Islands . .	508	135,976
Labuan . . .	31	6,000			
Sarawak . . .	45,000	300,000	Trinidad & Tobago	1,868	228,757
Brunei . . .	3,000	21,000	Barbados . .	166	182,322
Hong-Kong . .	30	221,441	The Bermudas .	20	15,884
			British Honduras .	7,560	31,471
			British Guiana .	109,000	284,837
			The Falkland Islands . .	6,500	1,789
In Africa :—					
The Cape Colony .	233,430	1,527,224			
Natal . . .	21,150	543,913	**In Australasia :—**		
Basutoland . .	9,720	218,902			
Zululand . .	9,000	150,000	New South Wales .	310,700	1,134,207
Bechuanaland .	222,000	—	Victoria . . .	87,884	1,140,411
Zambesia . .	500,000	—	Queensland . .	668,497	393,718
British East Africa	1,000,000	13,500,000	South Australia .	903,690	319,145
The North Somali Coast . .	30,000	240,000	Western Australia .	1,060,000	90,000
			Tasmania . .	26,215	146,667
The Gambia . .	2,700	50,000	New Zealand . .	104,471	714,000
Sierra Leone . .	15,000	180,000	British New Guinea	88,000	350,000
The Gold Coast .	40,000	1,500,000	Fiji . . .	7,740	123,000

Government of the British Empire: "Our Government, like other Governments, has a great deal to do besides making laws. It has to administer the affairs of an enormous empire. It has to conduct the foreign relations of an empire not less powerful than any empire in the world. It has to conduct the relations of that empire with every foreign State, civilised or uncivilised, all over the world. It has to administer the affairs of a great Indian Empire— a task the like of which does not devolve upon any other Government in the world. It has to regulate the relations of the mother country with vast self-governing colonies, to provide to a certain extent for the government of those great communities, to watch over their interests, and to regulate their relations with each other, and with ourselves. Our Government has vast naval and military departments to administer, not to speak of such unconsidered trifles as the Post-Office and Telegraph Department, which are in themselves departments as great as any of those railway or industrial companies with whose affairs we are acquainted. Above all it has to superintend the collection and the expenditure of an enormous annual revenue."[1]

The commercial and political value to the Mother Country of her Colonies and Dependencies, acquired by *conquest* or *treaty, purchase* or *settlement*, is incalculable, and without such boundless *fields* for *emigration* "under the flag" as Australia and New Zealand, Canada and Southern Africa present ; such *markets for British goods* as we find in India and the Colonies generally ; such facilities for the collection of raw materials from, and the distribution of our manufactures to, surrounding countries as are at our command in great *entrepôts* like Singapore and Hong-Kong, or *trading stations* as on the West Coast of Africa, *coaling stations* for our merchant steamers as at Aden and other places, and fortified *stations for our men-of-war* on all the great ocean highways—without such possessions and dependencies in all parts of the world, Britain could never have acquired her present predominance either in the commercial or in the political world. **One-fourth of the entire trade of the United Kingdom is with India and the Colonies** ; and were the rest of the world closed to our commerce, there is no product which we now derive from foreign countries that could not be supplied by one or other of our trans-oceanic possessions, the development of which would vastly increase the requirements of what are, even now, the most valuable markets for British manufactures in the world. **Trade follows the flag, and "colonial trade is safer and steadier than ordinary foreign trade."**

---

1. Extract from Lord Hartington's speech, August 12, 1890.

BRITISH POSSESSIONS IN EUROPE.

GIBRALTAR.

GIBRALTAR, in the extreme south of Spain, is a possession of the British Crown. It was captured by an English squadron in 1704, and has ever since been retained in British possession, notwithstanding several desperate efforts to recapture it, especially in 1789-93.

The town of Gibraltar occupies the western declivity and base of a lofty rock, which advances about four miles into the sea, and terminates to the southward in Europa Point. A narrow and sandy isthmus connects this rock with the mainland of Spain. The natural strength of Gibraltar is increased by extensive fortifications, and a large British garrison is always maintained there. An impregnable position such as Gibraltar, commanding the entrance to the Mediterranean, is of immense strategical value, and, with Malta and Aden, enables us to control the Suez Canal Route to India. Gibraltar is also important from a commercial point of view, being a free port, and serving as a depôt, coaling station, and port of call. The resident *population*, mostly descendants of Genoese settlers, numbers about 20,000, exclusive of the garrison of between 5,000 and 6,000 men.

MALTA.

THE MALTESE ISLANDS—Malta, Gozo, and Comino—lie in the Mediterranean Sea, between Sicily and the north coast of Africa. Malta is 17 miles in length, with an area of 95 square miles ; Gozo has an area of 20, and Comino about 2, square miles. *Total area*, 117 square miles. With a *population* of 163,000, Malta is probably the most densely-peopled place in the world, there being on an average no less than 1,400 persons to the square mile, or nearly three times the density in England and Wales.

The interior of Malta is mountainous ; there is neither river nor rivulet ; naturally barren, by dint of persistent labour a soil has been formed, capable of producing *cotton*, *cereals*, and all the ordinary sub-tropical plants and vegetables. The land is highly cultivated, and 50,000 acres are actually under crops, out of a total of 75,000 acres. The *water supply* is derived from springs—the rainfall only averaging 17 to 20 inches a year. The climate is hot, but, when the necessary sanitary improvements have been made, Malta will be the healthiest spot in the Mediterranean as a winter resort.   Four-fifths of the population are Maltese—a mixed race of Arabs and Italians—a most industrious and frugal people and excellent seamen.

VALLETTA (65), the capital of Malta, has one of the finest harbours in the world, and is strongly fortified.   It has an extensive arsenal and dockyard, and is the headquarters of the British Mediterranean Fleet.

Valletta is built on a long ridge on the eastern coast of Malta, with a harbour on either side—*Quarantine Harbour* on the north, and *Valletta Harbour*, or the Grand Port, on the south.   The arsenal and dockyard are in the suburb of *Vittoriosa*, on the opposite side of the Grand Port.   Great forts defend the seaward approaches; and, on the land side, both Valletta and Vittoriosa are enclosed by strong fortifications, so that this important strategical position is practically impregnable.

Malta is of vast importance in a commercial, as well as a military point of view, for it not only serves as a *depôt for collecting and distributing goods*, but also as a *coaling station*, and as a rendezvous for *refitting our fleet* in the Mediterranean.

In fact, the possession of Gibraltar at the entrance to this great inland sea, of Malta in the Central Mediterranean, and of Cyprus in the Levant, virtually converts the Mediterranean into an English lake, and secures the Suez Canal Route to India.

The trade of Malta is mainly a transit one, the *actual imports*, in 1893, amounting to 1½ millions sterling, and the *actual exports* to £95,000 ; while the *imports, in transit*, amounted in the same year to 12¼ millions, and the *exports, in transit*, to 12 millions.

The exportable produce of the islands consists of *cotton, potatoes, oranges, figs, honey*, and *corn ;* and there are manufactures of *cotton, filigree*, and *matches*.   A railway connects Valletta with CITTA VECCHIA, a beautiful town, formerly the capital, in the interior.   West of Valletta is the PORTO DE SAN PAOLO, where it is supposed St. Paul was shipwrecked.

The government of Malta is to some extent representative, the *Governor* being assisted by a *Council of Government* of 20 members, 14 of whom are elected, and an *Executive Council*.   The annual *Revenue* and *Expenditure* each amount to about a quarter of a million sterling.   In religion, the Maltese are *Roman Catholics*.

Malta was first colonized by the Phœnicians, and was afterwards held by the Greeks, Carthaginians, and Romans.   In the 9th century it passed into the hands of the Saracens, and at a later period belonged to Spain.   In 1530, Charles V. of Spain granted Malta to the Knights of St. John, who held it until 1798, when they capitulated to Napoleon.   Two years later, the French were forced to surrender the island to the English, and the cession was formally confirmed by the Treaty of Paris in 1814.

x
(11)

BRITISH POSSESSIONS IN ASIA.

CYPRUS.

Cyprus, the third largest island in the Mediterranean, lies in that portion of it known as the *Levant*. It is about 40 miles distant from the coast of Asia Minor, 60 miles from Latakia on the Syrian coast, and 240 miles from Port Said, at the entrance of the Suez Canal.

The main portion of the island is 105 miles in length (*i.e.*, from **Cape Arnauti** on the west to **Cape Greco** on the east) and 60 miles in greatest breadth (from **Cape Kormakiti** on the north to **Cape Gatta** on the south). On the north-east, a narrow peninsula runs out for 40 miles. Its extreme point, **Cape Andrea**, is 72 miles from Cape Khanzir, at the entrance to the Gulf of Alexandretta on the Syrian coast. The total area is **3,584** square miles, or rather less than half the size of Wales. In 1891, the population amounted to about 210,000.

Cyprus consists, physically, of three distinct regions—a narrow **mountain belt** along the northern coast, a much more extensive and loftier **mountain region** in the south, and between them **a broad plain**, called the *Mesorea*.

The northern range culminates in the peak of **Pentedaktylos** ; the southern mass of mountains attains in **Mount Troodos**, the ancient *Mount Olympus*, an elevation of 6,600 feet. The eastern part of the Mesorea is watered by the **Pedias**, the western portion by the **Potamos**, but there are no navigable rivers. There are several bays and harbours, but not a single harbour capable of sheltering a fleet.

Except in certain places, the climate is salubrious, and the **Mesorean plain is so fertile that it might be readily converted into one huge cornfield, and might again become, as of old, the "garden and granary of the East."** Agriculture is, however, carried on in the most primitive manner, and little more than one-tenth of the island is actually under cultivation. The productions include corn, cotton, wine, fruits, carobs, and olives, and these, with wool and hides, are exported, but the total trade only amounts to about half a million a year, and the imports from, and exports to, the United Kingdom are under £100,000 in value, although the population, consisting mainly of Greeks, is not much under **200,000**. The capital is the inland town of **NIKOSIA** or Lefkosia (12), in the centre of the island ; the other principal towns are **Larnaka** (8), Limasol (6), and **Famagusta** (2), on the coast.

Cyprus still nominally forms part of the Turkish Empire, but it was placed under the control of England in 1878.[1] The government is administered by a High Commissioner, aided by a Legislature of 18 members, 12 of whom are elected—3 by Mohammedan, and 9 by non-Mohammedan, voters. Native judges sit in all the courts of law, except in the Supreme Court for the whole island, which consists of two English judges. The island is divided, for administrative purposes, into six districts, namely, Nicosia, Larnaca, Limasol, Famagusta, Papho, and Kyrenia.

ADEN.

ADEN is a *British Dependency*, on the south-western coast of Arabia, about 120 miles from the entrance to the Red Sea. The colony includes a small peninsula, on which the strongly-fortified town of Aden stands, and another smaller peninsula to the west with a narrow strip of the adjacent mainland.

The total area of the colony is about 70 square miles, and the population, including that of the new town of *Shaikh Othman*, on the mainland, about five miles from Aden itself, with the troops and followers, numbers about 40,000. The coast tribes outside the limits of the colony, from Perim to Ras Sair, are under British protection.

Aden is an important military position, a great centre of trade with the surrounding countries, and a much frequented coaling-station on the route to India *via* the Suez Canal.

Aden itself stands in a deep hollow, the crater of an extinct volcano, and surrounded by high bare rocks. The colony is, in fact, absolutely non-productive, and nothing is manufactured in Aden except salt and water. The imports of coal, cotton and silk goods, &c., and the exports of coffee, feathers, gums, hides and skins, &c., are mainly in transit from or to the neighbouring countries.

The extensive and valuable trade with India, China, and Egypt, which Aden had during the 11th, 12th, and 13th centuries, was ruined by the discovery of the route to the East by the Cape of Good Hope. The main peninsula was taken in 1839, and the town then had 6,000 inhabitants—now it has five times the number. The ancient water-tanks above the town can hold 30 million gallons, but the rainfall is so scanty that the water supply is mainly derived from condensers, and partly from the mainland. The climate, though intensely hot, is healthy.

---

1. On June 4th, 1878, a "Convention" was secretly signed between the English Government and the Sultan—England undertaking to protect the Sultan's Asiatic Empire by 'force of arms' if need be; the Sultan promising to introduce all necessary reforms in the government of his Asiatic vilayets, and to assign the island of Cyprus to be occupied and administered by England as long as Russia should continue to hold Batum, Ardahan, and Kars. The promised reforms are yet to come: the British occupation and administration of Cyprus bid fair to revive the ancient glory and productiveness of the island, which under Venetian sway had a million inhabitants—more than six times its population when transferred to us.

Aden is governed by a *Political Resident*, who is also Commander of the troops, and is subject to the Government of Bombay. Also attached to the Aden Residency are Perim, a small island at the entrance to the Red Sea; Socotra, a large island in the Indian Ocean, 150 miles east of Cape Guardafui, the extreme eastern point of Africa; the Kuria Muria Islands, off the south-eastern coast of the peninsula; the Arabian Coast Protectorate, from Perim to Ras Sair; and the Somali Coast Protectorate, on the African side of the Gulf of Aden. Kamaran Island, in the Red Sea off the west coast of Arabia, is also British. It is 15 miles long and 5 miles broad, and affords a good anchorage. The Bahrein Islands, in the Persian Gulf, off the east coast of Arabia, are also under British protection. The great industry is the pearl fishing. The capital, Manameh, has 8,000 inhabitants.

OUR INDIAN EMPIRE.

OUR INDIAN EMPIRE comprises the central and by far the most important of the three great peninsulas of Southern Asia, together with large territories on the eastern side of the Bay of Bengal. The *total area* of these vast dominions, most of which are under direct British rule, and the rest subject to British control, is upwards of 1,800,000 square miles, or more than 30 times as large as England and Wales, while the *population*, according to the census of 1891,[1] is 286,000,000 (or more than 9 times the population of England and Wales).

The Asiatic Empire of Russia is very much larger in extent than our Indian Empire, but while it has an area of 6½ million square miles, its population amounts to not more than 18 millions, as against 286 millions on what is actually or virtually British territory.

The Chinese Empire, the only other great Asiatic power, is estimated to have an area of 4 million square miles, or 2¼ times that of our Indian Empire, and a population of 400 millions.

Within the limits of, or bordering on, our Indian Empire, are several independent Native States and a few unimportant French and Portuguese possessions. British India is divided into twelve Provinces under direct British rule. The Native or Feudatory States of India, some of them of considerable extent, are to all intents British dependencies.

---

1. The recent census of our Indian Empire was taken on February 26, 1891, and the totals, as the *Times* says in its review of the preliminary Census Report, are so huge that it is difficult to grasp their meaning. In the United Kingdom about 40,000 enumerators were employed in taking the census; in India no less than a million enumerators were employed, and in one province alone, Bengal, the staff exceeded 369,000.

According to the census of 1891, of the 286 millions of people in all India, 220½ millions are in British territory, and 65½ millions under feudatory rule.

In the densely-crowded British territories, after all deductions for new acquisitions have been made, the returns of the 1891 census show an increase of 11 per cent. during the ten years, and of 14½ per cent. in the less thickly-peopled Feudatory States.

This means a gross increase of 29 millions during ten years, an increase equal to the whole population of England. Deducting territorial acquisitions during the decade, no fewer than 26 millions of human beings have been added to the soil of India.

Although the government of India was transferred from the East India Company to the Queen in 1858, it was not until 1877 that Her Majesty formally assumed the title of **Kaisar-i-Hind**, or EMPRESS OF INDIA.

INDIA.

INDIA PROPER or HINDUSTAN extends from the Himalaya Mountains on the north to Cape Comorin in the south, and from the Sulaiman Mountains on the west to the head of the Bay of Bengal in the east.

"From Peshawar, the northern frontier station, to Cape Comorin, the distance is **1,900** miles, and the same distance separates Karachi, the port of Sind, from Sudiya, the frontier post on the eastern border of Assam."

BOUNDARIES : India is bounded by the **Himalaya Mountains** on the *north;* by Burma and the **Bay of Bengal** on the *east;* and by Afghanistan, Baluchistan, and the Arabian Sea on the *west.* To the *south* it terminates in **Cape Comorin,** a conspicuous headland which fronts the waters of the Indian Ocean.

In shape, India Proper is triangular, the vast range of the Himalayas forming the base, the Malabar and Coromandel Coasts the sides, and Cape Comorin the apex. It is worthy of notice that the boundaries of India are, for the most part, formed by strongly-marked natural features. Thus the Hala and the Sulaiman Mountains on the north-west, the Himalayas on the north, and the Naga, Khasia, and Tipperah Hills on the east, form an almost continuous "wall," enclosing the continental portion of India. The strictly peninsular portion south of the Tropic of Cancer is bounded on its eastern and western sides by the sea. The *political* importance of a naturally strong frontier, instead of merely artificial boundaries, is evidently very great, especially when an immense territory like India is held by a distant foreign power.

EXTENT: India Proper embraces an area of over **1¼** million square miles, a magnitude **14** times greater than that of the British Islands, and which exceeds by more than **20** times the area of England and Wales.

If we include Kashmir, with an area estimated at 80,900 square miles, **Manipur** at 8,000, Upper Burma at 90,000, and the British Shan States at 90,000 square miles, the total area of British India may be taken at 1,650,000 square miles, of which a million square miles are under British, and the rest under native, administration.

COASTS: The coasts of India are, on the whole, regular and unbroken, deficient in good harbours, and so exposed and surf-beaten as to be in many parts extremely dangerous to approach. The length of the coast-line is about 3,600 miles, equivalent to one mile of coast to every 416 square miles of area.

Various portions of the Indian coasts are distinguished by special names, such as—the Orissa Coast, between the mouths of the Hugli and the Godaveri ; the Golconda Coast, between the Godaveri and the Kistna ; the Coromandel Coast, between the Kistna and Cape Comorin ; and the Malabar Coast, between Cape Comorin and Goa.

Of the few *headlands* on the coast of India, the principal are Cape Comorin, the most southerly point of the peninsula, Cape Monze (near Karachi) and Diu Head (south of Guzerat) on the west coast, and Points Calimere and Palmyras on the east coast.

The chief *Inlets* are the Gulf of Cutch or Kach, leading into the *Runn of Cutch* (a vast salt marsh, flooded only during the rainy season), and the Gulf of Cambay, on the *west* coast ; and, on the *south*, the Gulf of Manaar, divided from Palk Strait by the remarkable ridge of sandstone known as Adam's Bridge, between Ceylon and the mainland.

ISLANDS : The principal islands are **Ceylon**, near the *south-east* coast ; **Bombay** and **Salsette**, close inshore on the *west* coast ; the **Laccadives** and **Maldives**, off the *south-west* coast ; and the **Andaman** and **Nicobar** Islands, in the Bay of Bengal, to the south of Lower Burma.

Ceylon,[1] a large island lying to the south-east of India, is not under the Government of India, but forms a distinct Colony, under the authority of the British Crown.

The groups of the Laccadive and Maldive Islands lie in the Indian Ocean, to the south-west of the Indian Peninsula. The Laccadives are surrounded by coral reefs, and the Maldive Islands are wholly composed of coral, and scarcely rise above the level of the surrounding waters. The *cocoa-nut* is the chief article of produce of these islands.

In the Andaman Islands, in the eastern part of the Bay of Bengal, the chief settlement is PORT BLAIR,[2] on Great Andaman Island.

The Nicobar Islands lie to the south-east of the Andaman Islands, and consist of two large islands and several smaller ones.

Of the Bombay Group, the principal islands are *Bombay, Salsette, Colaba,* and *Elephanta,* the latter being especially famous for its temples and idols excavated in the solid rock. The island of Bombay, near the southern extremity of which lies the town of BOMBAY, is connected with Salsette by an artificial causeway.

NATURAL FEATURES: The more noticeable of the great natural features of India are the vast range of the Himalayas, the loftiest mountains in the world, on the north, whose exterior ranges rise abruptly from the Great Plain of Hindustan, which is watered

---

1. For an account of Ceylon *see* page 31. | 2. Here Lord Mayo, the Viceroy of India, was assassinated by a convict in 1872.

by the **Indus** and the **Ganges**. The peninsular portion of India, to the south of this plain, forms a series of **tablelands**, crossed transversely by several considerable chains, and buttressed on the east and west by the **Ghats**, between which and the sea is a narrow plain. The highlands of Central and Southern India are everywhere seamed by irregular valleys and drained by numerous rivers.

India thus embraces two great divisions : the *north*, which is an extensive lowland plain ; and the *centre and south*, which form a plateau, bordered by mountains of moderate altitude. The plain of Northern India is specially distinguished as **Hindustan**, a name which is also commonly given to the whole country ; the centre and south constitute a region known as the **Deccan**.

MOUNTAINS : The chief mountain-chains of India are the Himalayas, the Western and Eastern Ghats, the Aravalli Hills, the Vindhya and the Satpura Mountains, and the Nilgiri Hills. The Himalayas are by far the most important, and they include the highest elevations on the surface of the globe.

The **Himalaya Mountains** extend for 1,500 miles in a well-defined line along the northern border of India, dividing that country from the tableland of Tibet. Like other great mountain-ranges, the **Himalayas** consist of several parallel ranges. The outer range, bordering on the great plain of India, rises abruptly to a height of 3,000 or 4,000 feet. The inner chains gradually increase in elevation and culminate in the main ridge containing the lofty summits of *Gaurisankar* or *Mount Everest*, 29,002 feet above the sea, the highest mountain in the world ; *Dhwalagiri*, 28,078 feet ; and *Kanchinjanga*, 28,177 feet. All the higher parts of the Himalayas are covered with perpetual snow. The **passes** over the Himalayas are lofty and extremely difficult. The best known passes are the *Karakoram Pass* (18,550 feet), and the *Mustagh*, leading from Kashmir into Eastern Turkestan ; and the *Seylub Pass*, leading into Tibet. All the loftier valleys are filled with vast glaciers, from which the great rivers of India derive a never-failing supply. The lofty **Karakoram Range**, on the northern frontier of Kashmir, contains *Mount Godwin-Austen* (28,250 feet)— the second highest elevation on the globe. The Karakoram Mountains are separated from the Himalayas by the valley of the Upper Indus.

The **Western Ghats** extend along the Malabar Coast of India. Their highest summits do not exceed 8,000 feet. A succession of detached portions of high ground, which extends along the eastern side of the peninsula, is called the **Eastern Ghats**. These have an average height of about 1,500 feet.

The **Aravalli Hills** lie along the western border of the tableland of Malwa, between the basins of the Ganges and the Lower Indus. The average elevation is inconsiderable, but *Mount Abu* rises to a height of 5,000 feet above the sea. Farther south is *Girna*, 3,000 feet high.

The **Vindhya Mountains** lie in the direction of east and west, along the north side of the peninsular portion of India. Their height is moderate, seldom exceeding 3,000 feet. The Satpura Mountains run almost parallel to the Vindhya range for 200 miles, between the Narbada and the Tapti. These mountains are prolonged eastward, almost to the banks of the Ganges, as the *Rajmahal Hills*, one peak of which, *Mount Parisnath*, reaches a height of 4,530 feet. Still farther east, beyond the Brahmaputra, are the minor *Garrow* and the *Khasia Hills*. Between these and the coast are the *Tipperah Hills*.

The Nilgiri Hills form a connecting link between the Eastern and the Western Ghats, and rise abruptly from the remarkable valley or "gap" of Coimbatore, which extends right across the peninsula. The highest point is *Mount Dodabetta*, 8,760 feet. To the south of the "gap" are the Aligherries or Cardamum Mountains, remarkable as containing the highest mountain in India south of the Himalayas, *Anamalli*, which is 8,837 feet above the level of the sea.

TABLELANDS: India contains two important tablelands, one in the *north* and a far larger one in the *south*.

The *Northern Tableland*, or the Plateau of Malwa and Bundelkhand, is in Central India, and is bounded on the north-west by the Aravalli Hills, and on the south by the Vindhya Mountains.

The *Southern Tableland*, or the Deccan, occupies nearly the whole of peninsular India, and is bounded on the east and west by the Eastern and the Western Ghats, and on the north by the Vindhya and the Satpura Mountains and the valleys of the Narbada and the Tapti.

PLAINS: The **Great Plain of Northern India** extends across the country between the northern tableland and the Himalayas. Its *south-eastern slope*, towards the Bay of Bengal, is drained by the Ganges, and its *south-western slope* by the Indus ; hence its division into the "Plain of the Ganges" and the "Plain of the Indus."

Within the latter is comprehended the fertile Punjab, the Great Indian Desert (or *Thar*), and the Runn of Cutch. In the north of the plain of the Ganges is the malarious swamp called the " Tarai ; " with this exception, the plain is fertile and productive, supporting an exceptionally dense population. The East and West Coast Plains lie between the Ghats and the sea ; the eastern is much wider and more fertile than the western, the extreme breadth of which nowhere exceeds 50 miles.

RIVERS: The rivers of India are naturally divisible into two great sections, namely, those draining the **south-eastern slope** into the *Bay of Bengal*, and those draining the **south-western counter-slope** into the *Arabian Sea*.

The principal rivers are the **Brahmaputra, Ganges, Mahanadi, Godaveri, Kistna,** and **Cauveri,** *draining the south-eastern slope into the Bay of Bengal ;* the **Indus, Narbada,** and **Tapti,** *draining the south-western counter-slope into the Arabian Sea.*

The **BRAHMAPUTRA** (1,680 miles) rises in the vast glaciers on the northern slopes of the Himalayas, and flows east for several hundred miles, but turns south through *Assam*, and unites in its lower course with the eastern outlet of the Ganges.

The **GANGES** rises on the southern slope of the *Himalayas*, and, after a south-eastern course of 1,500 miles through the great plain, finally enters the *Bay of Bengal* by numerous channels, of which the Hugli, on which stands CALCUTTA, the capital of India, is the most important. The Ganges is navigable for the largest vessels to CHANDERNAGORE, while light steamers can go up to CAWNPORE, and thence by canal to HARDWAR, more than 1,300 miles above its mouth. The chief tributaries are the Jumna and the Sone on the right, and the Goomti, Gogra, Gundak, and Coosy on the left, bank.

The Mahanadi, though notorious for its destructive floods, is yet navigable by boats for 400 miles. Its length is 520 miles, and the area of its basin is 70,000 square miles. Its extensive delta formed the old province of *Cuttack*.

The Godaveri rises in the *Western Ghats*, not far from the *Gulf of Cambay*, and flows south-east for 900 miles, entering the *Bay of Bengal* by two large channels. The navigation of this river is impeded by several rapids.

The Kistna also rises on the eastern slopes of the *Western Ghats*, and has a rapid and unnavigable course of 800 miles.

The Cauveri rises in the *Western Ghats* and enters the eastern coast-plain by two magnificent falls, of which the upper is 370, and the lower 460, feet high. It enters the sea by two branches, which enclose a delta 80 miles long. The south-eastern branch, the Colerun, is extensively used for irrigation.

Of the minor streams that flow into the Bay of Bengal the principal are the Bramini, between the Mahanadi and the Ganges ; the North Pennar and South Pennar, and the Palar, all of which rise in the hills of Mysore.

The INDUS (1,800 miles) rises in the tableland of Tibet, and flows first north-west through *Kashmir*, and then south through the *Punjab* and *Sind*, entering the Arabian Sea by numerous mouths. About 470 miles above the sea it is joined by a stream called the Panj-nad, which brings the collected waters of five tributary rivers. The district through which these rivers flow is called the Punjab, that is, the country of the five rivers. The names of these are the Jhelum, Chenab, Ravi, Bias, and Sutlej. The Indus is navigable from its mouths near KARACHI to its confluence with the Kabul River at ATTOCK, 900 miles from the sea.

The Narbada rises in the highlands of *Central India*, and flows west between the Vindhya and the Satpura Mountains, into the *Gulf of Cambay*. It is 800 miles long, and is throughout rapid and unnavigable.

The Tapti rises in the *Satpura Mountains*, and flows west through the valley formed by them and the northern edge of the Deccan. Both the Narbada and the Tapti are subject to sudden and destructive floods.

Of the smaller streams draining the south-western slope the chief are the Luni, which rises in the *Aravalli Hills* and flows into the *Runn of Cutch ;* and the Mahi, which rises in the tableland of *Malwa* and enters the *Gulf of Cambay*.

LAKES : The lakes of India are small and unimportant.

Among these are the Chilka and Palicat Lagoons on the *east coast ;* the Lagoons of the *Malabar coast ;* Lake Kolar, formed by the expansion of the Kistna and the Godaveri ; and Lake Wular, similarly formed by the Jhelum, in *Kashmir*.

CLIMATE : The climate of India is hot, excepting only in the higher mountain regions, where a cool temperature results from elevation above the sea. These elevated districts are, accordingly, much resorted to in the hot season.[1]

1. The chief hill stations, or sanatoriums, which are resorted to by Europeans during the hot season, are Simla, in the Himalayan hills, near the banks of the Sutlej; Darjeeling, on the borders of Sikkim, and Ootacamund, or Utakamand, in the Nilgiri Hills. The heat of the Indian summer is so intense and enervating, that there is a regular exodus, even of Government Departments, from Calcutta, Bombay, and Madras to the hills and the editor of this work has, more than once, received an intimation that certain matters would be attended to "*next cold weather*."

The lower slopes of the Himalayas, in the north of India, the Ghats, off the western coasts of the peninsula, and the region of the Nilgiri Hills, in the south, are well-known for their cool atmosphere and their refreshing breezes. In like manner, the mountain-districts in the interior of Ceylon, though only a few degrees distant from the Equator, enjoy a cool and invigorating temperature. The seasonal changes in India are those from *rain* to *drought*, and the reverse, and are intimately connected with the Monsoons, or periodical winds, which prevail throughout Southern Asia. The monsoons bring rain or drought alternately to the plains of India, according as they have passed over the ocean or over inland regions. On the Malabar Coast, the South-West Monsoon (which blows from April to September) is accompanied by rain, which falls in torrents along the whole seaward face of the Western Ghats. On the Coromandel Coast, on the other hand, the North-East Monsoon (October to March) is accompanied by rain. But the eastern side of India is generally hotter and more arid than the western coasts of the peninsula. These changes of the monsoons regulate, in great measure, the habits of life of the Indian population.

PRODUCTIONS: The natural productions of India are rich and varied; there are large areas of waste and unproductive land, but the alluvial soil of the great river valleys is **extraordinarily fertile.** Dense forests still cover extensive districts.

MINERALS : The gold and gems for which India is traditionally celebrated are of less real value than the coal and iron which are found extensively diffused throughout large portions of the country. Good coal is worked to the north-westward of Calcutta, and there are iron-works in Malabar and other localities. Some gold-mines, also, are being successfully worked in Mysore. Tin, copper, and other metals also occur, and large quantities of salt, the manufacture of which is a *government monopoly*, are obtained from the rich deposits of rock-salt in the Punjab.

PLANTS : The *vegetable products* are of high value. India supplies all, or nearly all, the fruits and other plants mentioned as belonging to Southern Asia in general. Millions of acres are devoted to the growth of millet, rice, and pulse —the staple foods of the people; wheat is largely grown for export, and a large area is under cotton and jute, or is utilized for the production of oil seeds, indigo, sugar, tea, and coffee. Vast forests of teak and other trees clothe the seaward face of the Ghats, and dense forests extend from the plains of Northern India far up the declivities of the Himalayas.[1] The least productive parts of India are the region known as the Great Indian Desert and the low-lying tract called the Runn of Cutch. The latter is alternately an arid and sandy waste, or a vast swamp, according to the season of drought or moisture.

ANIMALS : Of the *animals*, the principal are the domestic and wild elephant, the maneless lion (in Guzerat and Rajputana), the tiger, leopard, wolf, hyena, rhinoceros, buffalo, wild ass, deer, and other game, and monkeys. Besides the ordinary *domestic animals*, there are the elephant, the camel, the humped ox, the yak, and the Kashmir goat.[2]

INHABITANTS : According to the Census of 1891, the population of India amounts to 286,000,000—an average, for the whole

---

1. The forests of India are equal in extent to the entire area of Great Britain. About three-fourths of the forests have been demarcated and reserved by the Government.

2. According to recent Returns, there are in India nearly 90 million *cows* and *bullocks*, 13 million *buffaloes*, 2 million *horses* and *donkeys*, and about 28 million *sheep* and *goats*.

country, of more than 200 persons to the square mile. Upwards of 220 millions are under direct British administration; and the great bulk of the remainder, though with various Native sovereignties, are under the controlling power of Great Britain.

Density of Population: But vast as is its population, India is yet, relatively, less populous than several countries of Europe. The average density, according to the last Census (1891), is about 100 persons per square mile less than is the case in the United Kingdom. But some parts of the country are much more populous than others. The provinces on the Lower Ganges are the most densely populated; those in the north-west of India, the least so. In Bengal, there are now on an average no less than 474 persons to the square mile, or 745 to every square mile of the estimated cultivable area.

The annual increase in the population of India has reached the enormous number of 2¼ millions, and as the struggle for life in the densely crowded British territories becomes more and more fierce, the responsibilities of the Rulers of India grow daily more onerous and difficult to fulfil. The intensity of the pressure on the soil in the congested districts in Bengal is such that the lowest limit of subsistence, namely, half an acre to each person, has been reached. Happily, however, the extension of the railways and the general improvement in the means of communication all over the country are powerfully aiding an evident movement of the population from over-crowded to under-peopled areas. "The improvement in the means of communication is tending not only to relieve the purely agricultural districts by diverting the population to centres of trade and manufacture, it is also tending to relieve them by drawing off a portion of their surplus population to districts where land is still abundant and rents are low."

Race: The great mass of the people of India (six-sevenths of the whole) belong to the Hindu race, the various families of which, however, exhibit many points of difference. The inhabitants of the provinces that border on the Lower Ganges are of small stature and slender frame; those of the more inland and upland provinces are of larger proportions and greater strength. There are, besides, settled in various parts of India and intermingled with the Hindu population, descendants of *Arabs, Armenians, Afghans, Turks,* and other races, together with *Parsis,*[1] *Jews,* and people of various European nations (principally British).

Religion: The Hindus are uniformly followers of the Brahmanical religion, worshipping the Hindu trinity, of which Brahma, Vishnu, and Siva are the members. The division into *castes* is one of their most characteristic social usages. Of that portion of the population of India which is not included among the worshippers of Brahma, by far the greater number are Mohammedans. The total number of Mohammedans in India is upwards of 57 millions. There are also 7 million Buddhists (mostly in Burma) and 2 million Sikhs. Christianity is making rapid progress, and there are devoted missionaries in nearly all the large towns. The native Christians number about 2 millions, and are most numerous in Madras, Travancore, and Bombay.

Education: A very small percentage of the children of school age are under instruction, and only about one-thirtieth of the population not under instruction are able to read and write. Some progress has, however, been made, and

---

1. The Parsis, who are found chiefly in the city of Bombay and a few places in the immediate vicinity, are descended from the ancient fire-worshippers of Persia. They are among the most successful merchants and bankers of India. At the last census (1891) they numbered 89,877.

about 4 million pupils now attend the **Primary Schools**, while there are a large number of **Secondary Schools** and **Colleges**, and **5 Universities**—Calcutta, Bombay, Madras, Allahabad, and Lahore.

INDUSTRIES: Agriculture has always been, and still is, the chief industry in India ; but there are also important native **manufactures** of fine textile fabrics and metal wares, and the internal and foreign **trade** is very extensive.

Rice, millet, and *pulse* are the *staple foods* of the great mass of the population, who live mainly upon a vegetable diet—not less from its superior economy and from the natural influences of the climate, than from religious prejudices in its favour.[1]

The culture of the *poppy*—for the purpose of extracting opium—is very extensively pursued in some of the provinces within the valley of the Ganges, and also on the plateau of Malwa, to the northward of the Vindhya Mountains. Indigo, cotton, the sugar-cane, the **coffee** plant, and the mulberry, are objects of culture in various parts of India.

The *tea-plant* is very largely cultivated in Assam and Bengal, and more **tea** is now imported into Great Britain from India than from China.

Fine silks and muslins, with shawls and other articles of ornamental attire, and cotton fabrics, constitute the chief produce of Indian manufacturing skill.

The cotton goods produced in the cotton mills of **Bombay** compete with the products of the looms of Lancashire, not only in the Indian market, but also in those of China, Japan, and the Straits Settlements. Bombay also has large silk factories, and there are jute factories in Calcutta, &c.

Nearly 40,000 persons are already employed in the coal-mines of India, and the deposits of iron-ore are being gradually developed. In almost all the towns there are many skilled **metal-workers**, whose productions are often in the highest degree beautiful and artistic.

COMMERCE: The *import* of **manufactured goods**, principally from Great Britain, and the *export* of **raw produce** and native **manufactures**, are the distinguishing features of the extensive foreign commerce of India.

The annual value of the **Foreign Sea-borne Trade** of British India is about 120 millions sterling ; the imports amounting to about 50 millions, and the exports to 70 millions sterling. About thirty years ago the total foreign trade was only worth 40 millions.

The **Trans-frontier Land Trade** with the surrounding countries—Baluchistan, Afghanistan, Tibet, Nepal, Sikkim, Bhutan, Western China, Siam and the Shan States—amounts to about 5 millions a year.

The **Coasting Trade of India**, exclusive of Government stores and treasures, is very large—its annual value, imports and exports, amounting to 40 millions sterling.

[1] It is a mistake to suppose, as is commonly the case, that the Hindus abstain altogether from animal food. The ox is sacred, and its flesh is never touched, and the flesh of swine is regarded with horror both by the Brahman and the Mohammedan. But mutton is eaten without hesitation, and fish is largely consumed, whenever it is cheaply obtainable. In all hot countries, however, a vegetable diet is preferred by the mass of the people. The Greenlander, who consumes twelve pounds' weight of meat in a day, and the Hindu, whose chief nutriment is derive i from rice, act in each case upon the instinctive impulses that are always associated with climate and other conditions of physical geography.

The trade of India with the United Kingdom is considerably more than *one-half of its total trade*, and the Anglo-Indian fleet of London is second only to the Atlantic fleet of Liverpool.

The staple articles of export from India to the United Kingdom are *cotton*, *wheat, jute, oilseeds, tea, rice, indigo, leather, hides* and *skins, coffee*, and *wool*.

The chief articles of British produce imported into India are *cotton manufactures* and *yarn, iron* and *metal wares, copper, machinery*, and *woollen goods*.

About one-tenth of the total trade of India is carried on with China (through Hong-Kong and the Treaty Ports), nearly 5 per cent. with France, 4 per cent. with Germany, 3½ per cent. with the Straits Settlements, 3 per cent. with Belgium, and 3 per cent. with the United States.[1]

Italy, Austria, Turkey, Egypt, Persia, Ceylon, Australia, East Africa, Japan and other countries, all have a considerable trade with India.

The trade of India with China consists chiefly in the export thereto of *opium* and *cotton goods* and *yarn*, the principal imports from China being *raw silk* and *silk goods*.

The **seven great ports of India**, in the order of their importance, are *Bombay* and *Calcutta, Rangun* and *Madras, Karachi, Tuticorin*, and *Chittagong*. Of these, **Bombay** and **Calcutta** are by far the most important ports in India, and together do nearly *four-fifths* of the entire maritime trade of the country.

The foreign trade of **BOMBAY** has for many years slightly exceeded that of **CALCUTTA**, and the new *Nagpur-Bengal Railway* will form the *crux* of the rival claims of Calcutta and Bombay for commercial supremacy in India. Calcutta is a great emporium of home produce and an admirable distributing centre for foreign products, but Bombay has the advantage of being better situated for European commerce, and almost equally well placed with regard to trade with China, the Straits Settlements, and Australasia. Moreover, Calcutta is a river-port 75 miles from the sea, while Bombay is situated on an island, close in-shore, and with splendid docks and a sheltered harbour.

Madras has now a capacious artificial *harbour*. Rangun, on the Irawadi, in Burma, will be described in connection with that province. Tuticorin is a sea-port in the south-east of India, on the Gulf of Manaar. A railway connects it with Tinnevelli, an important town in the interior. Chittagong lies on the coast, east of the delta of the Ganges, and near the head of the Bay of Bengal. Karachi, at the mouth of the Indus, is the chief *grain port* of India, and Surat and Bombay are the principal *cotton ports*.

The means of internal communication in India were formerly very defective, the roads, except in the vicinity of the larger towns, being generally wretched; but, since the transfer of the government of the country to the British Crown, magnificent roads and an extensive system of railways have been constructed and are constantly being extended. These, with the navigable rivers and canals, have greatly augmented the internal trade of India, and have effectually prevented the possibility of a recurrence of the famines which inevitably follow bad

---

1. Nearly *seven-ninths* of the imports of merchandise into India come through the Suez Canal, and considerably more than *one half* of the exports and re-exports of India are sent by the same route.

harvests. Now, however, although the population has so largely increased during recent years, the Government of India, by the extension of railways, by the construction of irrigation and other public works, and by fostering the commerce, manufactures, and agricultural industries of the country, is able to cope successfully with the increasingly grave problem of subsistence in the congested districts of India. There is now a great network of main and district roads throughout British India, and nearly 19,000 miles of railways are already open for traffic.

The Bengal-Nagpur Railway, recently opened, is of vast importance, politically and commercially. It runs right across the heart of the peninsula, through a rich wheat-growing region, which will probably add largely to the grain exports of India, and forms a direct route for passengers and goods between the two great mercantile capitals of India, to which it will carry the mineral and agricultural wealth of the central regions. Both CALCUTTA and BOMBAY will profit by this new line, which will be, in a military point of view, one of the most important of all the railways in India.

GOVERNMENT : Nearly the whole of this immense country is directly or indirectly under British government. Three-fifths of the vast region lying between the Himalaya Mountains and Cape Comorin are included within the limits of British India, and are subject to the direct rule of authorities appointed by the British Crown.

The remainder is divided between various Native States, of which there are a vast number (many hundreds in all), attached to Britain by various ties, but all more or less dependent upon British power. These are sometimes called the "Tributary" or "Feudatory" States. Their rulers assume various titles. The Sovereign of Haidarabad, the largest of the Native States, is called the *Nizam ;* the ruler of the larger portion of Guzerat is known as the *Guicowar* or *Gaekwar.* More frequently, however, the title of Maharajah or Rajah is borne by the Native princes.

Prior to the year 1858, all the provinces of British India were under the rule of the East India Company—a body of merchants originally incorporated in the reign of Queen Elizabeth—subject only to a limited control on the part of the Crown. But in that year the political functions of the Company were terminated by Parliament, and the whole of their vast dominions brought under the direct authority of the British Crown. The Queen of England formally assumed the title of Empress of India (Kaisar-i-Hind), by an Act proclaimed at Delhi before the Princes of India, on January 1, 1877.

The Government of the Indian Empire is controlled by the Secretary of State for India,[1] aided by a consultative Council of not less than 10 members. The SUPREME GOVERNMENT IN INDIA is exercised by a Governor-General or Viceroy, who represents and is appointed by the Crown, assisted by an Executive Council of 6

[1]. The British Parliament is really the supreme ruler of India, and the Indian Budget is annually submitted to the House of Commons. The Secretary of State for India, like all other Cabinet Ministers, must have a seat in one of the two Houses of Parliament.

The apparently autocratic powers of the Secretary of State for India at home, and of the Viceroy in India, are limited by the responsibility of the former to Parliament, and the necessity for the Viceroy to act in all cases through his Council; and although the power of overriding his Council is reserved to him in "cases of emergency," every official Act must run in the name of the "Governor-General in Council."

members, including the Commander-in-Chief of the Imperial Forces in India. Other members are added to form a **Legislative Council** for making laws and regulations for the Indian Empire generally, and for those Provinces which have no **Local Councils.**

"The Governors of Madras and Bombay (including Sind) each have a Council of their own, as well as an army and civil service of their own. The Lieutenant-Governors of the North-West Provinces have each a Legislative Council only; the other Governors of Provinces have no Councils and no Legislative powers."

The **Annual Revenue** and **Expenditure** each amounts to about **55** millions sterling. The total **Debt** is about **140** millions sterling.

The three chief sources of **Revenue** are the *land*, *opium*, and *salt*, while the largest branches of **Expenditure** are those for the *army* and the *civil services.*

The **European Army** consists of about 74,000 men, and the **Native Army** of about 145,000 men. The armies of the various Feudatory and Independent Native States of India number collectively upwards of 350,000 men. Contingents of these native armies are being drilled and disciplined so as to render them able to co-operate with the Imperial forces.

The naval force consists of 2 armour-plated ships, a despatch-vessel, and 2 torpedo gunboats, belonging to the Government of India. The **British East India Fleet** consists of 10 war-vessels, and the **China Fleet** numbers 20 vessels.

DIVISIONS: India is politically divided into (1) British Possessions, (2) Native States, and (3) Foreign Possessions.

The Territories under direct British administration were formerly divided into the *three* "Presidencies" of Bengal, Madras, and Bombay, and the term "Presidency" is still applied to these three Provinces or Governments. But British India is now divided, not into *three* "Presidencies," but into *twelve* "Presidencies and Provinces," each with its own separate civil Government, subject to the Supreme Government at Calcutta, which derives its authority from, and acts under the orders of, the Secretary of State for India, who, as a Cabinet Minister, is directly responsible to the British Parliament. The minor divisions of Ajmir-Merwara, Berar, Coorg, British Baluchistan, and the Andaman Islands, are under the direct authority of the Governor-General.

Each of the British Provinces is divided into *Commissionerships* and *Districts,* termed "Regulation Districts," in contradistinction to the "Non-Regulation Districts," *i.e.,* those Districts—protected and semi-independent Native States—which are not under *regular* British rule.

The **Native States** of India are all governed by Native Princes with the help and under the advice of a British Resident or Political Agent, stationed at each of their Courts by the Viceroy.

The rulers of the Native States have "no right to make war or peace, or to send ambassadors to each other or to external States: they are not permitted to maintain a military force above a certain specified limit; no European is allowed to reside at any of their courts without special sanction; and the Supreme Government can exercise the right of dethronement in case of misgovernment. Within these limits the more important chiefs possess sovereign authority in their own territories. Some of them are required to pay an annual tribute; with others this is nominal, or not demanded."[1]

TOWNS : India has no less than 212 towns with over 20,000 inhabitants, and 27 of these have a population of more than 100,000. According to the Census of 1891, Calcutta, the political capital of the Empire, contains nearly a million inhabitants, and Bombay, the commercial capital of India, has over 800,000.

Twenty other towns have a population of between 100,000 and half a million. These are Madras, Haidarabad, Lucknow, Benares, Delhi, Patna, Agra, Bangalore, Amritsar, Cawnpore, Lahore, Allahabad, Jeypore, Mandalay, Rangun, Poona, Ahmedabad, Bareilly, Surat, Baroda.

Nagpur, Gwalior, Trichinopoli, Peshawar, and Dacca, had, in 1891, between 80,000 and 100,000 inhabitants.

PROVINCES OF BRITISH INDIA.

The provinces of British India are BENGAL, the NORTH-WEST PROVINCES AND OUDH, the PUNJAB, the CENTRAL PROVINCES, BURMA, ASSAM, MADRAS, and BOMBAY, together with AJMIR-MERWARA, BERAR, COORG, BRITISH BALUCHISTAN, and the ANDAMAN ISLANDS.

Bengal, the North-West Provinces and Oudh, and the Punjab, are under *Lieutenant-Governors;* Madras and Bombay, under *Governors;* and the Central Provinces, Assam, and Burma, under *Chief Commissioners.* A large number of the Native or *Feudatory States* are attached to Bengal, the North-West Provinces, the Punjab, the Central Provinces, Madras, and Bombay

Ajmir-Merwara, Berar, Coorg, British Baluchistan, and the *Andaman Islands,* are under the direct administration of the Governor-General. Berar is only provisionally under British administration. *Mysore* was restored to the Native Government in 1881.

BENGAL.

BENGAL, the most populous and productive of all the British Provinces in India, has an *area* of 150,000 square miles, or three times that of England, and a *population* of 71 millions (more than twice that of the United Kingdom) or over 760 persons to the square mile of the estimated cultivable area.

Bengal includes, besides the lower portions of the Ganges and the Brahmaputra valleys, the former province of Cuttack, at the mouth of the Mahanadi. The greater part of the presidency forms a vast alluvial plain, which is by far the most fertile and closely cultivated part of India. The principal industry is agriculture, and immense quantities of rice are grown. Besides rice, wheat, maize, and barley are also grown. Among its other products, the most important are opium, indigo, and jute. The coal mines in the hills are now largely worked, and there are important native manufactures.

Before 1853, Bengal was under the administration of the Governor-General. In that year, however, it was placed under a Lieutenant-Governor, who is assisted by a Legislative Council.

---

1. The Statesman's Year Book, 1894, p. 118.

All the great cities of this part of India are situated either on the Ganges or its various tributary streams, and the great lines of communication with the interior follow the course of the river and its tributaries. The principal towns in Bengal are CALCUTTA, MOORSHEDABAD, PATNA, CUTTACK, and DACCA.

CALCUTTA, the chief city of Bengal and the capital of British India, stands on the east bank of the River Hugli, the principal arm of the Ganges, at a distance of a hundred miles from the sea. Including Howrah and Báli, on the other side of the river, it has over a million inhabitants. The navigation of the Hugli is dangerous, but its channel is traversed by the largest sea-going vessels, and an immense trade is carried on, chiefly with England and China.

Plassey, the scene of Clive's great victory in 1757. lies to the northward of Calcutta. Moorshedabad (132) has important native *manufactures*. Patna (167), on the right bank of the Ganges, is the principal town in Bahar and the centre of the rice trade. The largest town in the maritime district of Cuttack is Cuttack (49), which is situated on an arm of the Mahanadi. Dacca (84), on the Buri Ganga, an arm of the Brahmaputra, is connected by railway with Calcutta ; it is noted for its manufactures of *muslin*.

THE NORTH-WEST PROVINCES AND OUDH.

The NORTH-WEST PROVINCES embrace the upper portion of the Ganges valley (including the *Dooab*, as the tract between the Ganges and the Jumna is called), and enclose OUDH on all sides but the north, which is bounded on that side by the independent State of Nepal. The whole Province has an *area* of 106,000 square miles, or twice that of England, and a *population* of 47 millions, an average of 442 to the square mile.

The chief industry in this division is *agriculture*, and large crops of wheat, rice, and other grains are grown. Indigo, opium, cotton, and sugar are also successfully cultivated, and much tea is now grown in the sub-Himalayan districts.

The North-West Provinces were separated from Bengal in 1833. Oudh was annexed in 1856, and until 1877 formed a distinct government under a Chief Commissioner. Since then, the North-West Provinces and Oudh have formed one Province under a Lieutenant-Governor.

The principal towns are ALLAHABAD, BENARES, CAWNPORE, AGRA, MEERUT, and HARDWAR, in the North-West Provinces, and LUCKNOW and FAIZABAD in Oudh.

ALLAHABAD (176), at the junction of the Jumna and the Ganges, and Benares (222), on the north bank of the Ganges, are two of the largest among the inland cities of India, and are among the sacred cities of the Hindus, their numerous temples being the crowded resorts of the devotees of Hindu worship. Cawnpore (182), on the right bank of the Ganges, is memorable for the massacre of its English residents during the mutiny of 1857. Agra (168) is on the right bank of the Jumna.[1] Meerut, notorious as the place where the great mutiny of 1857 broke out, is an important military station, 95 miles north-east of Delhi. Hardwar lies on the banks of the Ganges, where it issues from the Himalayas ; it is a sacred city of the Hindus.

[1] Near Agra is the *Taj Mahal*, a magnificent | gems, erected by Shah Jehan as a tomb for himself building of white marble, and inlaid with precious | and his favourite wife.

The capital of *Oudh* is **LUCKNOW** (273), memorable for the defence of the British Residency during the Sepoy insurrection of 1857-58.   Lucknow is on the river Goomti, one of the many affluents of the Ganges.   **Faizabad** (80), on the Gogra, was the former capital of Oudh.

THE PUNJAB.

The PUNJAB[1] embraces the north-western part of the great plain of India, and is so called from the "five rivers" which water it. Three-fourths of this immense territory are under direct British rule, the rest belongs to the 36 Dependent or Feudatory Native States attached to the Province.

The *area* of the "Regulation Districts" of the Punjab is 111,000 square miles —rather more than twice that of England.   The *population* in 1891 amounted to 21 millions, an average of 187 per square mile.

About a third of the land is cultivated, and large quantities of wheat, rice and other grains, and cotton, are produced.   The principal mineral product is salt, which is found in abundance in the hills—the Salt Range—in the north-west, between the Jhelum and the Indus.   The principal towns are LAHORE, DELHI, AMRITSAR, RAWAL PINDI, MULTAN, and PESHAWAR.

LAHORE (176), the chief city of the Punjab, stands on the River Ravi, one of the five tributaries of the Indus.   It is celebrated as the former capital of the Sikhs, or native inhabitants, of this part of India.   **Delhi** (193), on the right bank of the Jumna, is historically noted as the former capital of the Mogul Empire (which in the 16th and 17th centuries embraced nearly the whole of India), and has acquired more recent fame from its siege by the British in 1857. Here, on the 1st of January, 1877, the Queen of Great Britain was proclaimed *Kaisar-i-Hind*—EMPRESS OF INDIA.   **Amritsar**[2] (136), to the north-east of Lahore, is the holy city of the Sikhs.   **Rawal Pindi** (73) is a great *military station* on the north-western frontier.   **Multan** (75) is on the River Chenab. **Peshawar** (84) is situated to the west of the Indus, not far distant from the entrance to the Khyber Pass, and forms a strong military frontier post.   It is now connected by rail with Calcutta *via* Lahore and Allahabad, and with the rising port of Karachi *via* Lahore and Multan.

∵ The town of **Simla**, situated a few miles south of the Upper Sutlej, in the eastern extremity of the Punjab, is a much-frequented health-resort, and is the usual residence of the Governor-General of India during the hot season. It lies at an elevation of 7,800 feet above the sea, and enjoys an atmosphere which is free from the heat experienced in the lower plains.

THE CENTRAL PROVINCES.

The CENTRAL PROVINCES, which have an *area* of 86,500 square miles (nearly as large as Great Britain) and a *population* (which includes a large proportion of the aboriginal races of India) of 10 millions, include the interior districts enclosed between the upper courses of the Narbada and the Mahanadi, and traversed from east to west by the Satpura Mountains.

---

1. The Punjab was proclaimed British territory in 1848, and was placed under a Board of Adminis-tration until 1853, then under a Chief Commissioner | until 1859, when a Lieutenant-Governor was ap-pointed.
2. *i.e.* the 'pool of immortality."

This division was formed in 1861, previous to that year the various provinces were attached to the North-West Provinces and the Punjab. The natural productions of these provinces, notwithstanding their great areas of mountain and jungle, are rich and varied. There are large coal-fields, and valuable deposits of iron-ore. Cotton, rice, wheat, and opium are largely grown and exported. The principal towns are JABALPUR, NAGPUR, and SAUGOR.

NAGPUR (118) was until 1854 the capital of the Mahratta kingdom so named. It is connected by rail with BOMBAY and CALCUTTA, and is not only the capital of the Central Provinces, but also an important commercial centre for the richly-productive region between Bengal and Bombay. Jabalpur (85) is an important commercial town, the traffic which passes through it being "larger than that of any other town in India except Bombay." Saugor (45), to the north-west of Jabalpur, is an important military station.

BURMA.

BURMA belongs geographically to, and is therefore described under, "British Indo-China" (p. 36, *et seq.*).

ASSAM.

ASSAM was ceded by Burma in 1825, and was included in the Province of Bengal until 1874, when Lord Northbrook placed it under a Chief Commissioner. It has an *area* of **46,000** square miles, but although the soil is fertile, the province is thinly peopled, the *population* in 1891 amounting to $5\frac{1}{2}$ millions, an average of 117 per square mile, only about one-fourth the density in Bengal.

The **tea** plantations, for which Assam is chiefly famous, are in the hands of English capitalists. The climate is tropical, and the rainfall, especially in the Khasia States, excessive.[1]

The only considerable towns in the *Brahmaputra Valley* are **Gauhati** and **Goalpara**, both on the banks of the river. The largest towns in the *Surma Valley*, to the south of the hill region (which includes the Garrow, Khasia, and Jaintia Hills) are **Sylhet** (14) and **Cachar**, both on tributaries of the Brahmaputra, and the centres of the most important tea-producing district in India. The annual production of tea in Assam amounts to about 70 million lbs. The province is also rich in **coal** and **iron**.[2]

∴ Bengal, the North-West Provinces and Oudh, the Punjab, the Central Provinces, Burma, and Assam, although virtually under distinct Governments, are subordinate to the Governor-General in Council in a greater degree than Madras and Bombay. The Governors of Madras and Bombay can address the Secretary of State on all ordinary affairs directly ; the other Governors and Chief Commissioners can only do so through the Supreme Government at Calcutta.

---

1. At *Cherrapunji* the rainfall in some years reaches the enormous amount of 610 inches.

2. The Native State of *Manipur*, which the events of 1891 brought so prominently into notice, is on the eastern border of Assam.

BOMBAY.

The BOMBAY PRESIDENCY, which lies wholly on the western side of India, is about 1,000 miles in length, and has a *population* of 19 millions. The Native States attached to the Province, of which the largest are *Cutch* and *Baroda*, occupy a third of the total area, which is about 195,000 square miles.

The principal productions of the Bombay Province are cotton, rice, salt (in the Runn of Cutch), sugar, and indigo. Much of the cotton grown in the province is now worked up in the large cotton-factories of Bombay itself. The large province of SIND, which extends over both banks of the Lower Indus, forms part of this Presidency. The following are the largest towns :—BOMBAY, SURAT, BAROCHE, POONA, SATTARA, HAIDARABAD, and KARACHI.

The city of BOMBAY (804), the capital of the Presidency, is situated upon the island of Bombay, which closely adjoins the coast. Bombay has an excellent harbour, one of the best in India. It is rapidly rising in importance as the chief commercial centre of the Indian Empire, and already commands a larger amount of foreign trade than Calcutta. Bombay is historically noteworthy as one of the earliest English possessions in the East, having been part of the wedding dowry given to Charles II. with his Portuguese bride, Catherine of Braganza, in 1661. Surat (108), to the north of Bombay, is at the mouth of the Tapti River. Baroche, further north, is on the Narbada. Poona (160), an important military station, and Sattara lie to the eastward of the Ghats, on the tableland of the Deccan. The most important place in the province of Sind is the rising port of Karachi (104), a short distance west of the mouths of the Indus. Haidarabad (58), also in Sind, is on the east bank of the Indus ; near it is the village of *Miani*, where Sir Charles Napier gained his famous victory in 1843.

MADRAS.

The MADRAS PRESIDENCY embraces a large part of Central and Southern India, including both the eastern and western shores of the peninsula, besides an extensive portion of the interior plateau. Its *area*, 140,000 square miles, is nearly three times that of England, while its *population* numbers 35½ millions, an average of 253 per square mile.

The "Presidency" of Madras includes the old provinces of the *Carnatic*, the *Circars*, *Coimbatore*, *Malabar*, and *Canara*. Its principal towns are MADRAS, TANJORE, TRICHINOPOLY, MADURA, TINNEVELLY, TUTICORIN, and CALICUT.

The city of MADRAS (450), the capital of the Presidency, is on the Coromandel Coast. It is destitute of any natural harbour, the sea in front being merely an open roadstead. Its commerce is, nevertheless, very considerable and is increasing, especially since the construction of the new *pier* and *harbour of refuge*.[1] Masulipatam (39) and Coringa are to the north of Madras—the former near the mouth of the River Kistna, the latter at the mouth of the Godaveri. Tranquebar is a seaport near the mouth of the River Cauveri, to the south of Madras ; Arcot, Tanjore (54), Trichinopoly (90), Madura, and Tinnevelly (25) are inland cities. Tuticorin (25) is a seaport on the Gulf of Manaar. The old provinces of Malabar and Canara are upon the western side of the peninsula, stretching along the seaward face of the Western Ghats.

---

1. The eastern coast of India is nearly devoid, | for shipping, while the western, or Malabar Coast, throughout its entire length, of any natural shelter | abounds in good natural harbours.

Calicut (66), Cananore (27), and **Mangalore** (40) are flourishing seaports upon the Malabar Coast, but the first-named of these has fallen from the importance which belonged to it in a former day.

The small district of **Coimbatore**, which is wholly inland, has on its northern border the group of the Nilgiri Hills, which (like Simla in Northern India) are resorted to for the sake of their cool and refreshing breezes. The sanatorium of Ootacamund or Utakamuni, founded in 1822, is on the eastern side of the hills.

THE NATIVE STATES OF INDIA.

The Native States of India, including the Frontier States, are no less than 800 in number, and have a total *area* of 766,000 square miles, or 13 times the size of England and Wales, and a *population* of 65½ millions, or nearly twice the population of Great Britain. The Armies of the Native States number altogether about 300,000 men, while the Revenues of the Native Princes and Chiefs amount to about 20,000,000 tens of rupees.[1]

Of the 800 distinct States, about 200 only are of any considerable size, and but few of these are of any great political importance. Many of the Native States are attached to the various British Presidencies and Provinces, others are governed by their own Princes or Rajahs, but are more or less controlled by Residents or Political Agents appointed by the Governor-General, while two of the Himalayan States are independent. We may therefore class the Native States of India under three heads :—(1) Dependent States, (2) Tributary States, and (3) Independent States.

DEPENDENT NATIVE STATES.

The Dependent Native States are those attached to the various Presidencies and Provinces. They have an *area* of nearly 190,000 square miles, and a *population* of about 22 millions.

These States are 82 in number, 4 of them being attached to the Province of Bengal, 2 to the North-West Provinces, 36 to the Punjab, 15 to the Central Provinces, 5 to Madras, and 5 to Bombay.

Of these States, the most important are **Bahar, Chota Nagpur,** and **Orissa,** attached to *Bengal;* **Patiala** and other **Sikh States,** attached to the *Punjab;* **Garwhal** and **Rampur,** attached to the *North-West Provinces;* **Travancore** and **Cochin,** attached to *Madras;* and **Cutch, Kathiáwár** and **Palanpur,** attached to *Bombay.*

TRAVANCORE extends along the south-western coast of India, from Cape Comorin to the frontiers of Cochin on the north, and is limited on the east by the Cardamum Mountains. The greater portion of this State is covered with forests, but the coast districts are well cultivated and productive. The capital is Trivandrum, a seaport about 60 miles north-west of Cape Comorin.

COCHIN is a small district immediately north of Travancore, and is similarly bounded on the east by the Cardamum Mountains. The town of Cochin is situated near the mouth of a large inlet, on the Malabar Coast.

GUZERAT is the general name for the territories east and west of the Gulf of Cambay. Of the numerous Native States attached to the Presidency of Bombay, the most important are Cutch and Kathiáwár.

---

1. A "ten of rupees," written R, is nominally [rupee fell below is. 3 l. A "lac of rupees=10,000 equal to £1 sterling; but in 1893 the value of the [Rs.; a "crore =10,000,000 Rs.

CUTCH (Kach) is a small peninsula separated from the adjoining peninsula of *Kathiáwár* on the south by the *Gulf of Cutch*, and from the mainland on the north by the vast salt-marsh known as the *Runn* or Rann of Cutch. The *Rao*, as the Native ruler is termed, resides at Bhooj, an inland town. The productions of this peninsula comprise *cotton*, &c., and are mostly exported from the port of Mandivi, on the south coast.

Besides these, **Kashmir** and **Manipur** must now be classed among the Dependent Native States. **Baluchistan** and **Sikkim** are also dependent on, or feudatory to, British India.

KASHMIR.

KASHMIR includes the famous valley of that name, which lies between some of the highest ranges of the **Himalayas** and the **Karakoram Mountains**, watered by the upper course of the **River Jhelum**. It has an *area* of 80,000 square miles, and a *population* of about 1½ millions.

The **new railway** that is being constructed from the North-Western Railway near RAWAL PINDI, to SRINAGAR, the capital, is of the highest *political, commercial*, and *strategical* value, for Kashmir commands some of the most important **trade routes** into the heart of Central Asia, and is renowned for the beauty and magnificence of its **scenery** and its bracing and health-giving **climate**.

The new railway will turn Kashmir into the favourite health-resort and playground to India, as Switzerland is to Europe, and, besides attracting a large tourist immigration, it will also preserve the valley against famine. Happily, the Maharajah and his Council are working heartily under the advice and direction of the British Resident to improve the condition of the people and to develop the industries of this rich, but much-neglected and impoverished valley. The soil is extraordinarily fertile, but, as rain falls mostly in winter (the average is only 16 inches a year), cultivation is entirely dependent on irrigation, which is easily obtained from the Jhelum and the numerous tributaries that descend from the giant ranges of snow-clad mountains which environ the valley. The large herds of *goats* afford the fine wool which is woven into the celebrated **Cashmere Shawls**. SRINAGAR (150), the capital, lies on the banks of the *Jhelum*, at an elevation of about 5,000 feet above the sea.

SIKKIM.

This small **Himalayan State** lies between the Independent Native States of **Bhutan** on the east and **Nepal** on the west, and is bounded on the north by **Tibet**, and on the south by the British district of **Darjeeling**.

Sikkim is about 70 miles in length, with an extreme breadth of about 50 miles. The Maharajah acts under the direction of a British Agent at Tumlong, the capital, and is bound by treaty to keep open the great trade route between Bengal and Tibet, which passes through his territories. A treaty with China was signed in 1890, by which the entire control of Sikkim is assumed by the British Government.

MANIPUR.

MANIPUR is a small Native State in North-Eastern India, in the heart of the mountainous region which lies between Assam and Upper Burma.   It is about 8,000 square miles in extent, and has a population of over 250,000.

The hill ranges of Manipur are clothed with dense forests of *teak*, *indiarubber*, and other valuable woods, and the region generally is rich, but undeveloped, while *coal*, *iron*, and *gold* have been found, and the *tea-plant* grows wild in various parts of the country.   The State is separated from Cachar, in Assam, by seven ranges of hills, each from 2,000 to 7,000 feet in height, and four large rivers, unfordable during the greater part of the year.   In 1887, arms and ammunition were presented to the Maharajah by the Indian Government for assistance rendered to our troops in Burma, and it was, doubtless, these very arms which were used against us in 1891, when the Chief Commissioner of Assam and the British Resident were massacred.   Relief troops were immediately sent from Burma and Assam, and occupied the Manipuri capital. Order was promptly restored, and the State is now controlled as a Feudatory State.

BRITISH BALUCHISTAN.

British Baluchistan includes the north-eastern portion of Baluchistan, and extends from the Zhob Valley on the north to the Bolan Pass on the south, and from the Sulaiman Mountains on the east to the Pishin Valley on the west.

British Baluchistan, of which QUETTA is the administrative centre, was placed under a Chief Commissioner in 1887.   The Sind-Pishin Railway passes through the Bolan Pass to Quetta, and may be continued through the Khojak Pass towards Kandahar.

TRIBUTARY NATIVE STATES.

Of the tributary Native States, the principal are those of *Rajputana, Central India, Haidarabad, Mysore*, and *Baroda*.

RAJPUTANA, in the north-western part of India, is the name given to an extensive tract of country, within which are embraced numerous small States ruled by native Rajahs.   The southern part of this territory is traversed by the Aravalli Hills, to the north and west of which extends the **Thar** or Great Indian Desert.   Although measuring about 450 miles from north to south, and over 500 miles from east to west, with an area of fully 130,000 square miles, the total population in 1891 only amounted to 12¼ millions.   The ruling people in all the States are the proud and warlike *Rajputs* (hence the name of the country). Of the larger Rajput Principalities the most important are **Udaipur** (Oodeypore) and **Jaipur** (Jeypore), but **Jodhpur** is the largest.

The government of all the States is more or less under the control of **British Political Agents**, subordinate to the principal Agent, who resides at **Ajmir**, the chief town of a British District in the region of the Aravalli Hills.

Between the north-west frontier of Rajputana and the Indus and Sutlej lies the Native State of **BHAWALPUR**.   The town of Bhawalpur is situated on the banks of the *Sutlej*, about 60 miles south of MULTAN.

The State of **BARODA** is under the rule of a Native Prince styled the *Gaekwar*, whose territory lies on the north-eastern side of the Gulf of Cambay. The capital, Baroda, is about 250 miles north of Bombay, with which it is connected by rail. *Area*, 8,570 square miles ; *population* (1891), 2½ millions.

The States of **CENTRAL INDIA** lie between Rajputana and the Central Provinces, and have an *area* of about 75,000 square miles, with a *population* of over 10 millions. The largest State is that of Gwalior, governed by the Maharajah Sindia, but the British Agent resides at Indore, the capital of the dominions of the Maharajah Holkar. Sindia's territories embrace the greater part of the tableland of Malwa. The famous hill-forts of the capital, Gwalior, were taken by the British in 1842. Bhopal is a small but important Mohammedan State in the Vindhya Mountains.

HAIDARABAD (Hyderabad), with an *area* of 82,000 square miles, and a *population* of 11½ millions, is the most extensive of the Native States, and is under a ruler who bears the title of the *Nizam*. It is wholly inland, comprising a great part of the tableland of the Deccan. Not far distant from Aurungabad, in the north-west part of this territory, is the little town of *Assaye*, where the Duke of Wellington (then Sir Arthur Wellesley) gained one of his splendid victories in 1803. The capital, Haidarabad, on a tributary of the Kistna, is strongly fortified. The Nizam is advised on all important affairs by the *British Resident*, who has also the charge of the fertile province of **BERAR**, which lies to the north of Haidarabad, and the surplus revenues of which go to the Nizam's Government. The *finest cotton* grown in India is produced in Berar. The chief city is Ellichpur.

The State of **MYSORE** (Maisûr), which is also inland, is surrounded by the territories of the Madras Presidency. *Area*, 27,000 square miles ; *population* (1891), 5 millions. The city of Seringapatam, seated on an island in the River Cauveri, played a distinguished part in the wars of the last century, when it was the capital of Hyder Ali's extensive dominions. Under his son and successor, Tippu Saib, it was stormed by the British in 1799. Owing to the misgovernment of the Native ruler, this State was placed under a British Commissioner in 1832, but in 1881 it was restored to the native Rajah. The present capital, Mysore, lies about 20 miles south-east of the former capital, Seringapatam. The only territory in Mysore now held by the British is the fort and cantonment of Bangalore, near the eastern border.

INDEPENDENT NATIVE STATES.

The only Independent Native States are **Nepal** and **Bhutan**, on the southern slopes of the Himalayas.

NEPAL.

NEPAL lies between the Feudatory States of Sikkim on the east and Kumaon on the west, and is separated from the British provinces of Oudh and Bahar by the pestilential region of the Tarai.

Nepal has an *area* of nearly 57,000 square miles (and is thus nearly as large as England and Wales), and a *population* of about 2 millions.

Though bounded on the north by the lofty ranges of the Himalaya Mountains, the Nepalese carry on a considerable trade with Tibet. Until the British invasion of 1815, the country was virtually a dependency of the Chinese Empire.

In that year, however, a *British Political Resident* was placed at Khatmandu, the capital, which is connected by important **trade routes** with Bengal. The natural products of Nepal are rich and varied, and there are some manufactures of coarse woollen cloth and metal wares. The Maharajah is merely the nominal sovereign—the real power is in the hands of the Prime Minister or head of the dominant people, the Ghoorkas.

BHUTAN.

BHUTAN extends east of Sikkim, and comprises the mountainous region lying between the main ridge of the Himalayas and the British provinces of Bengal and Assam.

The inhabitants, who number about 200,000, apparently of Mongolian origin, profess *Buddhism*, and are under the rule of the *Deb Raja,* or the secular head, and the Dharma Raja, or the spiritual head. The Deb Raja has only a nominal power, the great chieftains being practically independent. The capital is **Punakha,** a place of great natural strength. There is no standing army, and the trade with the British provinces is not very large.

FOREIGN POSSESSIONS.

Two other European nations—the French and the Portuguese—possess a few stations in India, but they are of little importance either in extent or commercial value.

To the FRENCH belong—Pondicherry, a seaport town lying to the south of Madras ; Mahé, a few miles north of Calicut, on the Malabar Coast ; and **Chandernagore,** a small town on the River Hugli, north of Calcutta. These are the remains of a power which long contested with Britain the sovereignty of India. Their total area is but 196 square miles, while the population is under 300,000.

The PORTUGUESE Possessions, which together have an area of less than 1,300 square miles, and a population of scarcely half a million, consist of Goa, a small territory lying on the west coast of India, between the limits of the Bombay and the Madras Presidencies ; the port of **Daman,** to the north of Bombay ; and the town and port of Diu, situated on an island off the south coast of Guzerat. The city of Goa was long a splendid emporium of commerce and the chief mart of the Eastern world, but its importance has wholly passed away.

CEYLON.

CEYLON, the "Garden of India," is a Crown Colony, entirely independent of British India. It lies to the south-east of India, from which it is separated by the *Gulf of Manaar* and *Palk Strait,* and is one of the most valuable of the British Possessions in the East.[1]

Ceylon has an *area* of 25,364 square miles, and is therefore about three-fourths the size of Ireland. The *population* is now upwards of 3 millions, two-thirds of whom are *Singhalese ;* the rest being *Tamils, Moormen* or de-

1. Ceylon was first settled by the Portuguese in 1505, but they were dispossessed by the Dutch 150 years later. The Dutch settlements were taken by the British in 1772 and annexed to the Presidency of Madras, and two years later Ceylon was formed into a separate Colony. In 1815, the interior districts, which had remained independent under the King of Kandi, were occupied and annexed.

scendants of Arab immigrants, *Malays, Veddahs*—probably the aboriginal inhabitants—and *Europeans*, 4,000 of whom are *British.*

The interior of the island is a high mountain region, the loftiest summits of which exceed 7,000 feet in height. A broad belt of lowland extends around the coast. Numerous rivers water the fertile valleys and productive plains, and the climate, though tropical, is on the whole much healthier and more pleasant than that of the adjoining mainland.

The natural productions of Ceylon are varied and valuable. No part of our Eastern Empire is more richly endowed than this "pearl of the Indian Seas." Heber used no poetic licence when he wrote—' What though the spicy breezes blow soft o'er Ceylon's Isle, and every prospect pleases;' and the natives, under the stimulating influence of European energy, skill, and capital, the labour of devoted missionaries, and the attention paid by the Government to education, are prosperous and contented. Nearly one-half of the people are *Buddhists;* the *Hindus* number about half-a-million, and there is an equal number of *Mohammedans* and *Christians.*

The characteristic plant-products of the island are tea, coffee, cocoa, rice, cinchona, cinnamon, and tobacco. Elephant-hunting is a favourite sport in Ceylon. The most abundant mineral products are plumbago, and precious stones, especially *rubies* and *cats' eyes.* Fine pearls are obtained from the pearl-fishery in the Gulf of Manaar.

The trade of Ceylon is carried on mainly with the United Kingdom and India. Annual value about 8½ millions sterling—Imports, 4½ millions ; Exports, about 4 millions.

The principal articles of export, in order of value, are *tea, coffee, coco-nut products, plumbago, cinchona,* and *areca nuts.*

The chief articles of import are *rice, coals, cotton goods, salt fish, wines* and *spirits, metals* and *machinery.*

The trade with the United Kingdom—annual value 1¾ millions sterling—consists chiefly in the export of *tea, coffee,* and *cinchona,* and in the import of *cotton goods, coal,* and *machinery.* Disease has in recent years greatly reduced the production of coffee, the export of which is now scarcely a tenth of what it was 20 years ago, but the export of tea from the island has correspondingly increased.

The Government of Ceylon is in the hands of a *Governor*, aided by an *Executive Council* and a *Legislative Council*—the latter including representatives of the different races and interests in the Colony.

The political and commercial capital of the Colony is COLOMBO (120), a flourishing port on the western coast. Now that its harbours are protected by breakwaters, the "letter-box of the East," as Colombo is called, has superseded the south coast port of Galle as a coaling and steamship station. It is connected by rail with Kandi, the old native capital, in the interior. *Adam's Peak*, a lofty mountain (7,420 feet), with a Buddhist Temple on its summit, is to the south-west of Kandi. Trincomali, a busy port with an excellent harbour, is on the north-eastern coast of the island.

BRITISH INDO-CHINA.

BRITISH INDO-CHINA includes the Province of Burma, the Straits Settlements, and the Native Protected States of the Malay Peninsula.

The total *area* of the British Possessions and Protectorates is about 300,000 square miles, while the *population* is fully 8 millions.

BURMA.

BURMA, which is, politically, a Province of British India, includes the western division of the Indo-Chinese Peninsula.

Burma is bounded on the *north* and *north-east* by China ; on the *east*, by Siam ; on the *north-west*, by Assam, Manipur, and Bengal ; and on the *west* and *south*, by the Indian Ocean.

The *area* of the entire province, including the tributary Shan States, is estimated at 200,000 square miles, and the *population*, according to the recent Census (1891), numbers 9½ millions, or 47 to the square mile. The population is much more dense in the lower division of the province than in Upper Burma.

The government of the Province is vested in a *Chief Commissioner*, subordinate to the Governor-General and Council of India. The seat of Government is Rangun, a large town on the eastern delta-mouth of the Irawadi.

LOWER BURMA consists of three districts along the west coast of the peninsula. The northern and southern districts—Aracan and Tenasserim—were annexed in 1825, at the close of the first Burmese war. The central district—Pegu—was annexed in 1852, after the second Burmese war. These three divisions, which have a total *area* of 87,220 square miles, and a *population* (1891) of 4½ millions, were formed into the Province of British Burma in 1862.

ARACAN is a narrow strip of country lying along the east side of the Bay of Bengal. Its moist climate and marshy plains enable it to furnish a vast quantity of rice, which is exported from Akyab, the capital. Aracan, the old capital, is on a river of the same name about 50 miles from the sea.

PEGU, the most important division of Burma, includes the delta of the Irawadi, a fertile but unhealthy region. Although the area of this division is only twice that of Aracan, it has five times the population. The staple product is *rice*. There are also vast forests of *teak* and other valuable tropical woods. RANGUN (181), on one of the branches of the river, is an important seat of trade, and is the commercial as well as the political capital of the whole province. It was little more than a collection of huts, when the British took it in 1852 ; but now it is a large city, with good streets and buildings. Further north, near the northern border of Lower Burma, is the large town of Prome, on the left bank of the Irawadi.

The districts known by the general name of TENASSERIM extend along the eastern side of the Gulf of Martaban. The climate is tropical, and the productions include *rice*, *cotton*, *indigo*, &c. Most of the land, however, is covered by vast forests, and *teak* and other woods are largely exported. Amherst, Tavoy, and Mergui¹ are small seaports, but with considerable trade. The town of

¹ Large quantities of edible birds' nests, so Chinese, are exported from the islets of the *Mergui* highly esteemed as an article of luxury by the Archipelago, whence also pearls are obtained.

Maulmain (58), prettily situated near the mouth of the River Saluen, is the chief port and the chief town, and is an important seat of trade.

UPPER BURMA, which was annexed to British India in 1886, after the third Burmese war, occupies the north-western portion of the peninsula. Its *area*, including the tributary Shan States, is estimated at 180,000 square miles, or about three times that of England and Wales ; and its *population* at 5 millions.

Upper Burma is physically divisible into the three great valleys of the Irawadi, Saluen, and Mekong. That of the Irawadi forms, in fact, an extensive plain, bounded on the west by the Yoma Mountains. The climate is tropical, and the productions include *rice* (the staple crop), *wheat, maize, tobacco, cotton, indigo, teak*, &c. The mineral wealth is also considerable, *iron, lead, copper, petroleum,* and *coal* being widely diffused ; some *gold, silver,* and *precious stones*—especially *rubies*—are also found. The Irawadi is the main channel of communication, and is now regularly navigated by steamers as far as Bhamo, near the Chinese frontier, 700 miles from the sea. The chief town and capital is Mandalay (188), on the left bank of the Irawadi. Almost all the larger towns of Upper Burma have been at one time or other the seat of government of the former Burmese Empire. Mandalay was the capital of Theebaw, the last native ruler of the country. Further south are the old capitals of **Amarapura** and **Ava**, also on the banks of the same river. About one hundred miles south of Ava are the ruins of the ancient city of *Pagan*, with its numberless temples.

THE STRAITS SETTLEMENTS.

THE STRAITS SETTLEMENTS,[1] on the western side of the Malay Peninsula, form a distinct Dependency of the British Crown. They comprise Penang, Wellesley Province and the Dindings, Malacca, and Singapore,[2] and have an *area* of nearly 1,500 square miles, and a *population* of about 600,000, an average of no less than 400 per square mile.

Penang, or Prince of Wales' Island, is a small but beautiful and fertile island, off the west coast of the Malay Peninsula. It was ceded to the East India Company in 1785 by the Rajah of Kedah (or Queddah). **Georgetown**, on the east coast of the island, is the chief town.

Wellesley Province, on the mainland opposite Penang, acquired in 1800, and the Dindings, a group of islands 80 miles south of Penang, and a strip of the mainland cut out of the Protected Native State of Perak, are dependencies of Penang.

Malacca, the largest as well as the oldest of the Straits Settlements, comprises a strip of territory on the west coast of the peninsula about 240 miles south of Penang. The chief product is *tapioca*. The town of **Malacca** has

---

1. So called from their position on the Strait of Malacca.
2. The *Cocos*, or *Keeling Islands*, in the Indian Ocean, were annexed to the Straits Settlements in 1886, and *Christmas Island* in 1889.

about 5,000 inhabitants, but its formerly important trade has declined since the purchase[1] and settlement of Singapore.

The great emporium of **SINGAPORE** (140) is upon an island of the same name, at the extremity of the Malay Peninsula. The island, which is about 27 miles in length and 14 miles in breadth, is divided from the mainland by a narrow strait. Singapore is the seat of the general Government of the Straits Settlements, and the centre of an immense trade with the surrounding countries and the United Kingdom and America. The harbour of Singapore is defended by strong batteries, and there is a permanent British garrison.

The great **commercial importance** of the Straits Settlements may be inferred from the fact that the *imports* amount to about 150 million dollars[2] a year, and the *exports* to more than 140 million dollars. The direct trade with the United Kingdom alone amounts to 57 millions sterling.

These values are only approximate, as there is no Custom House at any of the Straits ports, which are entirely free from duties on imports or exports. The trade is to a large extent a transit trade, and centres at Singapore, which, in "the extent of its shipping, is one of the greatest ports in the world, being a port of call for vessels trading between Europe or India and the far East, the north of Australia and the Netherlands Indies."

Reclus says—"In the busy shipping quarter, the magnificent docks and the extensive quays are crowded with vessels from every part of the globe, whilst the bazaars and warehouses are stocked with the manufactures of Europe and America, and with the spices, cereals, tea, coffee, sugar, oils, gums, gutta-percha, and other produce of the surrounding regions."

The **Government** of the Straits Settlements is in the hands of a *Governor*, appointed by the Crown, and assisted by an *Executive* and a *Legislative Council*.

The **Revenue** amounts to about 3½ million dollars, while the **Expenditure** is about 4¼ million dollars. There is practically no Debt —the assets of the Colony amounting in 1893 to 2½ million dollars, and the liabilities to 783,000 dollars.

The harbour of Singapore is defended by strong batteries, and there is a permanent British garrison.

THE NATIVE PROTECTED STATES OF THE MALAY PENINSULA.

Nearly the whole of the Malay Peninsula, south of the territories conquered and annexed in 1821 by Siam, is now under British protection or included in the Colony of the Straits Settlements. The entire region has an *area* of about **35,000** square miles, and has immense **agricultural resources** and great **mineral wealth**.

---

1. By Sir Stamford Raffles in 1819, from the native Malay prince of Johor.
2. The only legal tender in the Colony is the *dollar* issued from the mint at Hong-Kong, or the silver dollar of Spain, the American trade dollar, or the Japanese dollar or yen. Nominal value, 4s.

All tropical products—*rice, tapioca, sugar, gambier, pepper, coffee, cinchona*—are grown ; while *tea*, and the ordinary *fruits* and *cereals*, can be successfully cultivated in the well-watered valleys and uplands of this richly-endowed region. *Tin* is as yet the most valuable mineral product,[1] but *gold* and *lead* mines are also being worked.

The Native States of **Pĕrak, Sĕlángor, Sungei Ujong,** the **Negri Sembilan, Pahang,** and **Johor,** are closely connected with the Straits Settlements—the *British Residents* or *Political Agents*, under whose advice and direction the native Sultans or Rajahs govern, being directly subject to the Governor of that Colony.

Pĕrak, Sĕlángor, and **Sungei Ujong** lie on the western side of the Peninsula, between Malacca and Province Wellesley ; the **Negri Sembilan** or "the nine States" are a confederation of small States in the interior ; Pahang is a large State on the east coast, its chief town is PĂKAN on the Pahang River ; and Johor is in the extreme south of the Peninsula, and its capital, JOHOR, is only about 30 miles distant from Singapore.

BRITISH EAST INDIES.

The richest portion of Borneo and the little island of Labuan belong to Britain. **Labuan** is a *Crown Colony*, and the territory of the **British North Borneo** Company, the Sultanate of **Brunei,** and the province of **Sarawak,** are *British Protectorates.*

LABUAN[2] lies on the northern side of the Bay of Brunei, on the north-western coast of Borneo, about 6 miles from the mouth of the Brunei River. It has an area of 30 square miles, and a population of about 6,000, mostly Malays, with some Chinese traders and a few Europeans. The climate is hot and unhealthy, but a considerable amount of produce is collected from Borneo, for export to Singapore. There are extensive coal measures, but the annual production of coal is only about 9,000 tons. The capital, VICTORIA, has a splendid harbour.

BRITISH NORTH BORNEO[3] includes the northern part of the island, and has an area of **31,000** square miles, a coast-line of nearly 1,000 miles, and a population of about **170,000**. The Company's territory, which has been placed under British protection, is described as a **magnificent country,** possessing the

---

1. " Half the tin of the world is exported from the Malay Peninsula, where mining is carried on mainly in Pĕrak State, and almost entirely by Chinese. The mining is that of stood tin, not rock, and the metal is taken from the lowlands near the mountains, where it is found in pockets 10 feet to 20 feet or more below the surface, in appearance like coarse, black sand, with here and there a mixture of tin and small particles of gold dust. To obtain the metal involves a great upheaval of the soil, pumping water from the pits, washing the tin, and finally smelting it. In most places the machinery employed is of the most primitive and simple, yet ingenious, description. The ore is smelted into slabs of irregular shape at the mines, and is then sent to Penang and Singapore to be purified and resmelted into slabs or blocks for the market. The ore is found in Larut, in Pĕrak, in

large quantities in a stratum of whitish clay, which is washed in long, open troughs, water passing through it and carrying off the soil, leaving the ore lodged against cleats nailed on the bottom of the trough. The mining companies now engaged in the work do not smelt the ore at the mines, but ship the sand direct to Singapore, where large smelting works have been established."—*United States Consular Report.*

2. Labuan was ceded to Britain in 1845, and erected into a colony in 1847. It was placed, in 1889, under the government of the British North Borneo Company.

3. Obtained by concession from the Sultan of Brunei in 1877, and from the Sultan of Sulu in 1878, and confirmed to the British North Borneo Company by Royal Charter in 1881.

only good harbours in the whole of Borneo, with a salubrious climate, and form-ing, in a mineral and agricultural point of view, the richest portion of Borneo.

The soil is fertile, and much **tobacco, rice, sago, coffee,** and other tropical products are grown, and **coal** and **gold** have been found.   Over a million acres have been already taken up for cultivation, and a brisk trade in *tobacco* and *sago* is carried on through Singapore (1,000 miles distant from Sandakan) with Great Britain, while large quantities of *timber, biche de mer, edible birds' nests,* &c., are sent to China.

The State is ruled by a *Governor* and *Council* in Borneo, subject to the *Board of Directors* in London.   SANDAKAN (7), the capital, on the north coast of the island, has an excellent natural harbour.

BRUNEI, or Borneo, is a Native State, on the north-west coast of Borneo, which formerly included Sarawak on the south and North Borneo on the north. The present area of the Sultanate is about 3,000 square miles.   The capital, BRUNEI (15), is situated at the mouth of a navigable river, which flows into the only considerable inlet on the north-western coast of the island.   The products are the same as those of North Borneo.

SARAWAK is a large territory to the south-west of Brunei, with a sea-board of about 400 miles, and an area of fully **45,000** square miles, but the population scarcely numbers **300,000.**   It was acquired from the Sultan of Brunei by the late Sir James Brooke in 1841, and the present Rajah is his nephew.   The *plant-products* are similar to those of other parts of Borneo, and among its great mineral resources are **coal, gold, silver, diamonds,** &c.   The capital is KU-CHING (Sarawak), a busy port on the extreme south-western coast.

HONG-KONG.

The British Crown Colony of Hong-Kong includes the island of Hong-Kong, off the south-eastern coast of China, at the mouth of the Canton River, and the adjoining peninsula of Kowloon, from which it is separated by a narrow strait—the Ly-ee-moon Pass—not more than a quarter of a mile in width.

The island of **Hong-Kong** is about 11 miles long and from 2 to 5 miles broad, and, with the Kowloon Peninsula on the mainland, has an area of about 30 square miles, and a population of **250,000,** about two-thirds of whom are Chinese.

The surface of the island is irregular and hilly, and rises in **Victoria Peak**— a favourite place of residence in the hot season—to a height of 1,800 feet. Though well watered, it is naturally barren, and the climate[1] is rather insalu-brious and oppressively hot in summer, but cooler and more invigorating in winter.

---

1. " The thermometer ranges from a *minimum* | inches, of which not less than 70 inches are received of 42° F. in February, to a *maximum* of 91° F. in | between May and September, when the south-west August.   The *average annual rainfall* is 85 | monsoon prevails."—*Whitaker's Almanack.*

Hong-Kong, which half a century ago was a bare rock, with a fisherman's hut here and there as the only sign of habitation, its great sea-basin but very rarely disturbed by a passing keel, has become one of the most important commercial, military, and naval stations of the British Empire. The capital—Victoria—on the northern side of the island, has a magnificent natural harbour, and is the main artery of British commerce in Chinese waters. Hong-Kong is, in fact, the third port in the British Empire, and therefore, with the possible exception of New York, the third port in the world. The tonnage frequenting its harbour is greater than that of the whole of the British possessions in the American continent, or than that of the four principal colonies of Australia.

Nowhere, perhaps, is to be found crowded into the same space such a swarming mass of humanity as in the Chinese quarter of Victoria, and alongside this eastern hive are found all the most characteristic and highly developed features of western civilization. Long lines of quays and wharves, large warehouses teeming with merchandize, shops stocked with all the luxuries, as well as with all the needs of two civilizations, line for four miles the island shores. Behind these, interspersed with tropical foliage, and rising, tier on tier, up the mountain-side, are handsome streets and stately public buildings, and higher still, solidly constructed roads, lined with bamboos and other delicately fronded trees, climb up to and over the heights behind, and are so studded with houses as to give almost an urban aspect to the higher elevation of the island, and indeed it seems as if, at no distant period, every available corner of Hong-Kong will be covered with houses. Kowloon, also, until recently an uninhabited waste of red rock, is becoming covered with houses, docks, warehouses, and verdure.[1]

More than half the foreign trade of Hong-Kong is with the United Kingdom, and a great deal of the British trade with China is done through Hong-Kong. The staple exports are *silk* and *tea*, and the chief imports, *cotton* and *woollen goods*, and *metals*.

Over 50,000 junks belong to Hong-Kong, and about 4,000 foreign vessels call annually at the port. There is no Custom House, but the actual trade of the colony, exclusive of the transit traffic, is estimated to amount to over 20 millions sterling a year.

The Governor of Hong-Kong, who is aided by an Executive and a Legislative Council, also controls the British trade with all the "Treaty Ports" of China. The Revenue, now about a million dollars a year, has, since 1855, generally exceeded the Expenditure. There is a small Public Debt, incurred for waterworks, &c.

1. Sir G. W. des Vœux, K.C.M.G., Governor of Hong-Kong. 1887.

BRITISH POSSESSIONS IN AFRICA.

BRITISH AFRICA includes:—

1. **British West Africa**—The Gambia, Sierra Leone, the Gold Coast, Lagos with Yoruba, the Niger Territories, and the Niger Coast Protectorate.

2. **British South Africa**—the Cape Colony, Natal, Basutoland, Zululand, Tongaland, Bechuanaland, and Southern Zambesia.

3. **British Central Africa**—Northern Zambesia and Nyassaland.

4. **British East Africa**—Ibea, Zanzibar and Pemba, with the Northern Somali Coast.

5. **British Insular Africa**—Mauritius, the Seychelles and Amirante Islands, Socotra, St. Helena, Ascension, and Tristan d'Acunha.

Other European Powers also have colonies and dependencies in Africa. Of these, the more important, from their contiguity to the British possessions on the continent, are the following :—

1. The **Congo State**, which adjoins British East Africa, on the north-east, and British Central Africa, on the south-east. British East Africa adjoins **Italian East Africa**, on the east, and **German East Africa**, on the south.

2. British Central Africa or Northern Zambesia lies between **Portuguese East Africa** and **Portuguese West Africa**, and is bounded on the north by the Congo State and German East Africa.

3. Southern Zambesia lies between Portuguese East Africa and **German South-West Africa**. The 20th meridian divides the latter from the British Colony and Protectorate of Bechuanaland, which are bounded on the east by the South African Republic.

4. In Western Africa, the British Colony of the Gambia is enclosed by the **French Colony of the Senegal**. Sierra Leone lies between the French territory, the Rivières du Sud, and the Republic of Liberia. The Gold Coast Colony lies between the **French Protectorate of the Ivory Coast** and the German Protectorate of **Togoland**. Lagos and the Niger Protectorates are enclosed on the west and north by the **French Sphere of Influence**. The Niger Protectorates are bounded on the south-east by the **German Protectorate of the Cameroons**.

BRITISH WEST AFRICA.

British West Africa includes the colonies of the **Gambia, Sierra Leone, the Gold Coast, Lagos,** the **Niger Territories,** and the **Niger Coast Protectorate**—all in Upper Guinea.

THE GAMBIA.

The small, but important, British Crown Colony of the Gambia includes St. Mary's Island, on which is the town of BATHURST, British Combo, Albreda, the Ceded Mile, McCarthy's Island, and various other islands and territories on the banks of the Gambia.

The River Gambia, which was frequented by English traders in the time of Queen Elizabeth, is navigable for vessels of 300 tons as far as the Rapids of Barraconda, a distance of 300 miles from the sea, but only the *lower river*, as far as McCarthy's Island, 180 miles above Bathurst, is regarded as British.[1] The staple export of the Colony is ground nuts, which are sent chiefly to Marseilles. Bees'-wax, indiarubber, and hides are also exported, but the total trade does not amount to half a million a year.

SIERRA LEONE.

The British Crown Colony of Sierra Leone includes the whole of the coast region between the Rivières du Sud on the north and Liberia on the south, together with the island of Sherbro, the Isles de Los, and other islands.

The coast-line is about 200 miles in length, and the Colony has a total area of about 4,000 square miles, with a population of perhaps 200,000. Sierra Leone Proper consists of a peninsula terminating in Cape Sierra Leone, and bounded on the north by the Rokelle or Sierra Leone River. This peninsula was ceded to Great Britain by the native chiefs in 1787, and was shortly afterwards formed into a place of refuge for liberated negroes. The colony is thickly-peopled, but there are only about 270 whites, the climate being simply pestilential to Europeans. The "White Man's Grave," as Sierra Leone is sometimes called, is, however, an exquisitely beautiful country, and its undulating hills are clad in an evergreen mantle of the most luxuriant vegetation, while all kinds of tropical fruits grow in abundance on the richly fertile and well-watered soil. And yet agriculture scarcely exists; there are no food resources, and if the colony were "cut off from England and America for three months, it would be in a semi-starving condition." The people of Sierra Leone are born traders, and are almost all engaged in trading European goods for the products of the interior—palm-oil, palm kernels, ground nuts, indiarubber, gums, hides, bees'-wax, kola nuts, &c. Three-fourths of the imports come from, but only one-third of the exports goes to, Great Britain. The trade, which amounts to about £800,000 a year, centres at FREETOWN, which is picturesquely situated on a slope of the "Sierra Leone" or *Lion Hill*, and has an excellent and strongly fortified harbour. There is frequent and regular steam communication with Liverpool, Hamburg, Havre, and Marseilles. The Rokelle River is navigable for 40 miles inland, and the Sherbro River for 20 miles.

There are numerous elementary and several higher-class schools in Freetown and a Training College at Fourah Bay, affiliated to Durham University. About 40,000 of the Sierra Leone Negroes are Protestant Christians, 8,000 are Mohammedans, and the rest are pagans.

1. The natives are Jolof and Mandingo Negroes; there are only 60 Europeans in the Colony.

THE GOLD COAST—ASHANTI.

The British Crown Colony of the Gold Coast comprises the harbourless coast between the Assinie River and the German Colony of Togoland, with a **Protectorate** extending inland to, and virtually including, the native State of Ashanti.

The Gold Coast Colony thus includes the whole of the low and unhealthy coast-plain between the Assinie River on the west and a point some 50 miles east of the entrance to the lagoon through which the **Volta River** enters the sea. The **Black Volta** rises in the Kong uplands, and is joined by the **Red Volta** and the **White Volta** in the Gonja country, and thence to a point about 70 miles from the sea it forms the boundary between the British and the German Protectorates. The Volta is navigable within the colony for small boats, as also are the **Ancobra** and the **Prah** Rivers, which flow through the gold mining district of Wassaw.

The Colony and Protectorate have an area of about **40,000** square miles and a population of about 1½ millions, of whom not more than 150 are Europeans. The natives are chiefly **Fantis**, who are akin to the fierce and more warlike **Ashantis** in the interior.

Of the numerous forts and factories established on the Gold Coast by various European nations, since the Portuguese built the Castle of Elmina[1] in 1481, the most important are **ACCRA** (20), founded by the Danes and purchased from Denmark in 1850, and now the capital and chief port of the colony ; **Cape Coast Castle**, the former capital, whose great church-like fort stands close to the water's edge, and the terminal port of the Great North Road which leads through Coomassie into the interior ; **Chama**, near the mouth of the Prah River ; **Axim** and **Elmina**, the most important of the Dutch settlements transferred to Great Britain in 1872 ; **Adda**, at the mouth of the Volta, and **Quitta**, further east on the coast near Cape St. Paul. Several trade routes from the interior converge at Accra, which is united by submarine cable with Europe, and by branch cables with Grand Bassam, Kotonu, Lagos, the Niger, &c., and by a land line with Cape Coast Castle.

Palm oil and **palm kernels** are the staple products, **india-rubber** abounds in the interior forests, and the whole region is rich in **gold**, which is now being worked by the aid of modern appliances.

Ashanti, before the power of its merciless tyrant was crushed by the British in the sanguinary war of 1873-4, was one of the most powerful of the Negro kingdoms of West Africa.

The Ashanti country is one continuous forest with small clearings around the native villages, which are usually perched on hills and always near water. The Ashantis are vigorous and well-formed Negroes, and belong to the same stock as the Fantis of the coast region. In the last war, the capital, **Coomassie**, was burnt, and the fearful massacres by which the Ashanti king had maintained his " reign of terror " happily came to an end, and the country is now being gradually brought under British influence.

---

1. In the 17th century, Elmina alone is said to have annually exported 3 million pounds.' worth of gold.

LAGOS—YORUBA.

The British Crown Colony and Protectorate of Lagos include the lagoon islands of **Lagos** and **Iddo**, and the adjoining lagoon-covered coast from **Kotonu** on the west to the **Benin River** on the east.[1]

Lagos thus lies between the French possessions on the Slave Coast and the British Protectorate of the **Niger**, from which it is separated by the Benin River, the most westerly of the numerous delta channels of the Niger. **Lagos Island** has an area of only 3¼ square miles—the whole Colony and Protectorate include about 1000 square miles, with a population of 100,000, only about 100 of whom are Europeans.

LAGOS itself is the largest town and most important port in all West Africa. It has a mixed population of about 60,000, and is fortunate in possessing the only safe harbour along 600 miles of coast, the 'river' of Lagos being one of the few navigable channels between the network of lagoons and creeks which extends along the coast. Lagos was formerly one of the chief slave marts on the Guinea Coast, but the trading stores or factories now exchange guns, cloth, tobacco—"anything from a fish-hook to a cask of rum"—for palm oil and kernels, cotton, ivory, gum copal, and other native products. A large trade is also carried on at Badagry to the west, and at Palma and Leckie to the east, of Lagos. The "Liverpool" of West Africa, as Lagos is called, is in regular communication with Liverpool itself by the mail steamers of the *African Steamship Co.* and by those of the *British and African Steam Navigation Co.*, and with Hamburg, New York, and Rio Janeiro by steamers and sailing vessels. The internal communication is chiefly by water along the network of lagoons.

The Ogun River affords an easy and direct passage from Lagos to Abbeokuta, the well-known and populous capital of **YORUBA**, an exquisitely beautiful and remarkably fertile country, bounded on the north and east by the Lower Niger. Northern Yoruba has been conquered by the Fulahs, and forms the riverain kingdom of Nupe, a vassal State of the empire of Sokoto, but directly attached to, and practically controlled by, the Royal Niger Company. South-east of Yoruba and extending thence to the Lower Niger and its delta, is the old kingdom of **BENIN**, now entirely within the British sphere. The Yoruba-Benin region, which measures about 250 miles from east to west, and some 200 miles from north to south, was split up into a number of more or less powerful native "kingdoms"—every petty chief in West Africa is a "king"—which were perpetually embroiled in little wars, and were active centres of the slave trade. The beneficent rule of the British authorities at Lagos and the energetic action of the Royal Niger Company, together with the remarkably successful missionary work carried on at Abbeokuta and other places, are gradually bringing order and security into this richly-productive tropical country, the climate of which, although not excessively unhealthy in the interior, debars the European from active labour, but does not injuriously affect the indigenous inhabitants. Here, as indeed throughout the verdant alluvial basin of the Niger, "millions might dwell in wealth and happiness were it not for incessant wars fomented by religious fanatics and petty kinglets," and these, wherever British influence extends, are rapidly becoming things of the past.

1. By an arrangement between England and France, a line intersecting *Porto Novo* at Appah Creek divides the French Territory from the British Colony of Lagos. The little Kingdom of *Kotonu*, which includes the narrow strip of Land between Denham Waters and the sea, and which had been attached to Lagos since 1879, was then also transferred to France in exchange for the 'Kingdom' of *Podrun*, between Badagry and Porto Novo.

THE NIGER PROTECTORATES.

The **Niger Protectorates** include the entire basin of the **Lower and Middle Niger**, with the whole of the coast from the **Benin River** to the **Rio del Rey**, and thus adjoin the British Protectorate of *Lagos* on the west and the German Protectorate of the *Cameroons* on the east.

On the north the British Sphere is limited by a line drawn (according to the Anglo-French Agreement of 1890) from Say on the Niger to Barrua on Lake Chad. On the west, to the north of Lagos, its limits are undefined, but are co-extensive with the territories belonging to Nupe, **Borgu**, and Gando, and on the east it includes all that fairly belongs to the empire of **Sokoto**. On the south-east British territory is divided from the German Protectorate by a treaty line drawn from the head of the **Rio del Rey** to the Ethiope Rapids on the Cross River, which debouches into the Old Calabar River, and thence continued in a north-easterly direction to the east of Yola, on the Benue, and thence to the southern shores of Lake Chad.

The **River Niger** and its great tributary the **Benue**, are the natural highways into the rich and populous countries of the Central Sudan, and the whole navigable course of both rivers is through **British territory**, while the enormous internal and coast trade is wholly in the hands of **British merchants**.

The dead level of the Niger Delta is broken by numerous channels, many of which, such as the **Benin** and the **Forcados** Rivers to the west, and the **Brass, New Calabar**, and **Bonny** Rivers to the east, of the main channel—the Nun River—are navigable by steamers of light draught, while vessels of 600 tons can ascend the Niger itself beyond the confluence of the Benue to RABBA, 600 miles from the sea. The Benue, also, which joins the Niger near LOKOJA, 300 miles from the sea, is navigable to RIBAGO, 70 miles above YOLA and 450 miles from its confluence with the Niger, and 750 miles from the sea.

The **Niger Protectorates**, which have an area of **500,000 square miles**, and a population of perhaps **20 millions**, include (1) the **Niger Territories**, governed by the Royal Niger Company, and (2) the **Niger Coast Protectorate** (formerly known as the **Oil Rivers District**), administered by an Imperial Commissioner.

The **Niger Territories** include the whole of the Middle and Lower Niger, except the portion of the coast-region under the Niger Coast Protectorate, and are governed by the ROYAL NIGER COMPANY (under the control of the Foreign Office), in virtue of treaties concluded with the native States and tribes (about 300 in number), including the empire of Sokoto and the vassal kingdoms of Gando, Borgu, Nupe, &c.

The **Niger Coast Protectorate** includes the whole of the coast-line between Lagos and the Cameroons, except that falling within the Niger Territories. British control is chiefly exercised in the estuaries between the Benin, Brass, and the Old Calabar Rivers.

The **staple** products of this region—*palm oil* and *palm kernels*, *vegetable butter, gums, ivory, hides, india-rubber,* &c.—are exchanged for *cotton* and *woollen goods, hardware, spirits, guns, gunpowder, salt,* &c., at the "factories," or trading stations, of which there are a large number on the coastal estuaries and on the Niger and its tributaries.

Nearly the whole of the **internal trade** of the Niger and the Benue is carried on by the **Royal Niger Company**, which has over 150 factories on or near the main stream and its tributaries, and a river fleet of 90 steamers, constantly employed in conveying the palm-oil and other native products collected at the various factories to **Akassa**, the port of entry at the Nun River (the main entrance to the Niger), where they are transhipped to the ocean steamers, most of which make Liverpool their terminal port. **ASABA**, a large town on the right bank of the Lower Niger, 150 miles from the sea and 50 miles above **Abo**, the very centre of the palm oil region at the head of the delta, is the seat of the local government (the general government is conducted by the Council in London), but the Company's military headquarters are at **Lokoja**, an important trading centre, near the confluence of the Niger and the Benue and nearly opposite the town of **Igbegbe**. Further south, on the same side of the river, is **Onitsha**, which marks "the northern limit of the palm oil trading region; higher up, ivory and shea-butter are the chief articles of trade."

In the **Niger Coast Protectorate**, formerly known as the Oil Rivers District and so called from the enormous amount of *palm-oil* brought down the **Benin, Brass, New Calabar, Bonny, Old Calabar,** and other "oil rivers" to the coast—most of the trade is in the hands of the *African Association* of Liverpool. The principal trading stations are at **Old Calabar**—Duke Town, Creek Town, &c.—**Opobo, Bonny, New Calabar, Brass, Warri,** and **Benin**. *Palm-oil* and *kernels* are the staple articles of trade at all the depôts in the district, but more *spirits, guns,* and *gunpowder* are imported into this district than into the territories of the Royal Niger Company, which imposes heavy duties on spirits[1] and on gunpowder and arms,[2] and absolutely prohibits the importation of spirits into the region north of lat. 7° N.

THE EMPIRE OF SOKOTO.

The Fulah Empire of **Sokoto**, which is practically a British Protectorate, is the largest and most populous of all the States of the Central Sudan, and includes all the former Haussa States between Lake Chad and the Niger, together with the tributary States of **Adamawa, Gando, Borgu, Nupe,** and other smaller "kingdoms" in the Niger-Benue region.

The agricultural resources of this fertile and well-watered region are considerable—rice and other cereals, dates and honey, are largely exported, and much cotton is grown, for the Haussa people are skilful in manufacturing it into durable material (which is coloured with indigo and other native dyes), and they also make excellent leather goods (shoes, sandals, harness, &c.).

---

1. 2s. per imperial gallon.
2. 100 per cent. ad valorem.
3. The student should read attentively the extracts given in the Appendix from Sir George Taubman-Goldie's admirable description of the | *International Struggle for the Niger,* which, owing to the energetic and patriotic action of the Royal Niger Company, resulted in the total elimination of foreign influence from the basin of the Lower and Middle Niger.

The Emperor of Sokoto is a direct descendant of the founder of the Fulah dynasty, and, as "Lord of the Mussulmans," has conferred on the Royal Niger Company full sovereign power throughout a large part of his dominions, and complete jurisdiction—civil, criminal, and fiscal—over non-natives throughout the remainder. He governs directly but a comparatively small part of the Empire ; the rest is ruled by vassal kings and chiefs who pay him an annual tribute. Most of the revenue is raised from these vassal States, over which the Sokotan army, estimated at about 90,000 infantry and 30,000 cavalry, is scattered. The armies and police forces of British West Africa are recruited from the Negroid Haussas of the Sokoto region, and render good service, being brave, faithful, and amenable to discipline.

Kano (35), the old Haussa capital, and still the commercial metropolis of the Central Sudan, lies about midway between Kuka, the capital of Bornu, and Sokoto, the former chief town of the Fulah conquerors of Haussaland. Sokoto is no longer the capital of the Empire, and its 100,000 inhabitants have dwindled to 10,000—scarcely half the population of Wurno, the present capital. Both towns are on the Sokoto River, an affluent of the Niger.

Gando, another large centre of population, is on the same river, about 90 miles south of Sokoto. It is the capital of the kingdom of GANDO, which extends on both sides of the Middle Niger Valley. Say, a town on the Niger, specified in the Anglo-French Convention as the starting point of the line of demarcation between the British and French spheres of influence in the Niger-Chad region, is within the territory of the King of Gando, with whom the Royal Niger Company has made an alternative treaty, by virtue of which his kingdom is practically a British Protectorate.

Niki, on the Oli River, a western affluent of the Niger, is the capital of BORGU, another vassal State of Sokoto. Borgu lies entirely to the west of the Niger, and to the south of Gando. The Oli River divides Borgu from the kingdom of NUPE, also a native State tributary to Sokoto, on the Middle Niger. The greater part of Nupe lies between the British Territories of Yoruba and Benin and the great bend of the Niger.

There are numerous smaller vassal States between the Niger and its great tributary the Benue. Yakoba, the capital of one of these, is a town of between 50,000 and 100,000 inhabitants, beautifully situated on a fertile plateau, at an altitude of 3,000 feet above the sea.

The basin of the Upper Benue is within the kingdom of ADAMAWA, the last conquest of the Fulahs to the south-east. This magnificent country is bounded on the north by Bornu, and on the north-east by Bagirmi. Towards the south and south-east its limits are undefined. The capital, Yola, is a busy trading centre on the Upper Benue, and is regularly visited by the Royal Niger Company's steamers, which ascend the river to Ribago, 70 miles above Yola. The line of demarcation between the British sphere in the Niger-Benue region and the German Protectorate of the Cameroons passes from the Rio del Rey on the coast to the neighbourhood of Yola and thence north-east to the southern shore of Lake Chad, and thus leaves almost the whole of Adamawa within the German sphere.

SOUTHERN AFRICA.

The Zambesi may be regarded as the natural limit on the north of this, the most important of all the great divisions of Africa. With the exception of the Algeria-Tunis region, in the extreme north, it is the only part of the continent suitable for permanent European colonisation.

Generally speaking, however, Temperate South Africa includes not only the vast region south of the Zambesi, but also the Nyassa Highlands to the north of that river and the adjoining uplands westward as far as the water-parting between the Zambesi and the Congo. On the other hand, the low-lying coastal zone on the east, from the Delta of the Zambesi as far south even as Delagoa Bay, belongs to tropical rather than to temperate Africa.

The whole of Southern Africa is not only ruled, directly or indirectly, by Europeans, but the richest and most productive lands are occupied by Europeans—mainly of Dutch or British origin—although, in all the South African States, the native races greatly outnumber the white population.

Politically and commercially, the British are the dominant people in Southern Africa, and, with the exception of the two Dutch Republics—the Orange Free State and the South African Republic—and of German South-West Africa, and part of Portuguese East Africa, the vast regions stretching from the Cape of Good Hope northward to the Zambesi and Lake Tanganyika, are included within the British Empire, either as colonies or protectorates. Even the two South African Republics may be regarded as British dependencies, inasmuch as they lie within the "sphere of British influence," and would certainly be included in any federation or union of the South African States.

BRITISH SOUTH AFRICA.

BRITISH SOUTH AFRICA includes the great self-governing colony of the Cape—the wealthiest and most important part of all South Africa—the colony of Natal, the Crown Colonies of Basutoland, British Bechuanaland, and Zululand, and the Protectorates of Tongaland, Bechuanaland, and Zambesia.

British South Africa, south of the Zambesi, is estimated to cover an area of not less than a million square miles, and to contain a population of not more than 4 millions, an average of only 4 persons per square mile.

THE CAPE COLONY.

THE CAPE COLONY, which derives its distinguishing name from the Cape of Good Hope,[1] is bounded on the *north* and *north-east* by

1. The *Cape of Good Hope* is important, both from its geographical position and from its place in the annals of discovery. It was discovered, in 1497, by Bartholomew Diaz, a Portuguese navigator, towards the close of a century which had been devoted by the Portuguese to the prosecution of maritime discovery along the western side of the African continent. Diaz succeeded in doubling the Cape in this voyage, and landed on the coast at some distance beyond. This he it was justly regarded as preparing the way for a passage to India by a maritime route. Ten years later, in 1497, Vasco da Gama (also a Portuguese) conducted the first fleet of ships to India by way of the Atlantic and Indian Oceans, passing, of course, round this famous headland. Diaz had called this long-sought headland (which he erroneously believed to be the extreme southern point of the African continent) by the name of "El Cabo Tormentoso," or the Stormy Cape; but the King of Portugal, on the return of Diaz, changed this name to *Cabo de Buena Esperanza*—the Cape of Good Hope—the name which it has ever since borne.

the Orange River, British Bechuanaland, the Orange Free State, Basutoland, and Natal ; on the *west*, by the Atlantic ; and on the *south*, by the Indian Ocean.

The first settlement was formed by the Dutch in 1652, in the immediate neighbourhood of Table Mountain.   The Dutch colonists gradually pushed inland to the Great Fish River on the east, and to the Roggeveld range on the north. In 1796, the colony was occupied by the British, but was given back in 1803, at the peace of Amiens.   Three years later, it was again taken by the British, and has since remained British territory.   But it was not until the year 1847 that the northern frontier of the Cape Colony was extended to the Orange River, an artificial line, drawn considerably to the southward of that stream, having previously marked its limits.   The area of country under British rule was gradually enlarged by the annexation of British Kaffraria in 1866, Basutoland[1] in 1868, Griqualand East in 1875, Griqualand West in 1876, and subsequently the Transkeian territories ; so that Natal and the Cape Colony are conterminous, the boundary being the River Umtamfuna.   Walfish Bay, on the coast of Damaraland, in German South-West Africa, also belongs to the Cape Colony.

EXTENT : The area of the Cape Colony, including the Transkeian Territories and Walfish Bay, is about 233,430 square miles, or more than 4 times that of England and Wales.   The *extreme length* of the colony is 600 miles, and the *breadth*, about 450 miles.

Griqualand West, to the north of the Orange River, and the whole region south of the Orange and west of the Kei River, are included in Cape Colony proper—the Transkeian Territories, between the Kei River and Natal (the *Transkei*, comprising Fingoland, the Idutywa Reserve, and Gealekaland, *Tembuland*, *Pondoland*, and *Griqualand East*), are dependencies of the Cape Colony, as also is the district of Walfish Bay, along with several small islands, on the West Coast.   The Transkeian Territories have an area of 15,283 square miles, and Walfish Bay 430 square miles, so that the colony proper has an area of nearly 218,000 square miles, or over 3½ times that of England and Wales.

COASTS : The coast-line of the Cape Colony is of the same regular and unbroken character as that of the African continent generally, and embraces but few harbours, although it is upwards of 1,300 miles in length.

Cape Agulhas and the Cape of Good Hope—the two most important headlands of South Africa—are both within the coast-line of the Cape Colony. The former is the most southerly point of the African continent and of the Old World.

INLETS : The most considerable inlets are St. Helena, Saldanha, and Table Bays, on the west coast ; False Bay with Simon's Bay, and Mossel and Algoa Bays, on the south coast.

Of the above inlets Saldanha Bay forms naturally the best harbour.   At Table Bay, which is somewhat unsafe during the north-west winds, extensive harbour works are in progress which will render the harbour permanently secure and accessible at all times.   Simon's Bay, which is entered through False Bay, is well sheltered and commodious, and is the principal South African

---

1. Basutoland, formerly attached to the Cape Colony, is now a British Crown Colony. *See* p. 60.

Station of the British fleet. Table Bay is so named from its vicinity to *Table Mountain*—an elongated mass of hill, with a flat top, which rises above its southern shore, and in the immediate vicinity of Cape Town. Table Mountain reaches 3,582 feet in height.

RELIEF : The surface of the colony rises from the sea-board to the interior, not regularly, but by a series of terraces or steps, the seaward edges of which are marked by the long ranges of mountains and hills that extend across the country from west to east.

The first high barrier between the coast region and the inland districts is formed by the **Drakenstein** and other ranges which extend from False Bay to the Orange River. Behind this range is another irregular series of elevations which bend eastwards parallel to the south coast, under the name of the **Lange Berge**. The land then rises another step to the lofty range known as the **Zwarte Berge**, which marks the seaward edge of the **Great Karroo** plateau—a vast plain of some 20,000 square miles—the clayey soil of which is parched and arid in summer, but covered with luxuriant vegetation whenever the scanty rains fall. It then affords excellent pasture for sheep, cattle, and horses, which thrive wonderfully on the aromatic herbage. The Karroo plateau is about 70 to 90 miles in width and between 2,500 and 3,500 feet in height, and is cut off from the scantily watered uplands which slope to the Orange River by the long and comparatively lofty range which, under various names, extends from the hills of Namaqualand on the west, to the still loftier range of the **Drakensberg** on the east. The central range, the **Nieuwveld**, is flanked by the **Roggeveld** on the west, and by the **Winterberge**, **Sneeuwberge**, and **Stormberge** on the east —the culminating point of this bold escarpment of flat-topped hills being **Compass Berg**, in the Sneeuwberge range, which rises to a height of nearly 9,000 feet. This long range forms the central water-parting of the colony— the drainage on one side running north to the *Orange River*, and on the other flowing south-eastward to the *Indian Ocean*.

RIVERS : Most of the numerous rivers of the Cape Colony are periodical streams, flooded to excess after the rains, and speedily drying up, or forming a mere chain of pools, in the dry season. Not one of them is of any considerable value for navigation.

The principal rivers are the **Orange** and the **Olifants**, flowing into the Atlantic ; and the **Breede**, **Gauritz**, **Gamtoos**, **Great Fish**,[1] **Keiskamma**, and the **Kei**, flowing into the Indian Ocean.

The **Orange River**, which forms the northern border of the Cape Colony, is about 1,200 miles long, but it has comparatively little volume of water, and is not navigable, except for small craft for about 30 miles above the bar at its mouth. The **Olifants River** is a Nile in miniature, and the coast-land, which it overflows and covers with the rich Karroo mud, produces splendid crops of grain.

The lower course of the **Breede** is navigable for small vessels for a short distance inland, but the **Gauritz**, **Gamtoos**, and **Great Fish River** are subject to sudden and violent floods, the water rising sometimes 20 or 30 feet in an hour or two after a heavy thunderstorm. The smaller streams further east have a more permanent flow.

---

1. There is another Great Fish River, a tributary of the Orange, in Namaqualand.

CLIMATE : The climate of the Cape is **temperate, dry, and healthy.** The occasional prevalence of **droughts** is its chief drawback. In the eastern province **rain** falls in summer, but in the western districts in winter.

The climate of the Cape Colony is **very healthy,** the air being buoyant, clear and dry. At the same time, it varies much in different districts. The coast climate is **warm, moist,** and equable ; the midland is **colder and drier** in winter and **hotter** in summer ; the mountain climate is **drier** still and **more bracing,** but with extremes of heat by day and cold by night. The hottest month is generally January (the average maximum temperature at Cape Town is 83° F.) ; July is usually the **coldest month** (the average minimum temperature at Cape Town is about 45° F.).

In the north-west province of the Great Karroo less than 6 inches of rain fall in the year. At Cape Town, the rainfall amounts to 30 inches ; at Kimberley, 16 ; and at Port Elizabeth, 25, as against 26 inches in London.

PRODUCTIONS : The extensive open plains of the interior are admirably suited for **pastoral pursuits ;** and the millions of live-stock reared upon them supply the most valuable products of the Colony.

The sheep, goats, cattle, and horses, supply the wool, mohair, skins, and hides, which, with diamonds, copper, and ostrich feathers, form the staple exports. The colony also produces excellent wheat and other cereals, with large quantities of grapes (principally used for making wine), oranges and other fruits. The heaths and other native plants are peculiar to this part of the world.

The chief mineral products are diamonds from the famous Kimberley mines, and copper from Little Namaqualand. But there are also rich deposits of iron ore and coal, and some gold has been found in the Knysna district.

INHABITANTS : The Cape Colony and its dependencies contain about 1½ million inhabitants, one-third of whom are **Europeans**—mainly of **Dutch, British, and German** origin. The rest are **Kaffirs** and other coloured races.

A large number of the white population is of British descent, but the bulk of the people in the western provinces are of Dutch origin, owing to the fact that the colony was originally established by the Dutch, and only came into the possession of Britain in the early part of the present century. The coloured races include Hottentots and Kaffirs—the two native races of this portion of the African continent. The pure Hottentots are now few in number ; the people of the Kaffir race are much more numerous, and perhaps amount to a third of the entire population of the colony. Malays are numerous in the larger coast towns.

INDUSTRIES : The chief pursuit is **sheep** and **cattle rearing,** for which the colony generally is better suited than for agricultural operations. Mining, especially for *diamonds* and *copper*, is an important industry, and furnishes by far the most valuable export.

The sheep farms are very large, but those in tillage are comparatively small. There are over 13 million sheep, 5 million goats, 1½ million **cattle,** and ¼ million horses in the colony. Large quantities of maize, **wheat, millet,** and other grains are grown, but the cultivable area is limited, owing to the deficient rainfall, and, except in some favoured localities, it is impossible to rely upon a regular return from the soil without *irrigation*. The vine is cultivated on a scale of some magnitude in the western provinces, and over 4½ million gallons of wine were produced in 1890. Almost all kinds of sub-tropical and temperate fruits grow well, but very little fruit is exported. Ostrich farming is an important industry, peculiar to, and practically a monopoly of, the colony. The birds are bred and reared like domestic poultry, but the value of the feathers fluctuates so much, that only large capitalists can risk the heavy losses caused by a sudden fall in prices, owing to some caprice of fashion.[1] There are about 114,000 domesticated ostriches in the colony, and, to prevent the industry spreading to other countries, there is an export duty of £100 on every ostrich, and £5 on every ostrich egg.[2]

Diamonds, however, form the principal and an apparently inexhaustible source of wealth to the colony. Since the gem was first discovered in South Africa in 1867, over 50 million pounds' worth of diamonds have been exported, chiefly from the Kimberley mines. Copper ore, the only other mineral exported, is obtained principally from the rich and productive mines of *Ookiep* in Little Namaqualand ; the ore is conveyed by a railway, worked by mules, to *Port Nolloth*, whence it is shipped to South Wales to be smelted.

Means of Communication : None of the rivers, as we have said, except a part of the lower courses of the Breede and the Orange, affords the means of access to the interior. The roads are generally good in the more settled districts. In the interior they are mere tracks. Railways are now being rapidly extended, there being 1,800 miles open for traffic, arranged in three systems—the Western System, starting from *Cape Town* and extending north to *Kimberley*, and thence through British Bechuanaland via *Taungs* and *Vryburg*, to *Mafeking*; the Midland System, from *Port Elizabeth* to *Colesberg*, and now extended to *Bloemfontein*, *Pretoria*, &c.; and the Eastern System, from *East London*, through *Queenstown*, *Cyphergat*, and *Burghersdorp*, to *Aliwal North*. Branch lines connect *Graaf Reinet* in the interior, and *Port Alfred* on the coast, with Port Elizabeth.

COMMERCE : Nearly all the trade of the Cape Colony is with the United Kingdom. Diamonds, wool, copper ore, hides and skins, ostrich feathers, gold[3] (from the Transvaal), with wine and brandy, form the chief exports. The imports are principally *textile fabrics, articles of food and drink, hardware*, and *machinery*.

The trade of the colony, owing to steady progress within, and the development of the countries beyond its borders, is rapidly increasing ; the total *imports*, in 1893, excluding specie, were over 11½ millions sterling. The *exports* have also increased from 5½ millions in 1885, to nearly 14 millions in 1893. Next to diamonds,[3] which average over 3 millions a year in value, wool is the chief product of the colony, but much of the wool exported from Port Elizabeth and other ports comes from the Orange Free State, the Transvaal, and Bechuanaland.

The chief ports are CAPE TOWN, on Table Bay ; Port Elizabeth, on Algoa Bay ; and Port Alfred and East London, on the south-east coast. Extensive harbour works are in progress at these four ports, and over 1½ millions have

---

1. For instance, the declared value of the 253,654 lbs. of the ostrich feathers, exported in 1882, was £1,093,989, but the value of the 229,137 l.s., exported in 1889, was only £365,884. In 1890, although only 212,00 lbs. were exported, the value was £561,000.

2. For a vivid description of the ostrich and of life on an ostrich farm in the Cape Colony, *see* Mrs. Annie Martin's *Home Life on an Ostrich Farm*. (London : George Philip & Son).

3. Gold, to the value of £7,000,000, was exported in 1893.

been already spent on the works at Table Bay. At **Port Nolloth,** on the west coast, the copper ore from the Ookiep mines is shipped.

GOVERNMENT : The Cape Colony possesses **responsible government** under a Governor appointed by the Crown.

The **Cape Parliament,** in which speeches may be made in Dutch as well as in English, consists of an elected **Legislative Council,** representing the provinces, and an elected **House of Assembly,** representing the country districts and towns of the Colony and the Transkeian Territories.

In 1892-3 the *Revenue* amounted to 6½ millions, and the *Expenditure* to 5¾ millions. The *Public Debt* of the Colony, which amounts to nearly 27 millions, has been incurred almost entirely in the construction of public works. About 20 millions have been spent on railways alone.

DIVISIONS : Cape Colony is divided into **eight electoral provinces,** which are subdivided into **75 magisterial districts.**

The **Western Province** includes the older settled and the most populous portion of the Colony.

CAPE TOWN, the chief place in the province, is the capital of the Colony and by far the most important town in the whole of British South Africa. The town, which derives its name from the famous promontory which it adjoins, stands on the shores of Table Bay, and, with its suburbs, has about 60,000 inhabitants. It was originally laid out by the Dutch, and presents in its general appearance a great resemblance to the towns in Holland. Its trade has largely increased within the last few years.

The CONSTANTIA district to the south, and the STELLENBOSCH and PAARL districts to the east of Cape Town, are famous for their rich vineyards, and *wine-making* is rapidly becoming an important industry in this province, which possesses a climate more favourable to the cultivation of the grape than that of any other country in the world.[1]

The **North-Western Province** extends from the wine district of MALMESBURY (north of Cape Town) to the Orange River. It has rich corn lands along the coast in the south, and productive copper mines in the north. The population of this province is scanty, and the few towns are small.

The **South-Western Province** extends along the coast between False Bay and Knysna Harbour, and stretches inland to the Zwarte Berge. On the coast are two small ports—MOSSEL BAY (Aliwal South) and KNYSNA (Melville). SWELLENDAM, ROBERTSON, LADISMITH, and OUDTSHOORN, are small towns in the wine districts between the Lange Berge and the Zwarte Berge.

The **Midland Province** includes the central portion of the Colony, from the Zwarte Berge on the south to the Orange River on the north. Here the chief place is GRAAF REINET, the oldest and the largest of the midland towns. This "gem of the desert" lies in a bend of the Sunday's River, and the avenues of trees which line the broad streets and vineyards, the gardens and orchards attached to nearly all the houses, contrast strongly with the arid appearance of the surrounding Karroo plains. The *Western Railway,* from Cape Town to Kimberley and Bechuanaland, traverses the eastern part of this province and crosses the Orange River near HOPE TOWN.

---

1. The grapes of the Cape are superior in quality | ber to March. The *Constantia wines* are equal to to European grapes, and I could, it is said, be shipped | those of any country, but Cape wines generally are in immense quantities during the season—Decem- | not properly prepared for export.

The South-Eastern Province includes the coast districts to the west and east of PORT ELIZABETH, the great *wool port* on Algoa Bay, now connected by rail with Bloemfontein, the capital of the Orange Free State, and with the Diamond Fields and Bechuanaland, and also with PORT ALFRED on the coast, and *Graaf Reinet* in the interior.

The Port Alfred branch railway passes through GRAHAMSTOWN, the metropolis of the eastern and frontier districts, which lies embosomed in green hills at an elevation of 1,760 feet above the sea. The city, with its broad streets lined with trees, and its houses interspersed with gardens, presents a thoroughly English appearance, and is a favourite health-resort and one of the most pleasant places of residence in the Colony.

The Eastern Province extends along the coast from the Great Fish River to the Great Kei River and inland to the Orange. The *Eastern Railway* runs through this province from the port of EAST LONDON, at the mouth of the Buffalo River, through QUEENSTOWN to ALIWAL NORTH on the Orange River. A short branch runs to KING WILLIAM'S TOWN, an important commercial centre between East London and the interior, and formerly the capital of the territory then called British Kaffraria.

The North-Eastern Province includes the interior districts between the middle course of the Great Fish River and the Orange, and is traversed by the *Midland Railway* from Port Elizabeth. The towns of SOMERSET and BEDFORD are near, and CRADOCK and COLESBERG on, this railway, which has been extended across the Orange River to Bloemfontein, the capital of the Orange Free State.

The Province of Griqualand West is the only part of the Colony to the north of the Orange River. Naturally bare and uninviting, except along the Orange and the Vaal, but with a dry and healthy climate, this province was inhabited, before the discovery of diamonds, by only a few thousand Griquas and a few Dutch and English settlers, until the "diggings" attracted thousands of fortune-hunters to it, and, now that diamond-mining has here become a settled industry, KIMBERLEY[1] has been transformed from a mushroom camp into a permanent mining centre. The town of BEACONSFIELD, which is built on the north-west side of the Bultfontein and the Dutoitspan mines, is rapidly growing in importance and ranks next to Kimberley as a mining centre.

The Transkeian Territories, which extend from the Great Kei River to the borders of Natal, are dependencies of the Cape Colony. The majority of the population are natives, who are all under British rule. These territories, which have a total area of over 20,000 square miles and a population of 800,000 (only 12,000 of whom are Europeans), are now grouped under three chief magistrates, with several subordinate magistrates. Pondoland was formally annexed to the Cape in 1894.

The Transkei includes *Gcalekaland*, on the coast between the Great Kei and the Bashee River, and *Fingoland* and the *Idutywa Reserve* in the interior.

Tembuland extends from the coast (between the Bashee and the Umtata rivers) inland to the Kathlamba Mountains.

1. Kimberley is 647 miles from Cape Town, and | express trains (Fridays only) take 32 hours (fare the third class trains take 44 hours (fare £2 3s. 10d.), | £8 1s. 9d.).

Pondoland, the last part of Kaffraria to be brought under direct British rule, includes the coast region between Tembuland and Natal—its limits being marked by the Umtata and the Umtamfuna rivers. There is a road from UMTATA in Tembuland to ST. JOHN'S, a small port on the Pondo coast belonging to the Cape Colony, and a good road has been constructed by the Cape Government from the St. John's River to KOKSTAD, in the fine territory of

Griqualand East, which includes the well-watered districts between Pondoland and the great range of the Drakensberg. Here sheep farming and the rearing of cattle and horses are successfully carried on. *Horse sickness*, which is so destructive in many parts of South Africa, is here unknown.

.. The Walfish Bay district on the West Coast, which has an area of 430 square miles, is attached to the Cape Colony, and is the principal outlet for the vast, and at present undeveloped, territories which form German South-West Africa.

NATAL.

The British colony of Natal[1] includes the territory lying between the Tugela River on the north and the Umtamfuna on the south, and extending inland from the coast to the Drakensberg Mountains, which separate it from Basutoland and the Orange Free State.

Natal has an estimated area of 21,150 square miles, nearly one-third that of England and Wales, with a seaboard of about 200 miles. The distance from the coast to the Drakensberg varies from 80 to 120 miles.

NATURAL FEATURES: The surface of Natal rises by a series of terraces from the sea coast to the Drakensberg Mountains, and may be roughly divided into the coastlands, the midlands, and the uplands, each with its characteristic climate and productions. The scenery is in parts picturesque in the extreme, and the climate, except along the coast at certain seasons, is on the whole extremely healthy.

The coastlands extend inland for about 25 to 30 miles, and are for the most part well wooded and watered. The climate and vegetation are almost tropical in character, and the soil is suitable for the production of *sugar, tea, coffee, indigo, arrowroot, rice, cotton, bananas, pineapples*, &c. *Maize*, or "mealies," is the chief crop throughout the colony, but *wheat* and other cereals, and green crops, are chiefly grown in the midland districts, the climate and soil of which are better adapted for European crops and stock. The higher uplands are suitable for *sheep-farming* and for the rearing of *horses* and *cattle*.

Much of the timber has been cut down, but there are dense forests in some of the upland "kloofs" and along the coast. There are extensive plantations of the magnificent Australian blue gum-tree *(Eucalyptus globulus)* in the more settled districts.

Stone and lime for building and other purposes abound, and some gold has been found at Umzinto (Alexandra County) and elsewhere; but more important than all is the coal found, and now extensively worked, at Newcastle and Dundee in Klip River County.

---

1. The coast of Natal was discovered by Vasco da Gama, a Portuguese navigator, on Christmas Day, 1497, hence its name.

INHABITANTS : Natal now contains over **550,000** inhabitants, of whom 40,000 are Europeans, and 35,000 Indian coolies. The rest are Zulu-Kaffirs.

The Europeans nearly all speak English, but half of them are of Dutch descent, and there are also some German and Norwegian settlers.

The occupations of the Kaffirs are almost exclusively pastoral, but the European colonists have brought a large portion of the land, especially along the coast, under cultivation. Sugar is the chief product of the coast region, but coffee, tea, tobacco, arrowroot, and the sweet potato are also grown. Wheat, barley, and oats are largely grown on many of the midland and upland farms. Maize and Kaffir corn (the staple food and drink material of the natives) grow luxuriantly everywhere.

Sheep-farming is the most important industry in the colony, and cattle and horses are reared in all parts, but principally in the midlands and northern uplands, where they are generally exempt from the peculiar sickness which is so fatal to stock in the low-lying districts.

Natal coal, mined at Newcastle and Dundee in the northern part of the colony, is now used on the Government Railways (the main line passes through Newcastle, and there is a branch line to Dundee) and is supplied to steamers in the bay at Durban. About 120,000 tons of coal are raised annually.

GOVERNMENT : The government of Natal is administered by a **Governor**, aided by an **Executive Council** of the chief government officers, and a **Legislative Council** of 11 members, and a **Legislative Assembly** of 37 members.

Before 1856, Natal formed part of the Cape Colony. In that year it was erected into a distinct colony, first under a Lieutenant-Governor, and, since 1879, under a Governor ; but, in 1893, the colony was granted full responsible government.

With the rapid increase in the trade of the colony, consequent on the extraordinary "rush" to the gold-fields in the Transvaal, the revenue and expenditure of the colony have increased by one-half in about six years. The actual revenue is over, and the expenditure under, 1¼ millions sterling. The Public Debt is over 7 millions, mainly for railways and other public works.

COMMERCE : Nearly all the trade of Natal is with the United Kingdom. Imports, 1893-4, 2¼ millions sterling. Exports, 1¼ millions. Three-eighths of the exports are from, and one-third of the imports is for, the neighbouring Dutch Republics.

The *chief exports* are wool, gold, sugar and rum, skins and hides. Much of the wool and hides comes from the Free State and the Transvaal, and almost all the gold from the Transvaal. Of the exports, about 60 per cent. go to Great Britain, and 75 per cent. of the *imports*—chiefly iron and iron goods and machinery, textile fabrics, and leather goods—are British.

DIVISIONS : Four of the counties—**Alfred, Alexandra, Durban**, and **Victoria**—are on the coast ; **Pietermaritzburg, Weenen,** and **Umvoti** occupy the central, and **Klip River** the northern, part of the country.

With the exception of **Pietermaritzburg**, the capital, which has a population of 16,000—of whom 9,000 are Europeans—and of **Durban**, the only port, with

a population of 24,000—of whom one-half are Kaffirs and Indians, the towns
are very small, the largest being **Verulam, 19** miles by rail north-east of Durban, **Ladysmith, 190** miles, and **Newcastle, 264** miles by rail north-west of
Durban.

. The whole of the foreign trade of Natal, and a great part also of that of the
Transvaal and the Orange Free State, passes through DURBAN or Port Natal,
and, now that the bar at the mouth of the bay is being removed and the railway has been extended beyond Ladysmith and Newcastle to Charlestown on
the Transvaal border, the port will soon rank with Cape Town and Port Elizabeth among South African ports. From Durban branch lines run north to
Verulam (19 miles distant) and south to Ispingo (12 miles). A branch line
from Ladysmith crosses the Drakensberg range by the *Van Reenen Pass* to
Harrismith, in the Orange Free State; and, midway between Ladysmith and
Newcastle, a short branch connects the **Dundee Coalfield** with the main line.
Post carts for mails and passengers, and *ox-waggons* for goods and produce,
run to Richmond and other towns to the west, and to Greytown, Weenen, &c.,
to the east, of the main line of railway. PIETERMARITZBURG, the capital
of the colony, is picturesquely situated on a tributary of the River Umgeni,
about 70 miles distant from Durban (by rail). It lies at an elevation of about
2,000 feet above the sea, and consequently has a much cooler and healthier
climate than Durban.

ZULULAND.

The **British Crown Colony of Zululand** lies to the north-east of
Natal, from which it is divided by the **Tugela River**, and includes
about two-thirds of the former Zulu kingdom.

The "New Republic," founded by the Boers in Western Zululand in 1886,
was united to the South African Republic in 1888, and now forms one of the
Transvaal districts under the name of **Vryheid**.

British **Zululand** has an area of **9,000** square miles and a population of about **150,000** (of whom only 500 are Europeans), or, including the adjoining territory of **Tongaland** or Amatongaland (as
far north as *Kosi Bay* and the *Usutu River*, which mark the
southern limit of *Portuguese East Africa*), **14,220** square miles,
with a population of over **180,000**.

The Zulus, from being a nation of warriors, always ready and eager for war,
have at last settled down to the more peaceful pursuits of agriculture and cattle-
rearing. Large crops of maize or "mealies"—the staple food of the Kaffirs
throughout South Africa—and Kaffir corn (principally used in making *pombe* or
native beer) are grown around the *kraals* or huts of the natives, who also possess
large herds of cattle.[1]

Oxen and maize are, in fact, the chief articles of trade, and are exchanged
for cotton goods and hardware, brought into the country by traders from Natal,
the absence of any good landing-place preventing communication by sea.
St. Lucia Bay, Sordwana Point, and Port Durnford are "nothing but exposed
and surf-bound beaches with no harbour."

Means of Communication: There is a main road from Eshowe to the borders
of the Transvaal and across **Rorke's Drift** (ever memorable for its gallant defence

---

[1]. No European can hold land in the colony except for missionary, mining, or trading purposes.

by Lieut. Chard and a few men, after the fearful disaster to the 24th Regiment at Isandhlana) into Natal, and good waggon roads lead from this to the magistrates' stations in the different *Districts* into which the territory is divided. ESHOWE, where the Commissioner and Chief Magistrate resides, is in communication with Natal by telegraph, and by daily post carried by native runners.

Ulundi, on the *White Umvolosi River*, was the chief kraal of Cetewayo, the last king of the Zulus, and here the Zulu army, which had for so long been the terror of South-Eastern Africa, was defeated and finally dispersed. The Zulu king was reinstated in 1883, but died the following year, and the British Government, finding that the Zulus could not form any orderly government, declared the country British territory in 1887.

SWAZILAND.

SWAZILAND is a small Native State enclosed on three sides by the Transvaal, and bounded on the east by Tongaland and Portuguese Territory. It has an area of about 8,000 square miles, and a population of about 60,000. The Lobombo Mountains extend along its eastern borders. The country is watered by the Usutu River and its tributaries.

Swaziland has long been known as a country rich in minerals, and particularly in gold, and the whole of it is now in the hands of concessionaries for mining, farming, grazing, &c. Its mineral wealth and agricultural resources caused it to be eagerly coveted by the Boers, who desired to absorb it and the adjoining strip along the coast, so that the Republic might have free access to the sea. The British Government, however, refused to recognise the Dutch claims, and, by the Convention of 1890, it was agreed to recognise the independence of the Swazi king and people, and to entrust the government of the country to a Government Committee of three members, representing the British, Dutch, and Swazi Governments respectively. By the Convention, the British Government recognised the concession, granted by the Swazi king to the South African Republic, to construct a railway through Swaziland to the sea, at or near Kosi Bay. The Boer Government, however, continued to press its claims, and, in spite of the protests of the natives and British settlers, the British Government has recently (1895) consented to allow the country to pass under the exclusive control of the Republic. The small town of Bremersdorp is the capital and chief centre of trade.

TONGALAND.

TONGALAND is a much smaller Native State on the coast to the north-east of Zululand, and was taken under British protection in 1887.

Tongaland is about 75 miles long and 40 miles broad, and has an area of about 5,000 square miles. It is bounded on the north by the Portuguese territory of Lorenzo Marquez (Delagoa Bay), and on the west by Swaziland. The country generally is level, and is watered by the Pongola and other streams. The Transvaal Government has the right to purchase land for a railway to connect the Republic with the sea at or near Kosi Bay, but by the completion, in 1895, of the Delagoa Bay-Pretoria line, the object of this concession has been already attained.

BASUTOLAND.

The British Crown Colony of Basutoland lies to the north-east of the Cape Colony, and forms an irregular oval, 10,000 square miles in extent, completely enclosed by the Cape Colony, Natal, and the Orange Free State. Except on the south-west, its boundaries are formed by strongly marked natural features, the great range of the Drakensberg separating the colony from Griqualand East and Natal, while the Caledon River, a tributary of the Upper Orange, divides it on the north-west from the Orange Free State.

Basutoland is really one continuous plateau, though broken by the Maluti Mountains and other rugged spurs of the Drakensberg. The eastern half of the colony is watered by the Upper Orange and its upper affluents; the western streams flow into its tributary, the Caledon River. But none of the numerous waterways are navigable, being flooded in summer and very low in winter.

This well-watered upland is in many respects the most favoured part of all South Africa. It has the best wheat-growing land in all Africa, and its rich pastures enable the natives to rear immense herds of cattle and large flocks of sheep, while the climate is simply delicious. The mountain scenery is grand, and in many parts the country is extremely beautiful.

The liberty-loving Kaffirs of the Basuto plateau have fought bravely against both the Boers and the Colonial forces, and when the Cape Government, which had annexed the country in 1871, attempted to carry out the Disarmament Act in 1880, nearly the whole tribe revolted. Strenuous efforts were made to reduce the Basutos to submission, but without success, and at a great national "Pitso" in 1883, they demanded to be taken under the immediate authority of the Crown. This was done in 1884, and since then the country has been quietly progressing under a Resident Commissioner at Maseru, the capital, and several Assistant-Commissioners or Magistrates. The districts or wards are presided over by hereditary chiefs, who settle all disputes between the natives, subject to a right of appeal to the Magistrates' courts.

The Basutos, who now number about 200,000, are the most industrious and civilized of all the Kaffir tribes of Southern Africa. European settlement is prohibited, and there are only about 500 Europeans (connected with the government, trade, or missionary work) in the country. There are good roads, and an active trade is carried on with the Cape Colony and the Orange Free State—grain, cattle, and wool being exported, and blankets, ploughs, saddlery, clothing, iron and tin wares, and groceries being imported.

BECHUANALAND.

BECHUANALAND lies to the north of the Cape Colony, and is bounded on the east by the Transvaal, and on the west by German South-West Africa. The total *area* is 170,000 square miles, or nearly 3 times the area of England and Wales.

Until 1884, Bechuanaland was an independent native territory, but a number of Boer freebooters from the Transvaal having intervened in disputes between two Bechuana chiefs, appropriated farms, and set up two Republics—Stellaland and Gosben—the British Government determined to protect the defeated chiefs —Mankoroane and Montsioa—and to keep open the great trade-route into the interior. A strong force of 4,000 men was therefore sent, under Sir Charles Warren, to occupy the country. This was effected without any armed opposition, and in 1885 the troops were withdrawn, after forming the territory south of the Molopo River into a British **Crown Colony** and declaring Northern Bechuanaland a British **Protectorate.**

The **Crown Colony of British Bechuanaland** is enclosed on the north and west by the Molopo-Hygap River, on the east by the Transvaal, and on the south by Griqualand West and the Orange River.

The **British Protectorate of Bechuanaland** includes the rest of the country north of the Molopo River to 22° of South latitude, and west to the 21° of East longitude. The colony has an area of 43,000 square miles, and the Protectorate about 127,000 square miles.

The country generally is about 4,000 to 5,000 feet above the sea-level, and is on the whole healthy, though very hot and dry in summer. Much of the colony is waterless, and even the Molopo or Hygap River (a tributary of the Orange River), with the Nosop River and other affluents, are dry in summer. A large proportion of the Protectorate lies within the arid Kalahari Desert.

In the more settled districts adjoining Griqualand West and the Transvaal, agriculture and cattle rearing are rapidly becoming important industries. The soil in many parts is extremely fertile, and with irrigation and slight culture will produce rich crops of grain, &c. Even now, considerable quantities of maize or "mealies," wool, hides, cattle, and wood are sent into the Kimberley market, in exchange for textiles, hardware, and other manufactured goods. Gold and other minerals have been prospected for, but without success. The railway has been extended from *Kimberley* to **Taungs** and **Vryburg** (774 miles from Cape Town) and is now open to Mafeking, in the extreme north-east of the colony, close to the western border of the Transvaal. VRYBURG is the capital of the colony, but MAFEKING is the largest European township and the chief centre of trade, which is carried on with the natives of the Protectorate and with the Boers of the Western Transvaal. The great events of the year 1890 were the opening of the railway from Kimberley to Vryburg (extended since then to Mafeking), and the passage through the country of the **British South African Company's Expedition** to Mashonaland. Both these causes largely increased the population and trade of the colony.

The Governor of the Cape Colony is also **Governor of British Bechuanaland,** and has power to legislate by proclamation. The actual administration is entrusted to an **Administrator** at Vryburg, and a Resident Magistrate at each of the five *districts* into which the Colony is divided, namely, Vryburg, Mafeking, Taungs, Kuruman, and Gordonia. Order is maintained in the Colony and Protectorate by the **Bechuanaland Border Police,** a well trained force of about 500 mounted men.

The dominant native ruler in the Protectorate is **Khama,** the paramount chief over the whole of the **Bamangwato** tribe, except a small section. The Bamangwato people are a peaceful race, feeding and clothing themselves by agriculture and hunting, and their king, Khama, is a man of great refinement and intelligence, a thorough Christian, an able ruler, and a shrewd diplomat. Travellers, traders, and hunters, all speak of him with respect. "Of Khama's splendid character," writes one, "I cannot speak too highly." The country under Khama's rule extends beyond the limits of the Bechuanaland Protectorate (22° S. lat.) to the Zambesi. Khama's country possesses very considerable natural riches, were they but developed.[1] Especially is this so with regard to the district around **PALAPYE**, the present capital, and to the north and north-east; while the mountain district behind Shoshong, the former capital, is said to be the very garden of the country. There are fine sites for farming and much well-watered land in other parts of Bamangwatoland, and Khama is very anxious for English settlers to come and settle in his country. But they must, he says, be "true settlers and workers, and not travelling traders." Khama has been most resolute in prohibiting the importation of spirits by traders and the manufacture of the stupefying native beer, with the result that at Shoshong, "it required no police to manage the native part of the town." Shoshong was deserted in 1889, and Khama, with out the slightest European assistance, laid out a new town at Palapye, covering some 20 square miles of ground and holding 30,000 people—a wonderful town for native South Africa, with broad avenues, and comfortable and neat cottages, numerous schools and places of worship.[2]

ZAMBESIA.

ZAMBESIA includes the vast region under British protection, or within the British 'sphere of influence' in South Central Africa, extending from **Bechuanaland** and the Transvaal on the south, to **Lake Tanganyika** on the north, and divided by the Zambesi into two great sections, which we may distinguish as *Southern Zambesia* and *Northern Zambesia*.

The direct distance between the great bend of the Limpopo and the south end of **Lake Tanganyika** is considerably over 900 miles, while the width of this enormous territory, both north and south of the great river by which it is bisected, is no less than 800 miles. The total area must be considerably over half-a-million square miles, or more than ten times the size of England.

British Zambesia includes the entire region between the Transvaal and Lake Tanganyika, with the exception of two portions of foreign territory—the Portuguese Tete-Zumbo wedge on the east, and the **German** Okavango-Chobe extension on the west. Further, the Limpopo River forms but a small part of the actual southern boundary, which, west of the junction of the Tuli River with the Limpopo, is marked by the parallel of 22° S. latitude, the common boundary of Zambesia and the Bechuanaland Protectorate. Strictly speaking, then, Zambesia is bounded by the 22° S. latitude on the south, the 21° E. longitude and the **Chobe Valley** on the *south-west;* by Gazaland, Manicaland, and the Tete-Zumbo region, and by the **Shiré River** with its tributary the Ruo, and Lake Nyassa on the *east;* by the **Upper Zambesi** on the *west;* by an irregular, and as yet undefined, frontier from the **Kabompo River** to **Lake Moero** on the *north-west;* and by a line joining the north end of **Lake Moero** with the south end of Lake Tanganyika and continued from thence to the north end of **Lake** Nyassa on the *north.*

1. The railway from Mafeking is now being constructed to Palapye (trader's road) and will thence be continued to Buluwayo and Gwelo.

2. See for the the intensely interesting account of Khama and his country, in "Zambesia, England's El Dorado in Africa," by E. P. Mathers, F.G.S., F.R.G.S., Editor of South Africa.

Southern Zambesia may be regarded as part of Temperate South Africa, but Northern Zambesia, except perhaps the Nyassa Highlands, is within Tropical Africa. The former is adapted for European settlement and colonisation in the true meaning of the term, the latter is not, although Europeans can live comfortably and in fairly good health for many years on the higher uplands.

Zambesia, especially the rich metalliferous districts in the south-east, is being rapidly opened up, and it is, in fact, marvellous that a region which, a year or two ago, was only known to a few adventurous sportsmen and travellers, has reached a point at which special postage stamps have become necessary for both the great divisions of the enormous territory entrusted to the British South Africa Company.

SOUTHERN ZAMBESIA.

SOUTHERN ZAMBESIA is bounded by the Bechuanaland Protectorate on the *south*, German South-West Africa on the *west*, Portuguese East Africa on the *east*, and Northern Zambesia on the *north*. It includes *Matabeleland*, *Mashonaland*, the greater part of *Khama's Country*, and other native territories which were, until 1888, absolutely independent.

The actual limits of Southern Zambesia are marked by the 22° S. latitude and the Limpopo River on the *south;* by the 21° E. longitude, and an irregular line along the Chobe River to its junction with the Zambesi, on the *west* and *north-west;* and by the middle Zambesi from the confluence of the Chobe to the junction of the Loangwa (near Zumbo), on the *north*. On the *north-east* and *east*, according to the Anglo-Portuguese Convention of 1891, the boundary is marked by a line drawn due south from the mouth of the Loangwa, a tributary of the Zambesi, as far as the 16th parallel of south latitude, then bending eastwards to the point where the Mazoe River is intersected by the 33° E. longitude; thence it turns south along the upper part of the eastern slope of the Manica Plateau to the centre of the main channel of the Sabi River, following that channel to its confluence with the Lunte, whence it strikes direct to the north-eastern point of the Transvaal frontier.

Matabeleland and Mashonaland are rich in *gold*, and the healthy and fairly fertile uplands are well adapted for *European settlement*. The Manica Plateau, especially, has immense mineral wealth, and the fertility of its soil and the salubrity of its climate make it capable of sustaining a large European population.

In Matabeleland, Englishmen have lived for 20 years without needing homeward journeys for health. Missionaries have reared families, and their children have married other missionaries and settled in the country. Traders and Dutchmen have reared families there, who have again married and reared children of the second generation, robust in health, but lacking, alas, in education. This proves that white men can live there and colonize on sound principles.[1] The country generally is capable of great farming development, and can support a large white population in addition to a much larger native population than it now sustains.

Some thousands of miners and farmers have already entered the country, and numerous gold reefs, which are easily found by the aid of the old workings,

---

1. Mr. Maund, in a letter to Sir Lambert Playfair.

are being worked, while sheep and cattle have been placed on the splendidly
grassed and well-watered Mashona uplands. The future prosperity of the
country does not, therefore, depend on gold alone, but, as was the case in
Australia and California, gold is invaluable, and indeed indispensable, in
attracting a white population, and in creating a demand for supplies, which
must sooner or later be met by the cultivation of the soil and the rearing of
sheep and cattle, and thus mining and agriculture will become settled industries.

The administration and commercial development of these territories have been entrusted to the British South Africa Company,
which, by its Charter, specially undertakes to develop the mineral
and other resources of the country, and to promote and aid immigration.

The progress of the British South Africa Company's work in opening up the country
has been extraordinarily rapid. The Charter was signed on October 29th, 1889. A few
days later, the extension of the railway northwards from Kimberley was begun, and, 13
months after, the section to Vryburg was opened, and has since then been completed to
Mafeking, the frontier town whence the Pioneer Expedition to Matabeleland and the
Mashona country was despatched north.

The *Times* says that, "immediately on the signing of the charter a police force of 500
picked men was raised to protect the interests of the Company in its immediate territories.
Part of this force was left at the camp on the Macloutsie River to keep open communications and protect the base, while the remainder escorted a body of 180 carefully-
selected miners and pioneers, many of them young men from English public schools and
universities, who had undertaken to cut a waggon road through the south and east of
Matabeleland up to Mount Hampden, in Mashonaland, about 150 miles from the
Zambesi. This pioneer force crossed the Macloutsie on June 25, 1890, and proceeded to
make a road, by which they took up 70 waggons, four machine guns, an electric search
light, engine, and other material; and they reached Mount Hampden, some 400 miles
from the base, on September 12, 1890 (about ten weeks), without the loss of a single
man. A fort was established at the Tuli drift, and another at Fort Victoria, 31° E.
by 20° 8′ S., on the edge of the high Mashona Plateau, which stretches from that point
for over 250 miles to the north, at an average height of upwards of 4,500 feet above the
sea. Two more forts, Fort Charter and Fort Salisbury, were also established, the
latter, near Mount Hampden, being made the headquarters of the administration for
Mashonaland, and around it a township quickly sprang up."

The " Pioneer Expedition " to Mashonaland, in 1890, founded Salisbury, the present
capital, Victoria, and other townships ; and considerable progress was being made in the
development of the mineral and other resources of the country, when in October, 1893,
war was declared against Lo Bengula, the Matabele king. The Company's forces, aided
by the Bechuanaland Border Police and Khama, the Bamangwato chief, defeated and
dispersed the Matabele impis. The king fled towards the Zambesi, and shortly after
died. Buluwayo, the capital, was occupied in November, 1893, and is rapidly becoming
an important centre of trade. About 120 miles of the East Coast Railway, from Beira, in
Portuguese East Africa, to Salisbury, the largest town in Mashonaland, are now open
for traffic. The Kimberley-Mafeking line is now being built to Gaberones, and will
ultimately be extended to Palapye, Khama's chief town, and to Buluwayo.

NORTHERN ZAMBESIA.

NORTHERN ZAMBESIA, or British Central Africa, as some prefer
to call the region—now definitely declared British territory—to the
north of the Zambesi, has a total area of not less than 300,000 square
miles, or 6 times the area of England.

Commissioner Johnston thus defines the boundaries of British Central Africa:—

" British Central Africa is the name given to a considerable area of South-Central Africa which lies for the most part within the northern basin of the Zambesia, but which also includes within its limits a considerable part of the watershed of the Upper Congo. It is bounded on the north by the Congo Free State, the waters of Tanganyika, and German East Africa ; on the east by the German and Portuguese East African possessions ; on the south by the Portuguese possessions on the Zambesi, by the middle course of the Zambesi river, and by German South-West Africa ; and on the west by the Portuguese Province of Angola."

By the Anglo-Portuguese Convention of 1891, the eastern frontier is formed by the eastern shore of Lake Nyassa as far as 13°30′ S. latitude, thence it runs in a south-easterly direction to the eastern shore of Lake Chiuta, and thence to that of Lake Shirwa, and from thence in a direct line to the River Ruo, which it follows to its confluence with the Shiré. It then runs along the centre of the channel of the Shiré to a point just below Chiwanga, and then along the watershed between the Zambesi and the Shiré and Lake Nyassa, until it reaches 14° S. latitude, from whence it runs in a south-western direction to the point where south latitude 15° meets the River Loangwa, and follows the mid-channel of that river to its junction with the Zambesi. The Zambesi itself, from the confluence of the Loangwa to the mouth of the Chobe, forms the southern boundary—the middle course of the Zambesi is thus through British territory—while the Upper Zambesi, from the Katima Rapids northwards, forms the western frontier, which is to be actually delimited by an Anglo-Portuguese Commission.

A British Imperial Commissioner and Consul-General [2] controls the administration of Nyassaland, the rest of Northern Zambesia being administered directly by the British South Africa Company.

The Shiré Highlands, and the rest of Nyassaland claimed by the African Lakes Company, were proclaimed a British Protectorate in 1889. The African Lakes Company is now a purely trading concern, and has no territorial rights or jurisdiction whatever. The entire region is practically taken over by the British South Africa Company, whose sphere of action, according to the Anglo-German Agreement of June, 1890, extends as far as the south end of Lake Tanganyika, and includes the Stevenson Road. The B. S. A. Co. (through the Lakes Co.) has established a fort—Fort Abercorn—on Lake Tanganyika, and another—Fort Fife—on the high plateau between Lakes Nyassa and Tanganyika, and grants £10,000 a year (in addition to the Government allowances) to the Imperial Commissioner, and £2,500 a year to the Lakes Company, to help it to develop the country, which is evidently the most favoured portion of Northern Zambesia, and, if possible, to divert the trade on both sides of Lake Tanganyika down to Nyassa, the Shiré, and the Zambesi.

The British Commissioner has a strong police force, and a corps of Sikhs from the Indian army, and there are also a number of British gunboats on Lake Nyassa, the Shiré River, and the Zambesi. Protection is thus afforded to all legitimate commerce, and the slave-trade, which was until recently carried

1. The *Chobe Valley* extension of *German South West Africa* also extends to the south-western boundary of this region.
2. Mr. H. H. Johnston, C.B.

on in the Nyassa country and adjacent regions, has been practically suppressed. The seat of administration is at ZOMBA, in the Shiré Highlands, but Blan-tyre, which lies to the south of Zomba, is a more important and better known place.

Physically, Northern Zambesia "belongs to the characteristic plateau region of Central Africa. Except for some distance on each side of the Lower Loangwa, it has an average height of from 4,000 ft. to 6,000 ft. ; and dotted all over it, especially to the south of Lake Tanganyika, are spots rising to over 6,000 ft., which in the future may become so many little Simlas. So far as rainfall goes, it is fairly well situated for Africa ; it averages about 50 inches annually. There is a network of permanent rivers ; indeed, in some regions the surface moisture is too much for health and comfort. As is the case everywhere in Africa, there are here and there districts which are practically desert, but, as a whole, it is among the most hopeful regions in Central Africa. There is ample room for industrial development in many directions,[1] and for a very much larger native population than at present exists. Under the peace and freedom from raids which, it is hoped, British rule will bring, this population is sure to increase. At present, the total population probably does not exceed three millions, or about ten to the square mile. Among this population there are probably at most not more than 300 Arabs, or rather, men with any proportion of pure Arab blood in their veins. There is, besides, a floating population of about 400 Swahilis. It is these who were the great disturbers of the peace in Central Africa, the real slave-raiders and ivory-stealers. The true Arabs in Northern Zambesia, even when they have a considerable proportion of African blood in their veins, are, with few excep-tions, respectable men, most of them having property on the Persian Gulf, at Aden, and at Zanzibar. As for the slave-raiding Swahilis, the British Commissioner has carried on active operations against them. Several military expeditions have attacked their strong-holds on both sides of Lake Nyassa, and they have, in general, been successful. Several forts—Fort Johnston, Fort Maguire, and Fort Sharpe—have been built, and Makan-jira, the most important of these Arab lake chiefs, was forced to take refuge in Portuguese territory ; the district is still, however, in an unsettled state, but the energy and skill of Mr. H. H. Johnston, with the additional military and naval aid placed at his disposal, will soon be able to entirely subdue and pacify this part of Nyassaland.

"As to European colonisation, in the true sense of the term, that is not to be thought of north of the Zambesi. Englishmen and other Europeans may spend many years of their life on the higher uplands, and move about just as they do in India and Ceylon, helping the natives to help themselves and developing the resources of the region. Here, as in every other part of Central Africa and of the tropics generally, the *maximum* of comfort is indispensable for the maintenance of health and vigour in the European. "Roughing it," when it can be avoided, is a fatal mistake ; with comfortable surround-ings no European constitution need be any the worse for a few years in Central Africa."

BRITISH EAST AFRICA.

BRITISH EAST AFRICA extends along the coast from the Umbe River in the south, to the Juba River on the north—a distance of 400 miles. Inland, the British "Sphere of Influence" extends westwards beyond the Victoria and the Albert Nyanzas to the borders of the Congo Free State—a distance of 900 miles. On the east, the British sphere is conterminous with the Italian sphere in Gallaland and Abyssinia. There are no definite limits on the north or north-west, but the Upper Nile Region and the Eastern Sudan and Nubia are open to British influence, and to it alone.

1. *Rice* of excellent quality is being grown more and more extensively in the low-lying lands about the lake and the rivers, and has a high reputation in the London market. It is believed that the soil | and climate at many points on the Nyassa uplands are well suited for *tea* culture. The *coffee* from the Shiré Highlands already commands a good price in the London market.

The boundaries of British East Africa, as defined by the Anglo-German Agreements of 1886 and 1890 and the Anglo-Italian Agreement of 1891, are as follows :—The line of demarcation between Ibea and German East Africa runs in a north-west direction from the north bank of the mouth of the River Umbe towards and round by the north of Kilimanjaro, to where the 1° N. lat. cuts the Victoria Nyanza; thence across the lake and westwards, along the same parallel, to the boundary of the Congo Free State, but deflecting south, so as to include Mount Mfumbiro and the districts covered by Stanley's treaties within the British 'Sphere.' Between the British and the Italian spheres on the east, the line follows, from the sea, the *Thalweg*, or mid-channel, of the Juba River up to 6° N. lat., Kismayu, with its territory on the right bank of the river, thus remaining to England. It then follows the 6° N. lat. as far west as the 35° E. long., which it follows north to the Blue Nile.[1] Further north, as we have said, the entire region up to the Egyptian frontier is within the British Sphere of Influence, and as Egypt itself is practically a British Protectorate, and Zambesia is definitely declared to be British, had it not been for the unfortunate cession of the Victoria-Tanganyika uplands to Germany, one might have travelled from Cape Town to Cairo through British territory.

The Protectorate has an area of some 250,000 square miles, and the rest of the British Sphere of Influence to the Egyptian frontier at least 800,000 square miles—a total area of over 1 million square miles, of which, however, but a comparatively small portion in the south has been as yet effectively occupied. The population, which consists of various Negro tribes, Arabs, and Banyan or Hindu traders, for so vast a territory is small, amounting to not more than 6 millions in the southern, and to 7 or 8 millions in the northern, division of a region, which measures over 1,600 miles from north to south.

The administration and development of all the territories included within the British sphere were, until 1895, entrusted to the Imperial British East Africa Company, incorporated by Royal Charter in 1888. The coast territory and ports, which formerly belonged to the Sultan of Zanzibar, were conceded to the Company by the Sultan in 1887-9, and treaties have been made with the native chiefs between the coast and the Victoria Nyanza, and also with the King of Uganda and the chiefs to the west of that lake. Exploring expeditions and trading caravans are rapidly opening up the country, and "trading stations have been established, and a trade road, protected by stockaded stations and suited for baggage animals, has been opened along the line of the Sabaki River for a distance of 250 miles in the direction of the lake." The great navigable waterways of the Tana and the Juba rivers afford excellent means of communication with the populous regions through which they flow. Steamers have been already placed on the Tana, and from Mombasa there is regular steam communication with other African ports and with Europe and India. The chief port and seat of government, MOMBASA, is situated on an island off the coast. It has a fine harbour, with piers and jetties; a new town is being built, and a railway is to be constructed from the coast to the Victoria Nyanza, by means of which the resources of the thickly-populated districts in the interior will be opened up to trade. Malindi, Lamu, and Kismayu are other important trading stations

1. Of course, these and other mathematical frontiers in Africa, may be subsequently altered in | accordance with the hydrographical and orographical conditions of the country.

on the coast. 'The British Company have been more fortunate and adroit in the management of their territories than their German neighbours and rivals, and consequently they have made steady progress unhampered by native disaffection of any kind. On the contrary, they have been well received everywhere, mainly because their administration is based on justice and common sense, and has been carried out with a firm hand.'

∴ The formerly independent states of **Uganda** and **Unyoro**, between the Victoria Nyanza and the Albert Nyanza, are important links in the chain of communications between the British stations on the East Coast and the rich provinces of the Upper Nile.

UGANDA.

The British Protectorate of UGANDA, formerly the largest and most powerful of the native states of Equatorial East Africa, includes the region on the north and north-west of the **Victoria Nyanza**, and, with Usoga and the vassal States on the eastern side of the **Nile**, has an area of **70,000** square miles and a population of perhaps **5** millions, mostly negroes of the *Bantu* race.

Uganda is a country of undulating plains, with rich and fertile valleys and well-wooded hills. The climate is hot, but not unhealthy—the country lying at an altitude of from 5,000 to 6,000 feet—and the rainfall is often heavy in spring and autumn. The Uganda people supply ivory, **tobacco**, **cattle**, **goats**, and other native products to Arab traders in exchange for fire-arms and ammunition, woven fabrics, and other goods. The chief caravan-routes pass through the great market town of **Mruli**, on the Somerset Nile, and from the opposite shore of the Victoria Nyanza to Mombasa and Zanzibar. The projected railway, from Mombasa to the Victoria Nyanza, if completed, will, of course, absorb the entire trade of Uganda and the adjoining regions.

The late King Mtesa was a powerful sovereign, and under his arbitrary rule the country became rich and, in some degree, civilised. His son, Mwanga by a treaty (1889) with the British East Africa Company, placed his country under British protection, but, owing to severe conflicts between the Mohammedan, Catholic, and Protestant parties, and the excessive cost of permanent occupation, the company threatened, towards the end of 1892, to withdraw their agents and abandon the country. This gave rise to such strong protests, that the Home Government delayed the evacuation, and sent out an Imperial Commissioner to report on the state of affairs and "the best means of dealing with the country." After considering Sir Gerald Portal's report, the British Government determined, in 1894, to establish a regular administration, and for that purpose declared Uganda to be under British protection. **MENGO**, on Murchison Bay, an inlet on the north-western coast of the Victoria Nyanza, is the royal residence, but **Port Alice** is the seat of administration.

UNYORO lies to the west of Uganda, and includes the fertile and well-watered plateau between Uganda and the **Albert Nyanza** and the **Semliki River**.

The Wanyoro people are, like the Waganda, of the *Bantu* stock, but the chiefs and sovereigns of both States are the descendants of the Wa-huma (Galla) conquerors and founders of the great **Empire of Kitwara**, which was broken up into the States of Unyoro, Uganda, Karagwe, Ruanda, &c. Kabba Regga, the former king of Unyoro, was a lineal descendant of the "King of Kitwara." He opposed the British advance, but was defeated and driven out of the country in 1894, and a line of forts was then erected along the south-western shores of Lake Albert. The people are skilled metal workers, and cultivate the soil and rear large herds of **cattle**, which, with **ivory**, **salt**, and **gums**, are disposed of to Arab traders in exchange for guns and ammunition, cotton, and other goods.

ZANZIBAR.

The Sultanate of Zanzibar formerly included a long strip of the coast, from **Cape Delgado to Kipini**, and several *points* on the coast as far north as **Warsheikh**, besides the productive and populous islands of **Zanzibar and Pemba.**

The partition of the Zanzibar Sultanate was commenced in 1888, by the cession of the Mrima or mainland, between the Rovuma and the Umbe River, to the German East Africa Company for 50 years, and a similar lease of the coast-lands and ports north of the Umbe to the British East Africa Company. A further agreement was signed in 1890, by which Zanzibar and Pemba were placed under British protection, and the entire coast-land and islands between the Umbe River and the Juba were ceded to the British East Africa Company as well as the 'northern ports'—Brava, Merka, Magadoxo, and Warsheikh—which were afterwards transferred to Italy. In 1895, when the Home Government took over the administration of the British East Africa Company's territories, the coast-land was retroceded to Zanzibar.

The island of Zanzibar has an area of 625 square miles, and Pemba 360 square miles, with a population of 125,000 and 40,000 respectively. The town of **ZANZIBAR** has a population of 100,000, and is an active centre of trade in the products of the island and those of the adjoining mainland. Much of the *ivory, caoutchouc*, &c., formerly sent to the Zanzibar market is now shipped direct to Europe from Mombasa, Dar es Salaam, and other coast ports, but the " Liverpool" of East Africa is still the great emporium not only for the world-famous cloves of the islands of Zanzibar and Pemba, but also for the hides of Somaliland, the gums of the Swahili coast, and the ivory of East Central Africa. Now that it is a " free port " its trade is likely to increase rapidly.

BRITISH ISLANDS OFF THE COAST OF AFRICA.

MAURITIUS.

The beautiful and fertile island of **Mauritius**,[1] one of the most prosperous of all our colonies, lies in the Indian Ocean, about 500 miles east of Madagascar. With an area of only **705** square miles, it has a population of nearly **380,000**, or **530** to the square mile, and a trade of between 5 and 6 millions sterling a year.

Mauritius is "famous for the wondrous beauty of the landscape, surpassing even that of Tahiti in the Pacific." Its surface exhibits a succession of hills and valleys ; its volcanic soil was once covered by magnificent forests, which have been cleared and replaced by vast sugar plantations. The cultivation of the sugar-cane is the principal industry, and sugar forms fully three-fourths of the total exports. Rum, vanilla, aloe fibre, and cocoa-nut oil are also produced. The trade of Mauritius is carried on chiefly with the British Colonies of South Africa, Australia, and India, and with the United Kingdom.

The coast is fringed by coral reefs, in which there are only two permanent openings—one on the north-west coast, leading to the splendid harbour of

---

1. Mauritius was discovered in 1505 by the Portuguese, but was first settled in 1598 by the Dutch, who called it *Mauritius*, in honour of their Prince Maurice. Being abandoned by the Dutch in 1712, it was taken by the French, who renamed it "*Ile de France*," and held it until 1810, when it was captured by an expedition under Sir Ralph Abercromby, sent out by the East India Company. In 1874, and again in 1892, terrific hurricanes caused great damage on the island.

PORT LOUIS (70), the capital of the colony, and the other on the opposite
side of the island, leading to the harbour on which Mahebourg, the only other
large town, is built.

Dependencies of Mauritius : The twin groups of the *Seychelles*
and the *Amirantes*, about 600 miles north-east of Madagascar, and
1,100 miles east of Zanzibar, together with the scattered groups of
coral islands in the Indian Ocean, are dependencies of Mauritius.

The prosperous Seychelles, covered with a luxuriant vegetation, especially
cocoa-nut palms and date trees, are 940 miles north of Mauritius ; the coralline
Amirantes are about 100 miles south-west of Mahé, the largest island of the
Seychelles group. Of the numerous other islands administered by the Mauritius
Government, the most important are the granitic islet of Rodriguez, 300 miles
east of Mauritius, and the island of Diego Garcia, one of the Chagos Archipelago. The latter has an excellent harbour, and is an important coaling station on the direct route from the Red Sea to Australia. The French steamers
to and from Australia, however, coal at PORT VICTORIA, a fine harbour on
the north-eastern side of Mahé, in the Seychelles.

SOCOTRA.

SOCOTRA, a **British Crown Colony**, is a large island 150 miles
east of **Cape Guardafui.** Its surface is broken by granitic mountains
and limestone hills, and its sterile soil yields little beyond *aloes* and
a few *dates*.

The inhabitants consist of a few thousand Arabs, and the island formerly
belonged to the Sultan of Keshim, on the opposite coast of Arabia, by whom
it was ceded to Great Britain in 1876, but it was not formally annexed until
1886. Socotra is important from its position at the entrance of the Gulf of Aden,
on the direct route to India, and its consequent strategic value in case of war.

ST. HELENA.

ST. HELENA,[1] another isolated island in the South Atlantic, 800
miles south of Ascension, and 1,200 miles from the nearest point of
the African mainland, is a huge mass of rock, of volcanic origin,
rising steeply out of the waters of the Atlantic, and reaching in its
highest point—*Diana's Peak*, near the centre of the island—2,700
feet above the sea.

The coast of the island forms a perpendicular wall or cliff, the principal
opening in which is on the north-western coast, where Jamestown, the capital,
is situated. The interior exhibits a succession of hills and valleys, some of the
latter very fertile and capable of producing *vegetables* and *fruits* in abundance.
The climate is warm, but the position of the island, in the midst of a vast
ocean, preserves it from the intense heat of the Torrid Zone. It has an area
of 47 square miles and a population of about 5,000.

The chief interest attaching to St. Helena is derived from the fact of its
having been, during the last six years of his life, the place of exile of Napoleon

---

[1] St. Helena was discovered by the Portuguese | English took possession of it in 1651.
in 1502, on St. Helena's Day, hence its name. The |

the Great, who died at **Longwood** in 1821. His body was removed to Paris in 1840.

St. Helena is a British Crown Colony, and the strong fortifications which defend **James Bay** are always manned by a British garrison. The island, commercially unimportant since the opening of the Suez Canal and the use of large steamers, is yet of great strategic importance as a coaling station for the British fleet.

ASCENSION.

The British island of ASCENSION[1] is nearly 1,000 miles distant from Cape Palmas, the nearest point on the African coast. It is only 35 square miles in area, and consists of a mass of volcanic rock, rising to a height of nearly 3,000 feet above the sea.

This solitary oceanic island, rugged and barren, with only one cultivable spot—a veritable oasis surrounded by old lava streams and scoriæ—has yet the driest and most salubrious climate in the world, and only requires more water and vegetation to become a delightful invalid resort and sanatorium. A large number of turtles are caught during the season, and rabbits abound.

Ascension is under the control of the Admiralty; the governor is a naval officer, in command of a company of marines, which forms the garrison. **Georgetown**, on the north-west of the island, is the only settlement.

TRISTAN D'ACUNHA.

TRISTAN D'ACUNHA, which was garrisoned until the death of Napoleon I., is the largest of a group of three islands, lying to the west of the Cape of Good Hope, at a distance of about 1,700 miles.

Tristan d'Acunha is a barren volcanic rock, the highest point of which reaches 6,400 feet above the sea. The inhabitants, about 100 in number, have a few sheep and cattle, and grow potatoes, &c. Provisions are sent them, at intervals, by the British Government.

BRITISH POSSESSIONS IN AMERICA.

BRITISH NORTH AMERICA.

BRITISH NORTH AMERICA embraces the whole of **the northern half of the continent**, except Greenland and Alaska. This immense territory, which is nearly as large as all Europe, is, with the exception of *Newfoundland* and *Labrador* and the small islands of *St. Pierre* and *Miquelon* off the south coast of Newfoundland, included in the **Dominion of Canada**.

The Colony of **Newfoundland** and **Labrador**, its dependency, are thus the only parts of the British possessions in North America not included in the Dominion, and as they stand, as it were, at the threshold of the New World, and Newfoundland itself is the nearest land to Europe—St. John's, Newfoundland, being only 1,665 miles from Galway, in Ireland—we will describe them first.

---

1. Ascension was discovered by the Portuguese | mained uninhabited until 1815, when it was taken on Ascension Day, 1501, hence its name. It re- | possession of by the British as a naval station.

NEWFOUNDLAND.

NEWFOUNDLAND,[1] the oldest and, in many respects, the most peculiar British Colony, is a large island at the mouth of the Gulf of St. Lawrence. With an area of over 42,000 square miles, or one-sixth larger than that of Ireland, and with **inexhaustible fisheries and large tracts of rich agricultural, mineral, and timber lands,** the colony has a population of only 200,000, or less than that of Belfast.

The island is triangular in shape—the three extreme points being **Cape Norman** on the north, **Cape Ray** on the south-west, and **Cape Race** on the south-east. The coast-line, 2,000 miles in length, is deeply indented by several deep bays, such as Hare, White, Notre Dame, Bonavista, Trinity, and Conception Bays on the east coast; St. Mary's, Placentia, Fortune, and Hermitage Bays on the south coast; and St. George's Bay and the Bay of Islands on the west coast. The south-eastern part of the island—the Peninsula of Avalon—is almost cut off from the main portion by Trinity Bay on the north and Placentia Bay on the south. Hundreds of islands fringe the coast, two of them—Miquelon and St. Pierre—belong to France. One of them, Belle Isle, gives its name to the narrow channel—the Strait of Belle Isle—which separates Newfoundland from Labrador. Most of these inlets form excellent harbours, well sheltered and with safe anchorages.

The interior of Newfoundland long remained a *terra incognita*, and, until recently, the Avalon Peninsula was the only part of the island containing any settlement more than a mile from the coast. The greater portion appears to be an undulating prairie, watered by so many rivers and containing so many lakes that fully **one-third of the island is covered with water.** Grand Pond (185 square miles) is the largest lake, and the River of Exploits (150 miles) the longest river.

The climate is not so extreme as that of Canada, being much milder in winter and not nearly so hot in summer, and, in spite of the *fogs* which prevail on the 'banks' and occasionally extend to the island, the people are **exceptionally healthy.** Although the island could support an agricultural population numbered by millions, agriculture is practically unknown, and **cod-fishing** in summer, and **seal-hunting** in winter and spring, are the chief occupations of the people. The cod-fishery on the **Grand Banks** of Newfoundland is the most important in the world, but the islanders depend chiefly upon the less risky shore fishery.[2]

The **trade** of Newfoundland, which amounts to about 3 millions sterling a year, consists in the export of **codfish, seal oil, sealskins, herrings, lobsters,** and other products of the fisheries, with **copper and copper ore,** and the import of textile fabrics and food stuffs. About a third of the trade is carried on with the *United Kingdom*—the rest is principally with *Canada* and the *United States.* Enormous quantities of fish are sent to Catholic countries such as *Brazil, Spain,* and *Portugal.*

1. Newfoundland was first discovered by Norwegians about the year 1000, and was visited by Cabot in 1497. The first settlement was made in 1623, and the first governor was appointed in 1728.   2. The annual catch on the banks and shores of Newfoundland amounts to about 2½ million quintals.

Newfoundland possesses responsible government. The Governor is appointed by the Crown ; the House of Assembly is elected by the people every
four years.

The capital, ST. JOHN'S, is a town of over 30,000 people, beautifully situated
on a splendid harbour—" one of the very best on the Atlantic coast "—on the
eastern side of the Peninsula of Avalon. In 1892, a most disastrous fire
occurred at St. John's, by which the greater part of the town was destroyed.
It is now connected by rail with Harbour Grace (7), on the west side of Conception Bay, and Placentia, on Placentia Bay.[1] The railway is being extended
westwards across the island, so as to open up the rich timber and agricultural
lands in the interior and along the west coast. The *copper mines* on the eastern
side of the island have been very productive, and *coal, plumbago*, and other
useful minerals abound.

LABRADOR.—The adjoining coast territory of Labrador, from the Strait
of Belle Isle on the south, to Cape Chudleigh, at the entrance of Hudson
Strait, on the north, is politically attached to Newfoundland.

Some of the most valuable fisheries—cod, herring, trout, and salmon—in the
world, are found off the coast of this sterile region. There is a resident population of about 4,000, which is increased to 50,000 during the fishing season.
Nain and Hopedale are the chief settlements, and there are also a few posts of
the Hudson Bay Company.

BRITISH ARCTIC AMERICA.

BRITISH ARCTIC AMERICA is politically included in the vast
Dominion of Canada, but in truth "the entire Polar regions of
America, from Bering Strait to Baffin Bay and the whole territory
of Greenland, nominally owned by three different powers—Denmark, Britain, and the United States—are the domain of the one
race which can wrest a living from them ; one of vast antiquity,
which has adapted itself through ages of evolution to its terrible
environments—the Eskimo."

British Arctic America includes not only the apparently inextricable maze of
"lands" and islands, intersected by innumerable channels, straits, and sounds,
but also the inhospitable shores of the adjoining mainland, which stretches
poleward in the great Peninsula of Labrador, to the east of Hudson Bay. To
the north and west of this great inland sea, two small peninsulas—Melville
Peninsula and Boothia Felix—also run northward ; the one reaching the 70°
N. lat., and the other—the most northerly point of the American mainland—
attaining a point (lat. 72° N.) only 1,260 miles from the North Pole itself.

An attentive examination of a good map will show that the mass of "lands"
and islands to the north of this irregular and deeply indented coast is resolvable
into two great divisions, and that these again may be grouped into smaller

---

1. The *Atlantic Cables* terminate at Trinity Bay, on the east coast.

divisions, so that a clear and definite idea[1] may be gained of the absolute and
relative position of the various members of what may be termed, as a whole, the
Arctic Archipelago, a term which some geographers limit to the **Parry Islands**
and North Devon. The long and continuous channel—formed by **Lancaster**
Sound, Barrow Strait, Melville Sound, and **McClure Strait**—leading from
Baffin Bay westwards to the open ocean and Bering Strait, may be taken as the
main line of demarcation. To the north of it are the **Parry Islands** and **North**
Devon and the long coastland, of which the eastern and part of the northern
margin only has been traced, and which, under various names—**North Lincoln,**
Ellesmere Land, Grinnell Land, and **Grant Land**—stretches north along the
western side of **Smith Sound** and **Robeson Channel.**

To the south of the channel referred to, the Arctic Archipelago is resolvable
into three groups—the *eastern*, which includes the long **Baffin Land**, extending
along the western side of Baffin Bay and Davis Strait, from Lancaster Sound
to Hudson Strait ; the *central*, which includes the two large islands of **North**
Somerset and **Prince of Wales Land**, separated from each other by Franklin
Strait ; and the *western*, which consists of one large central mass—**Prince**
Albert and **Victoria Land**—and two islands, **Banks' Land**, separated from it by
Prince of Wales Strait on the north-west, and **King William's Land**, separated
from it on the south-east by **Victoria Strait.** The central mass is divided from
the adjoining mainland by the **Dolphin Strait, Coronation Gulf,** and **Dease**
Strait. Boothia Peninsula is divided from North Somerset by **Bellot Strait,**
while the adjoining Melville Peninsula is separated from Baffin Land by the
narrow **Fury and Hecla Strait** and the broad **Fox Channel**, which leads into
Hudson Strait and Bay. All these and other islands and channels are inseparably linked with the memory of the brave men who risked their lives in
the vain endeavour to open up a highway between the Atlantic and the Pacific,
and of those who perished in the search for the **North-West Passage**, the most
famous of whom, Sir John Franklin, with the *Erebus* and *Terror*, went north
in 1845, and was last seen by some whalers in Melville Bay. The search for
the missing vessels was energetically prosecuted, but it was not until 1850 that
a clue was found, namely, a cairn and three graves on Beechey Island. In the
same year, Capt. McClure, in the *Investigator*, entered the Arctic Ocean through
Bering Strait, but was frozen up in **Prince of Wales Strait**, whence Melville
Sound could be seen. Seven years later, Capt. M'Clintock took the *Fox* north
and wintered in **Bellot Strait**, whence sledge parties were sent out, and on **Cape**
Victory, on the north-western coast of **King William's Land**, the long-sought-for
record of the Franklin Expedition was found. It seems that the expedition had
wintered at **Beechey Island** and had pushed south towards King William's
Land, but the ships had been beset in the ice near **Cape Felix**, Franklin had
died, and the survivors had abandoned the vessels, and had started overland
for Back's **Fish River.** A few reached the mainland, but not one lived to tell
the tale. " Ninety more miles of open water and Franklin and his heroic
followers would have won the prize they sought." But the North-West Passage,
of which they were indeed the first discoverers, can never be utilized as a commercial highway between the two great oceans, the channels, by which the passage might under unusually favourable circumstances be made, being generally
ice-bound.[2]

1. Such an idea is absolutely necessary to an intelligent perusal of the deeply interesting annals of
Arctic Exploration, and especially those connected
with the search for, and the discovery of, the *North-*
West Passage.

2. The longer but more open North-East Passage
round the northern coast of Asia is almost as valueless for all practical purposes, although some attempts have been made to open up communication
with Siberia by sea.

THE DOMINION OF CANADA.

The magnificent **Dominion of Canada** embraces the whole of the mainland of North America to the north of the United States, except Alaska and Labrador, and all the adjacent islands, except Newfoundland, St. Pierre, and Miquelon.

This immense territory stretches right across the continent, from the **Atlantic** on the *east*, to the Pacific on the *west*, and from the **Great Lakes** and the 49th parallel of N. lat. (which divide it from the United States) on the *south*, to the Arctic Ocean on the *north*.

The Dominion of Canada is a confederation of **seven Provinces** —*Ontario, Quebec, Nova Scotia, New Brunswick, Prince Edward Island, Manitoba,* and *British Columbia;* **four Districts**—*Assiniboia, Saskatchewan, Alberta,* and *Athabasca,* which together form the **North-West Province ;** together with the District of *Keewatin,* and **two Territories**—the *North-West Territory* to the west, and the *North-East Territory* to the south and east, of Hudson Bay.

The formerly separate colonies of **Canada** (which included the present Provinces of Ontario and Quebec), **New Brunswick,** and **Nova Scotia,** were united in 1867, and, two years later, the vast **North-West Territory** was added to the Dominion by purchase from the Hudson Bay Company. **Manitoba,** a Province formed out of a portion of it, was admitted into the confederation in 1870. British Columbia joined it in 1871, and Prince Edward Island in 1873, so that *Newfoundland* alone has stood aloof, and still remains a distinct colony. The District of **Keewatin,** to the north of Manitoba, was formed in 1876 and is administered by the Lieutenant-Governor of Manitoba.

The student should carefully note the relative and actual position of the various Provinces of Canada. The *Maritime Provinces*—**New Brunswick, Nova Scotia,** and **Prince Edward Island**—are on the south side of the Gulf of St. Lawrence. **Ontario** curves round the northern shores of the Great Lakes, while the St. Lawrence runs through the Province of **Quebec. British Columbia** includes the broken country between the Rocky Mountains and the Pacific Coast, together with the great island of Vancouver, the Queen Charlotte group and other islands, while **Manitoba** and the immense North-West Territories, the southern portion of which has been divided into the four provisional Districts of **Assiniboia, Saskatchewan, Alberta,** and **Athabasca,** form the central or prairie section of the Dominion. The District of **Keewatin** lies to the north of Manitoba, and adjoins the western side of Hudson Bay. The **North-East Territory** embraces the desolate peninsula on the east side of Hudson Bay, and also the sterile tract on the west side of James Bay, north of the Albany River, which forms the northern boundary of Ontario. It should also be noted that British Columbia, Alberta, Assiniboia, Manitoba, Ontario, Quebec, and New Brunswick, touch or border upon the United States, and that the **International Boundary,** from the Gulf of Georgia to the Lake of the Woods, is formed by the *49th parallel of North latitude,* then by the *Rainy River, Rainy Lake,* and *Pigeon River* to Lake Superior, whence the line runs through the middle of *Lakes Superior, Huron, Erie,* and *Ontario,* and along the *St. Lawrence* to the *45th parallel of North latitude,* which it follows to the banks of the *Upper Connecticut River,* it then strikes north along the *Green Mountains,* and around *Maine* to the Atlantic Coast at the mouth of the Bay of Fundy.

Eastern Canada, in the following description, is taken as including the Provinces of Ontario, Quebec, New Brunswick, Nova Scotia, and Prince Edward Island ; Central Canada as embracing Manitoba and the provisional Districts of Assiniboia, Alberta, Saskatchewan, Athabasca, Keewatin, and the rest of the North-West Territory ; while Western Canada is limited to the Province of British Columbia.

Eastern Canada is the great WOODLAND REGION ; Central Canada is the vast PRAIRIE COUNTRY ; while Western Canada forms the MOUNTAIN REGION of the Dominion.

EXTENT : It is difficult to convey any idea of the vastness of the Dominion of Canada. From the Atlantic on the *east* to the Pacific on the *west* is a direct distance of more than **3,000** miles, while more than **2,000** miles of mountain and prairie extend between the International Boundary and the shores of the Arctic Ocean. With an area of **3½ million square miles**—not including the area covered by the Great Lakes—the Dominion is nearly as large as Europe, and 40 times the size of Great Britain.

COASTS : Bounded, as the Dominion is, by three oceans, it has, besides its numerous inland **seas**, many thousands of miles of **sea coast.** The older Provinces have **2,500** miles of sea coast and inland seas, while the sea coast of British Columbia alone is over **3,000** miles in extent, exclusive of minor indentations.

Hudson Bay is a vast inland sea, 1,000 miles long and 600 miles wide, running south into the heart of Canada, midway between the Atlantic and the Pacific, and communicating with the Atlantic by Hudson Strait, a broad channel 400 miles in length. James Bay, the southern extension of Hudson Bay, is 300 miles long and 80 miles wide at its mouth. The "Mediterranean of America," as Hudson Bay has been termed, is unfortunately *icebound* in winter. Ships enter it generally about the middle of July and leave about the middle of September. PORT NELSON, on the western side of Hudson Bay, is a hundred miles nearer to Liverpool than New York, and between Winnipeg and Liverpool, *via* Hudson Bay, there will be, when the new railway from WINNIPEG to this port is completed, a saving in inland carriage of about 2,000 miles as compared with the present routes *via* New York or Halifax.

The Arctic Coast, on the *north*, is broken by numerous inlets and is for the most part skirted by "lands" and islands, the channels between which and the mainland are generally ice-bound, and are consequently of no commercial importance.

On the *east*, are the great Gulf and Estuary of the St. Lawrence, opening into the Atlantic by the broad Cabot Strait, between Newfoundland and Cape Breton Island. The narrow Strait of Belle Isle, between Newfoundland and Labrador, and the still narrower Gut of Canso, between Cape Breton Island and Nova Scotia, are the only other entrances. The *cul-de-sac* of the Bay of Fundy is also an arm of the Atlantic. The tidal wave, as it moves swiftly up this funnel-like inlet, sometimes attains the extraordinary height of 70 feet.

The British Columbian Coast, on the *west*, is more deeply indented than that of Norway ; it is, in fact, a perfect maze of inlets and islands, and its total length cannot be far short of 10,000 miles. Vancouver Island is separated from the mainland by a long channel known as the **Juan de Fuca Strait** in the south, the **Strait of Georgia** in the middle, and **Queen Charlotte Sound** in the north. Queen Charlotte Islands are separated by the **Strait of Vancouver** from the mainland and by **Dixon Entrance** from the Alaskan portion of the archipelago.

RELIEF : By far the greater part of Canada is level—the only mountainous region is in the west, where the magnificent natural rampart of the **Rocky Mountains** forms the western boundary of the Prairie Region of Central Canada. Eastern Canada is in parts hilly; there are no extensive level plains and no high mountain ranges.

The portion of the **Rocky Mountains** in Canada is about 1,500 miles in length and culminates in **Mount Brown**, 16,000 feet, and **Mount Hooker**, 15,690 feet, about 250 miles north of the southern frontier of the Dominion, and, gradually decreasing in elevation to the north, finally merges into the Richardson Mountains and other hills, which divide the basin of the Lower Mackenzie from that of the Yukon River.

Between the Rocky Mountains and the Pacific Coast, and parallel with them, are other ranges—the **Gold Ranges** and the **Coast** or **Cascade Range**. The Gold Ranges, more particularly known as the **Selkirk**, **Purcell**, and **Cariboo** Mountains and the Gold Range proper, are about 80 miles in width. Between these mountains and the rugged Cascade Range is the interior plateau of British Columbia, with an average width of 100 miles and an elevation of 3,500 feet. The Cascade or Coast Range is also about 100 miles in width, and, as it receives on its seaward slope the moisture from the sea, it has a very luxuriant vegetation.[1]

The Pacific slope and the Atlantic slope of the Dominion are both heavily timbered, but the **Great Prairies** which extend from the Red River Valley to the foothills of the Rocky Mountains are almost *treeless*, though well grassed and with a soil of unsurpassed fertility. A remarkable feature of this great area is its division, along lines running generally north-west and south-east, into three distinct prairie steppes.

The First Prairie Steppe, according to Professor Fream, includes the *Red River* and *Lake Winnipeg*, with the adjacent lands on the west. The average altitude of the plain is 800 feet above the sea,[2] and it has an area of about **56,000** square miles. The *Red River Valley* itself is perfectly flat and undiversified, the most absolutely level prairie-region of America. This steppe is bordered on the east by the Laurentian plateau, and extends westwards as far as the *first escarpment*, marked by the *Duck, Porcupine,* and other hills which lie to the west of *Lake Winnipegosis*. When the summit of the first escarpment is reached in the neighbourhood of MACGREGOR, 80 miles west of WINNIPEG, a vast open country, called the Great Plains, forming the Second Prairie Steppe, is entered upon. This second steppe is about 200 to 230 miles in width, and its surface is not so even as that of the Red River Valley. Low hills rise from the prairie level in certain localities, such as *Turtle Mountain* and the *Touchwood Hills*, but neither of these rise over 500 feet above the prairie, which has an average elevation of about 1,600 feet, and is bounded westward by the remarkable *Grand Coteau of the Missouri*, which extends from south-east to north-west for a distance of about 800 miles, and is 30 miles wide under the 49th parallel. This coteau or hill slope of the Missouri, which is chiefly a great

---

1. " Agricultural Canada," by Prof. Fream. | 2. The surface of Lake Superior is 627 feet above the sea.

mass of glacial detritus and ice-travelled blocks, and is perhaps the most remarkable monument of the Glacial Period in the western plains, is about 400
miles west of Winnipeg, and fringes the eastern margin of the Third Prairie
Steppe, which extends with a gentle ascent westward to the base of the *Rocky
Mountains.* The average altitude of this third and highest prairie steppe is
about 3,000 feet, though its eastern edge is generally not over 2,000 feet, while
it attains an elevation of over 4,000 feet at the base of the Rocky Mountains.
Its area, including the highland and foothills along the base of the mountains,
is about 134,000 square miles, and of this by far the greatest part, or about
115,000 square miles, is almost devoid of forest.   It is nearly 500 miles broad
along the frontier, but it narrows rapidly towards the north.   The total area of
prairie country, including that of all the three Steppes, may be estimated at
192,000 square miles.

Eastern Canada, as we have said, has no extensive plains or high mountain
ranges.   It has, however, a "Height of Land," insignificant in elevation, but
of great physical and geological importance.   The Laurentian Mountains or
Laurentides, as this low and long range is called—their average elevation does
not exceed 1,500 to 1,600 feet, and they really extend from the Atlantic to the
shores of the Arctic Ocean—stretch from *Labrador* along the northern side
of the *St. Lawrence,* diverging north about 50 miles below Quebec and running along the *Ottawa* on the north side for about 100 miles, "sweeping round
thence to the Thousand Islands, near Kingston, from which they gain the southern extremity of Georgian Bay and continue along the eastern and northern
shores of *Lake Huron* and *Lake Superior.*   The range then turns up to the
north-westward, and, gradually diminishing in height to a tableland, ultimately
reaches the *Arctic Ocean."*   The distance of the whole course of the Laurentian range from Labrador is about 3,500 miles.   Its slopes are, in general,
thickly covered with pines and firs, and the region through which it runs, and
of which it forms the water-parting, is remarkable for an enormous number of
lakes and pools, more than 1,000 of them being marked on the maps already
published, although large portions of it are still unknown or but partially explored.[1]

This Laurentian chain forms the water-parting between the basin
of the St. Lawrence and those of the rivers flowing into Hudson
Bay, except the Churchill and Saskatchewan, which pierce it and
discharge into the Bay.   Further north, it forms the "divide" between the Mackenzie on the west and the Coppermine and other
rivers on the east.

To the south of the St. Lawrence, the Acadian Highlands, as the various hill
ranges of the Maritime Provinces may be termed, sweep through the centre of
Nova Scotia and the northern part of New Brunswick.   The Notre Dame
Mountains run along the south side of the St. Lawrence and keep close to its
margin as far as Kamouraska, 100 miles below Quebec, whence they diverge
southwards, being opposite Quebec about 30 miles, and opposite Montreal 50
miles, from the river.   The highest portion of these sub-Laurentian ranges are
the Shickshock Mountains (3,000 to 4,000 feet), at the base of the Gaspé
Peninsula, about 12 miles from the St. Lawrence.

With regard to the "Relief" of the Dominion, Onésime Reclus[2] says that, physically,
its regions are six.   British Columbia is the first, the heavily-timbered and well-watered

---

1. Professor A. R. C. Selwyn, F.R.S., Director | 2. A Bird's-eye View of the World (La Terre à
of the Geological Survey of Canada. | Vol d'Oiseau).

Pacific Coast and slope of the Rockies. East of this is the northern part of the great Central Plain of America, filled with a network of lakes and drained into the Arctic Ocean by the Peace and Mackenzie River system, and into Hudson Bay through Lake Winnipeg and the Nelson by the immense Saskatchewan system. This is the great wheat belt, rising in the centre and the west to a rather high latitude of habitability, owing to a dryness of air which makes the severe winters not quite unendurable ; it shades by infinite gradations to the Arctic wastes. Eastward, along Hudson Bay and the Great Lakes, is a stony tableland of utter sterility, despite its situation between great waters, severe and permanently uninhabitable. Southern Ontario and Quebec are a well-timbered region, with much good land, the former exceptionally favoured, and the St. Lawrence forming virtually an ocean frontage for the latter. Fifth, are the Maritime Provinces along the Atlantic ; and lastly, comes the sterile Peninsula of Labrador."

RIVERS and LAKES : The splendid rivers and huge lakes of Canada, so easily interconnected by a few canals, form an unrivalled system of inland navigation, and powerfully influence the climate, productions, and trade of the Dominion. No lake system in the world, except perhaps that of Equatorial Africa, approaches the Great Lakes of Canada in magnitude, which, with the mighty St. Lawrence and its tributaries, contain more than one-half of all the fresh water on the globe.

The drainage system of Canada is on the same extensive scale as the country itself. The valley of the St. Lawrence penetrates the continent by a navigable route to a distance of over 2,000 miles from the ocean. The rivers, which flow eastward into Hudson Bay, have their sources in the Rockies, 1,600 miles distant from their mouths. The northward-flowing rivers have a length of over 2,000 miles.

There are *four main river-systems* in the Dominion : (1) That of the St. Lawrence, in Eastern Canada ; (2) that of the Saskatchewan-Nelson, in South Central Canada ; (3) that of the Mackenzie-Athabasca, in North Central Canada ; and (4) that of the Fraser, in British Columbia.

Of the numerous *minor* river-systems, the principal are the St. John, on the Atlantic Coast, to the south of the St. Lawrence ; the Severn, Albany, East Main, and other rivers which discharge into Hudson Bay ; the Coppermine and Great Fish River, in the Arctic Coast region, to the east of the Mackenzie ; and the Stikeen and Skeena in British Columbia.

The Basin of the St. Lawrence includes not only the broad belt of country drained by the noble stream itself and its tributaries, but also the GREAT LAKES,¹ of which it forms the outlet. These lakes—Superior, Michigan, Huron, Erie, and Ontario—though of immense size (their united area is 90,000 square miles and exceeds that of Great Britain) have comparatively slight drainage areas, because of the nearness of the "Height of Land" and the southern

---

1. The following statistics, relative to the Great Lakes, are taken from the *Canadian Handbook* :—

	Length miles.	Breadth miles.	Depth feet.	Elevation feet.	Area sq. m.
Superior	420	170	1,000	600	31,500
Michigan	320	70	700	576	22,400
Huron	280	105	1,000	574	21,000
Erie	240	57	200	565	9,000
Ontario	180	55	600	235	5,400

In order to gain as adequate an idea as possible of the scale of the lakes and rivers of Canada, the student should compare the figures given with well-known areas at home. Thus, Lake Superior equals in length the distance from Berwick to Land's End, and has an area very nearly as large as that of Ireland, while the four smaller lakes would cover the whole of England and Wales and, with Lake Superior, the whole of Great Britain. The united length of the five lakes is 1,440 miles, which is almost exactly 200 miles less than the distance between Galway and St. John's, N.F.

water-parting, and thus streams gather almost at their very shores, and flow to feed the Mississippi or the Saskatchewan. In spite of this, the excess of rainfall over evaporation in the basin of the St. Lawrence is so enormous, that the river carries as much water to the sea as the Mississippi, whose drainage area is four times as large.[1] The river and the lakes together contain 12,000 cubic miles of water, or more than one-half the fresh water on the globe. The lakes diminish in size and decrease in altitude from west to east.

Lake SUPERIOR, the most westerly of the five great inland seas, with an area of 31,500 square miles, is the largest and also the most elevated of them all—its surface being about 630 feet above the sea. It is the largest expanse of fresh water on the globe, being 420 miles in length, 160 miles in greatest breadth, and having a circuit of 1,750 miles. It is fed by over 200 impetuous torrents, and is connected by the *St. Mary River* with Lake Huron. The shores and islands of Lake Superior are rich in minerals, especially copper, silver and iron. The picturesque rapids of *Sault Sainte Marie* have been turned by two canals, one on the United States side,[2] and another on the Canadian side.

Lake HURON is scarcely two-thirds the size of Lake Superior, but is quite as deep (mean depth 1,000 feet). Its shores are extremely irregular—*Georgian Bay*, a great inlet on the Canadian side, is almost cut off by a long peninsula and the chain of the Manitoulin Islands. Off *Saginaw Bay*, on the United States side, the depth is 1,800 feet, or 1,200 feet below the level of the sea. The waters of Lake Huron are remarkably pure, clear, and sweet, and the scenery in some parts is magnificent.[3] It receives on the north the surplus waters of Lake Superior by the *St. Mary River*, and those of

Lake MICHIGAN, the only one of the Great Lakes that is entirely within the United States, by the *Strait of Mackinaw*. This lake is about the same size as Lake Huron, but its shores are much less indented. The traffic on it—to and from CHICAGO and MILWAUKEE, the great grain and provision ports of the West Central States—is enormous. Lake Huron overflows into

Lake ERIE by a channel 100 miles long, formed by the *St. Clair River*, *Lake St. Clair*, and the *Detroit River*. Lake Erie, the fourth in size, is the shallowest of the Great Lakes, the maximum depth being only 270 feet; but the populous States to the south, and the prosperous Province of Ontario on the north, render it of immense importance as a commercial highway. The descent from Lake Superior, through Lake Huron, to Lake Erie, is very gradual and comparatively slight, but between the surface of Lake Erie and that of Lake Ontario it amounts to more than 300 feet, and the *River Niagara*, which connects them, is precipitated midway between them over the great Falls of Niagara—the grandest and most awe-inspiring of all the wonders of nature.[4] The Falls and the Rapids below them are avoided by the *Welland Canal*, by which vessels pass directly from Lake Ontario to Lake Erie.

---

1. The mean discharge of these great rivers is estimated at between 500,000 and 600,000 cubic feet per second.

2. Some idea of the magnitude of the trade and industry of the surrounding country may be gained from the fact that the tonnage and value of the products which pass through the *Sault Sainte Marie Canal* (compressed within 7 months of the season of navigation, e. jual, and probably exceed, those which pass through the Suez Canal in the entire year. Here, in the northern part of North America, between two inland lakes, with only one shore of these developed, a commerce has been created, which equals that between two oceans, whose traffic is almost as old as the world, and contributions to which are made from every country and clime on the globe. And this is so, in spite of the fact that the water communication of the lakes is competed with by the most perfectly-equipped railway systems of the age, while the commerce of Suez is practically without a competitor.—*Wiman*.

3. Lake Huron is said to contain above 3,000 islands, most of them well wooded.

4. The Cataract is divided by Goat Island into two Falls—the *Horse Shoe Fall* on the Canadian side, 178 feet high, and the *American Fall*, 180 feet.

Lake ONTARIO, the most easterly of the Great Lakes, is the smallest of them, its area being not more than 5,400 square miles. Ontario is much deeper and lower in level than Erie, and it is also safer, but all the Great Lakes are subject to fearful storms, and some of the most awful disasters have occurred on them. The shores of Ontario, like those of Erie, are dotted with flourishing villages and many large towns. On the Canadian side are TORONTO, the capital of the Province which owes its name to the lake, KINGSTON, HAMILTON, PORT DALHOUSIE, and NIAGARA ; on the American side are OSWEGO, CHARLOTTE, the port of ROCHESTER, &c. The shores of Ontario are bold and regular, except on the north-east, where Prince Edward Island and other islands and the adjoining coast are deeply indented. At Kingston, the waters of Lake Ontario flow into

The River St. LAWRENCE, which, between KINGSTON and BROCKVILLE, some 40 miles below, is a broad expanse, studded with innumerable clusters of the most lovely islets. This " Lake of the Thousand Isles " is the most exquisitely beautiful 'lake within a river' in the world, and below it are other beautiful lake expanses and some of the grandest and most extensive rapids imaginable. The "run" over the boisterous Long Sault, the picturesque Cedars and Cascades, and the terrible Lachine Rapids,[1] where it seems as if the vessel must be dashed to pieces among the breakers, is most impressive. A few miles above MONTREAL, the steamer gets into smooth water, and, passing under the magnificent *Victoria Bridge*, steams alongside the quay of this historic city, the commercial metropolis of Canada, and now accessible by ocean steamers. Montreal stands at the junction of the St. Lawrence and its noble tributary the Ottawa. Another great tributary, the Richelieu from Lake Champlain, joins the St. Lawrence just before it expands into Lake St. Peter, the last of the lakes in its passage to the sea. Below Three Rivers, where the excessively rapid St. Maurice falls into the St. Lawrence from the north, the banks gradually increase in height, rising at Quebec into the magnificent Cape Diamond, upon which the far-famed citadel is built. The scene from the citadel or from Pointe Levis, on the opposite side of the river, is said to be, in picturesque beauty, the finest in the world. Below Quebec are the Montmorenci Falls, which are higher than those of Niagara, and the beautiful Isle of Orleans. Thence the river broadens out, receiving from the north the Saguenay, an immense river, over a mile in width and 1,000 feet in depth, which rises in Lake St. John and is navigable for the largest vessels from TADOUSSAC at its mouth to HA HA BAY, a lumbering port 70 miles up the stream. The river is now 20 miles in width, and at CACOUNA, a bathing place on the southern side, the water is quite salt. From Tadoussac all the way down to the sea, the villages, churches, telegraph stations, mills, and farmhouses, all painted white, produce a dazzling contrast to the dark woods which clothe the rising grounds in the distance to their very summit. When we reach the island of Anticosti, at the entrance to the Gulf, we have covered the whole sweep of the mighty St. Lawrence, navigable for the largest ocean steamers as far as Montreal, 1,000 miles from the sea ; and by a system of canals, engineered to overcome the St. Lawrence Rapids and the difference in the levels of the Great Lakes (600 feet), there is uninterrupted navigation from the sea to the head of Lake Superior, a distance of 2,400 miles, of which only about 70 miles are canals.

1. To avoid the Lachine Rapids, the *Grenville Canal* was constructed, by which vessels can reach Ottawa, where they enter the *Rideau Canal* which extends to Kingston on Lake Ontario. Crossing the lake, they enter the *Welland Canal* near the mouth of the River Niagara, and thus escape the Niagara Falls and rapids, and reach Lake Erie through 27 locks. By the *St. Clair River* they pass from Lake Erie to Lake Huron, and thence by either the Canadian or the American *Sault Sainte Marie Canal*, they enter Lake Superior.

In the Great Plains, which extend from Lake Superior to the Arctic Ocean and westwards from Hudson Bay to the foot of the Rockies, there is a perfect labyrinth of lakes and watercourses, connected together by cross channels, or separated only by short portages.

In the Saskatchewan-Nelson River-system, *Lake Winnipeg*, an immense sheet of water 240 miles long and 55 miles wide, is, as it were, the central reservoir, as it receives the great navigable streams of the Red River, the Assiniboine, and the Saskatchewan, besides the overflow from Lakes Manitoba and Winnipegosis, and finds an outlet by the Nelson River into Hudson Bay.

The Saskatchewan is formed by the junction of two streams, both of which rise in the Rocky Mountains in Alberta—the *North Branch*[1] having a course of 772 miles, and the *South Branch* about 810 miles, before they unite. From their junction, the main stream has a course of 282 miles to Lake Winnipeg. Besides the enormous mass of water brought down by this great river, Lake Winnipeg receives the drainage of an immense region by the Assiniboine from the prairie-lands of the south-west, by the Red River from the south, and the Winnipeg River from the Lake of the Woods on the south-east.

The Assiniboine joins the Red River at WINNIPEG, about 40 miles from Lake Winnipeg. Both are navigable for hundreds of miles—steamers ascend the former as far as FORT ELLICE on the western border of Manitoba, and the latter, which rises within 10 miles of the Mississippi, forms an admirable waterway through North Dakota, Minnesota, and the southern part of Manitoba. The tortuous course of the Winnipeg River is broken by a succession of magnificent cataracts, tumultuous cascades, and foaming rapids, and it frequently expands into large lakes studded with islands.

The immense mass of water thus poured into the lake finds an outlet by the Nelson River, one of the great rivers of the world, which flows into Hudson Bay by an estuary 6 miles broad. A narrow tongue of land, on which stands YORK FACTORY, separates the mouth of the Nelson from that of the Hayes River, which, with the Severn, the Albany, and other streams, drains the desolate territory between Lake Winnipeg and James Bay. The navigation of the Nelson is unfortunately obstructed by rapids and shallows, but river-steamers can ascend it for about 100 miles from the sea. It is not likely, however, to be even partially utilized, the grain and other products of Manitoba and the North-West, shipped by the Hudson Bay Route to Europe, will be brought to the seaboard by the railway that is being built from WINNIPEG to PORT NELSON.

The Churchill, another river of great volume and length (over 1,000 miles), enters Hudson Bay to the north of the Nelson, and receives the overflow of an extensive system of lakes and channels in the belt between the Saskatchewan-Nelson and the Athabasca.

Further north, the Great Fish River and the Coppermine River inundate the dreary tundras of the Arctic Coast in summer, while their lakes and expansions are frozen over during the long winter.

---

1. The North Branch of the Saskatchewan has a greater volume of water than the Rhône, while the South Branch discharges more water than the Rhine.

The rest of the Great Central Plain, to the north and west of the rivers already named, belongs to the immense basin of the **Atha-basca-Mackenzie.**

The **MACKENZIE**[1] is by far the largest river in the American section of the Arctic river-system. Measured from the source of either of its main tributaries—the **Peace River** or the **Athabasca**—this great river has a length of not less than **2,500** miles, of which not less than 2,000 miles are navigable for steamboats, while its drainage area, estimated to embrace more than half-a-million square miles, is double that of the St. Lawrence. The **Peace River**, which rises in British Columbia, on the western side of the Rocky Mountains, and is noted for the boundless resources and wonderful fertility of its valley, is regarded as the main branch of the Mackenzie, but the **Athabasca**, the most southerly tributary, is not inferior to the Peace, either in length or volume. This great river rises on the eastern slopes of the Rocky Mountains, and flowing through the *Athabasca Pass*, west of *Mount Hooker* and *Mount Brown*, the culminating points of the main ridge of "the backbone of the North American continent," it enters the great Central Plain and flows northward to **Lake Athabasca**, an irregular expanse of water, 230 miles in length, the principal outlet of which—the **Stony River**—is joined, 30 miles north of the lake, by the **Peace River**. The united stream, now called the **Slave River**, runs north for 300 miles before it enters the **Great Slave Lake**, from which it issues as the **Mackenzie**; and, after receiving from the west the **Liard River**, which, like the Peace River, rises to the west of the Rocky Mountains, and from the east the **Great Bear River**, which brings the overflow from the **Great Bear Lake**, the mighty stream enters the Arctic Ocean by several channels. There are innumerable islands at its mouth, some of them well grassed and wooded, although they are within the Arctic Circle, and as far north of Manitoba as Florida is south of the "Prairie Province."

The **Pacific** slope of the Dominion is drained by the Upper **Columbia** and the **Fraser** in the south, and by tributaries of the **Mackenzie** and the **Yukon** in the north.

The **Columbia**, the largest of the Pacific rivers of America, rises in the Columbia Lakes, on the western slope of the Rocky Mountains. It has a total course of about **1,400** miles, nearly one half of which is through British territory. The upper courses of two of its tributaries—the **Kootenay** on the east, and the **Okanagan** on the west (both of which flow through, or discharge from, long lakes of the same names)—are also north of the International Boundary (49° N. lat.). The principal river of British Columbia is, however, the **FRASER**, which is entirely within the Province. Like the Columbia, the Fraser rises on the western side of the main ridge of the Rockies, and, curiously enough, both run north-west for about 180 miles, and then "both make a sharp turn, generally known as 'The Big Bend,' to the south, flowing almost due south and nearly parallel with each other for 350 miles—the Fraser to HOPE, and the Columbia to COLVILLE—whence they each reach the sea by almost rectangular bends to the westward through profound gaps in the coast range." In the case of the Fraser, this gap or gorge extends from LYTTON to HOPE (60 miles), but steamers can ascend the river from its mouth to YALE, 13 miles above Hope; above Yale

1. The Mackenzie River is so named after Sir Alexander Mackenzie, who discovered it in 1789, and followed it to the sea in 1793. He accomplished the whole distance, from Fort Chippewyan to the mouth of the river and back, a boat voyage of 3,000 miles, in 100 days (June 3—Sept. 17).

it is unnavigable even for canoes. The main western branch, the **Nechaco River**, and its tributary the **Stuart River**, bring down the waters of a long chain of lakes. On the east, the **Quesnelle** from the *Cariboo Gold Fields*, joins the Fraser at QUESNELLEMOUTH, and the **Thompson** from *Lake Kamloops*, at Lytton, 200 miles lower down. North of the Fraser, the largest rivers in British Columbia are the **Skeena** and the **Nasse**, both navigable for stern-wheel steamers for some distance inland. The immense and almost unknown area beyond the **Stikeen River** (which enters the sea 120 miles south-east of Sitka), between the Rocky Mountains and the eastern boundary of Alaska and extending north to the Arctic Ocean, is drained principally by the **Liard River**, a tributary of the **Mackenzie**, and by the **Pelly** and other tributaries of the **Yukon**. The **Pelly River** alone flows for 700 miles through British territory before it enters Alaska.

CLIMATE : The Dominion of Canada, extending from the latitude of the North Cape in Norway to that of Rome, naturally exhibits a **great variety of climate**. Generally speaking, it is severe and "formidable" in the north, but genial and temperate in the south. The summers in all parts of Canada are finer and hotter than those of England, but the winters are far colder. But in winter the air is dry, bracing, and exhilarating, and the climate of Canada is, on the whole, one of the healthiest and most favourable in the world to the highest development of man.

The Dominion, from its vast extent, has been truly said to possess all the climates of Europe, from the Mediterranean to the Arctic Ocean. The *Gulf Stream* in the Atlantic and the *Japan Current* in the Pacific are both singularly favourable to the maritime portions of Canada, while in **Ontario**, the portion enclosed between the Great Lakes, in particular, enjoys a temperate and a delightful climate. In **Quebec**, the winter is long and severe—the St. Lawrence is frozen over and closed to navigation for about 140 days every year. But during the greater part of this time the sky is clear and the cold healthy and invigorating. The Maritime Provinces—**New Brunswick**, **Nova Scotia**, and **Prince Edward Island**—have, of course, a milder and more equable climate ; and while **Manitoba** and the **North-West** have a "continental" climate—a hot summer being followed by a cold winter, and spring and autumn being of exceedingly brief duration—the coast region of **British Columbia** possesses an insular climate, having all the advantages of that of England without any of its disadvantages. East of the Cascade Range, the climate is drier and more extreme, and on the higher lands the winter is as severe as in Eastern Canada.

PRODUCTIONS : Immense forests, luxuriant pastures, fertile wheatlands, inexhaustible fisheries, and vast stores of mineral wealth—these are the most important of the rich and varied resources of Canada.

Canada possesses thousands of square miles of the finest forests, and forest products constitute one of her main sources of wealth. Canadian forests are rich in a great variety of the most useful and valuable trees, which yield lumber of many kinds for building purposes, for furniture and, in many parts of Canada, for fuel. Among the varieties are the *maple, elm, ash, cherry, beech, hickory, ironwood, pine, spruce, balsam, cedar, hemlock, walnut, oak, butternut, basswood, poplar, chestnut, mountain ash, willow, black and white birch,* and many more.

" These forest trees add a singular beauty to the landscape in many parts of the country, and also exercise a very beneficial influence on the climate in affording shelter and attracting rainfall. The beauty of the tints and the brilliancy of the colours of Canadian forest trees in autumn require to be seen to be appreciated."

With the exception of the prairie lands of Manitoba and the North-West, which are generally treeless, and the extreme Arctic Coast, Canada may be said to be covered with forests—many of the trees, such as the *Douglas pine* of British Columbia, the *Banksian pine* on the shores of Hudson Bay, and the *balsam poplar* on the Athabasca, Peace, and Mackenzie rivers, attaining a height of from 100 to 300 feet ! The forests on the banks of the St. Lawrence and Ottawa, notwithstanding the immense quantities of *timber* which they have yielded for so long, are still of vast extent, and the supply may be said to be, for all practical purposes, inexhaustible. And the same may be said of the forests of Central and Western Canada.

But, besides her magnificent forest trees, the fruit trees of Canada are unsurpassed, and fruit-growing is a very important industry in Eastern Canada, and will also be so, in the near future, in the southern portion of British Columbia and Vancouver Island. *Apples, pears, plums, peaches, apricots, grapes,* and *berries* of every description, grow luxuriantly, and attain a size and flavour unknown in Europe. All kinds of *vegetables* thrive everywhere, even in the far North-West and along the shores of Hudson Bay.[2]

The meadows and pastures of the Dominion are co-extensive with its vast forests ; wherever trees grow, there the cultivable grasses will thrive. The pasture lands of Canada (including the comparatively small portions now under actual cultivation) are estimated to extend over an area of more than 2 million square miles, or more than 12 hundred million acres.

The wheat lands of Canada, writes Mr. Wiman in the *North American Review* for January 1889, possess all the advantages of the regions to the south, but in richness, fertility, and extent, infinitely greater. The Dominion possesses a wheat area larger than do the entire United States, and the soil of this wheat area is richer, will last longer, and will produce a higher average of better wheat than can be produced anywhere else on the continent, if not in the world. Manitoba alone produced, in 1890, 20 million bushels of wheat, and at KIL-DONAN, on the Red River, wheat has been grown for 35 consecutive years, without rotation, without fertilization, and now produces crops averaging 30 bushels to the acre ! If one-half of that comparatively small portion of Central Canada which is drained by the Red River and its affluents were sown with wheat, the product, at an average yield, would be 500 million bushels, or more than the entire amount raised in the United States in 1890. But, besides the Red River Valley, there are the immensely larger Saskatchewan, Athabasca, and Peace River regions, and the extensive wheat areas of Eastern Canada and British Columbia. And not only wheat, but oats, barley, potatoes, and other vegetables come to perfection over the greater part of the settled and cultivable portions of the Dominion.

---

1. In some places on the coasts of British Columbia the Douglas pine or spruce—commercially called the "Oregon pine"—frequently exceeds 8 feet in diameter at a considerable height, and reaches a height of from 250 to 300 feet, forming prodigious, dark forests. The "Giant Cedar," another valuable tree, is also often found from 100 to 150 feet high and 15 feet thick.
2. There are *vineyards* in the Province of Ontario of 50 to 60 acres in extent, *peach orchards* of similar extent, and *apple orchards* almost innumerable. *Tomatoes* and *melons* ripen in the open air, as field or market garden crops. The *apples* of Canada are highly prized—620,000 barrels of apples went direct to Great Britain from Ontario, Quebec, and the Maritime Provinces in 1889. Berries and other fruits grow abundantly in Central Canada and in British Columbia, and, in fact, almost everywhere the settler can always obtain a supply of the healthful luxury of delicious fruits.

Animals and their produce are a greater source of wealth to Canada than even her immense forests. All the ordinary domestic animals thrive wonderfully on the rich pastures, and *live animals, meat, butter, cheese, eggs, hides, skins,* and *wool* form the largest items in the exports. Over 100,000 *cattle* are annually exported, and nearly four times that number of *sheep*, besides about 20,000 *horses.* The export of live cattle and horses, when the ranches established on the grasslands of Alberta and Athabasca, at the base of the Rocky Mountains, have been developed, will be enormous. Dairy-farming and stock-breeding are, in fact, rapidly becoming as important as wheat-growing, and Canada now exports over 3½ million lbs. more *cheese* than the United States.

Of wild animals, the *bison* or *buffalo*, millions of which, 50 years ago, roamed over the prairies and north-western plains, is now almost extinct; the *grizzly bear* is still found in the Rocky Mountains; the *moosedeer, sable,* and other *fur-bearing animals* are more widely distributed. West and north of Ontario and Quebec, the Hudson Bay Company has numerous 'forts' or trading-stations, and the traders in its employ collect enormous quantities of *furs* and *skins* from the Hudson Bay region, the North-West Territory, and British Columbia. Among wild birds, there are any number of *prairie chickens, wild geese, wild ducks, pigeons,* &c., besides *eagles, hawks,* and other birds of prey.

The fisheries of Canada are the largest in the world, embracing fully 8,000 miles of sea coast, in addition to inland seas, innumerable lakes, and a great number of rivers. Nearly 70,000 men and 30,000 boats are employed in the fisheries, and the annual value of the produce amounts to 3 millions sterling.

The sea-fisheries are well-nigh inexhaustible—a fact attributable to the fishes' food-supply being brought down by the Arctic currents from the northern seas and rivers. This consists of myriads of minute organisms which swarm in the Arctic seas and are deposited in vast and ever-renewed quantities upon the fishing grounds. Salt-water fishes of nearly every variety are to be found along the Canadian coast, but the marine fisheries of the greatest commercial importance are the cod, herring, mackerel, lobster, salmon, and seal.

The fresh-water fisheries are also of great importance, the immense lakes and rivers supplying an abundance of fish of great commercial value, both for home consumption and export, besides providing sportsmen with some of the finest salmon and trout fishing to be found anywhere.

Cod is the most abundant and valuable catch on the Atlantic Coast, and salmon on the Pacific Coast. Over 20 millions lbs. of salmon are canned every year on the Fraser River in British Columbia. Trout are everywhere abundant, but especially in Lake and River Nipigon in Ontario.

Canada is marvellously rich in minerals, and there are vast deposits of coal and iron, with copper, gold, silver, and other useful metals and minerals.

Little has been done, comparatively, to develop the undoubtedly great mineral resources of the Dominion, but the gold mines of British Columbia have yielded 60 million dollars' worth of gold, and the precious metal is also obtained from Nova Scotia, Quebec, and Ontario. Large quantities of native silver have been obtained from the Thunder Bay and Silver Islet Mines at Lake Superior, and silver lodes occur near Hope on the Fraser River, in British Columbia. The iron-ores of Canada, which include some of the richest ores yet discovered, occur in boundless abundance in Nova Scotia, New Brunswick, Quebec, Ontario, and British Columbia.   Native copper is extensively and profitably worked on the northern shores of Lake Superior, and there are rich deposits of copper ore in other parts of Ontario and also in Quebec.

The **coalfields of Canada** are of immense extent, and many mines on the Atlantic and Pacific coasts, and in the North-West and the Rocky Mountains, are actively worked.   The coal and lignite area of Canada is over 100,000 square miles in extent, of which 65,000 square miles are in the North-West. The coal of Nova Scotia and New Brunswick is sent inland by the St. Lawrence and by rail.   The Nanaimo coal of British Columbia commands the highest price in San Francisco.   Large quantities of lignite exist, and beds of true bituminous coal are being worked, in the North-West Territories.   Anthracite coal of the best quality is mined in the Rocky Mountains, and is conveyed east and west by the Canadian Pacific Railway.

Petroleum is known to exist in several parts of the Dominion, and the petroleum wells in Ontario are extremely productive.   It is also found in the North-West and in the Rocky Mountains.   There are large deposits of apatite or phosphate of lime, so valuable as a fertilizer, in the Ottawa River Valley, and extensive **salt-works** in Ontario and New Brunswick.

INHABITANTS: Canada had, in 1891, a total population of over 4¾ millions,[1] which is a small number for so large and richly-endowed a country.

The general average, in the habitable portion of the country, may be about 2 or 3 to the square mile, and the highest density, which occurs in Prince Edward Island, does not exceed 50 to the square mile—only one-tenth of the density in England—so that there is "ample space and verge enough" in Canada for millions more.

The great bulk of the population are of *British* descent, except in the case of Quebec, where the majority are of *French* origin—descendants of the settlers in Canada prior to its falling under the rule of Great Britain in 1763.   The *Indians* are comparatively few in number.   In the Provinces and the North-West Districts, they live on certain tracts of land reserved for them.   A few tribes of *Eskimo* live along the Arctic Coast from the mouth of the Mackenzie to Labrador.

There is no State Church in Canada, and complete religious liberty prevails. About 2½ millions of the Canadian people are Protestants, and nearly 2 millions are Roman Catholics.   A large number of the Indians and Eskimos are still heathens.

In no country in the world is education so generally diffused.   Primary education is free and compulsory, except in New Brunswick, and there is every facility for obtaining the highest education at a small cost.   There are 16 Universities and Colleges that confer degrees in divinity, arts, law, medicine, &c., besides several purely Theological Colleges.

INDUSTRIES: Canada is mainly an agricultural and pastoral country, largely covered with forests, the produce of which, until recently, formed the chief source of wealth and the most important item of export.   Fishing and mining are also important industries. but manufactures, which are chiefly connected with the main industries, are as yet in their infancy.

1. The actual number, in 1891, was 4,823,344—an increase, since 1881, of nearly half a million.

The industries of Canada include the cultivation of the soil and the growing of enormous quantities of wheat and other cereals, and of all kinds of fruits and vegetables; the rearing of cattle, sheep, and horses; and the manufacture of cheese and butter on a very large scale. But the "produce of the forest"[1] greatly exceeds in value the purely "agricultural produce," and is but little less in value than the "animals and their produce," which now form the chief source of wealth. The "produce of the fisheries" is less than a third in value of that of the forest, while the "produce of the mines" as yet amounts to less than one-third of the produce of the soil, although in the near future it must exceed in value that of all other resources. The manufactures of Canada include *shipbuilding*, the making of *agricultural implements, leather goods, furniture, musical instruments*, &c. The iron industry has also a great future, owing to the enormous deposits of excellent *iron-ore* within the coalfield areas in the various Provinces, and the rich deposits of *nickel* at Sudbury and along the shores of Lake Superior.

TRADE : The trade of Canada is larger than that of any other British Colony, and is mainly carried on with the United States and the United Kingdom. Annual value, about 46 millions sterling.

The principal exports, in order of value, are *timber* and *other forest products, cheese, grain and flour, live animals, fish, coal, eggs, canned lobsters* and *salmon*, &c. Value, 1892, over 22¾ millions sterling.

The chief imports, in order of value, are iron goods, woollen goods, coal, bread-stuffs, cotton goods, sugar, tea, coffee, silks, tobacco, &c. Value, 1892, over 23 millions sterling.

The chief exports from Canada to the United Kingdom are *timber, cheese, oxen, wheat* and *flour, maize, bacon* and *hams, skins* and *furs, fish* and *apples*. Value, 14 millions sterling.

The principal imports into Canada from Great Britain are *iron* and *iron goods, woollen* and *cotton goods, clothing*, &c. Value, 6¾ millions sterling.

PORTS : The chief *ports* of Canada are Halifax in Nova Scotia, St. John in New Brunswick, Quebec and Montreal on the St. Lawrence, Ottawa on the Ottawa River, Toronto on Lake Ontario, and Vancouver and Victoria in British Columbia.

SHIPPING: Canada stands fifth among maritime countries in tonnage of shipping owned and registered in the country. Over 65,000 vessels—about 70 per cent. of the whole under the British flag—enter and leave Canadian ports every year. Including the arrivals and clearances of coasting vessels, the total number is 160,000, and the total tonnage is over 35 million tons. Regular communication with England is maintained by the *Allan Line* mail steamers, and by the steamships of the *Dominion* and *Beaver Lines*, which carry passengers and cargo between Liverpool and Quebec and Montreal in summer, and Halifax and Portland in winter. The cargo-carrying steamers of the *Furness Line* sail every 10 days between London, Halifax, and St. John. Railways start from Halifax, St. John, and Quebec to all parts of the Dominion. The distance from Liverpool to Quebec is 2,661 miles, and to Halifax 2,480 miles, and the voyage takes from 9 to 10 days. The *Trans-Pacific steamships* of the Canadian Pacific Railway Co. run from Vancouver to Yokohama and Hong-Kong in from 12 to 14 days.

---

1. In the Canadian Trade Returns, the home produce of the Forest, (4) Animals and their Produce, (5) Agri-ducts are classified under seven heads :—(1) Produce cultural Produce, (6) Manufactures, and (7) Miscel-of the Mine, (2) Produce of the Fisheries, (3) Produce laneous.

CANALS : The canals of Canada and the **river improvements** are works of immense importance, which have largely increased the trade of the country.

From the sea to **Quebec**, the St. Lawrence is naturally navigable for vessels of any size. Above Quebec, the channel has been deepened, so that the largest ocean-going vessels can ascend as far as **Montreal**, which is thus a fresh water port, nearly 1,000 miles inland from the Atlantic, 250 miles above salt water, and nearly 100 miles above the highest limit of the tide. Above Montreal, there is a system of canals to overcome the *Rapids of the St. Lawrence* and the difference in the levels of the Great Lakes, which affords **uninterrupted navigation** from the **Strait of Belle Isle** to the head of **Lake Superior**, a distance of 2,384 miles, of which only 71¾ miles are canals. The depth of water is from 9 to 14 feet, and it is proposed to make the whole route available for vessels drawing 14 feet. The **Ottawa Canals** overcome the difficulties of the *River Ottawa* between **Montreal** and **Ottawa**. The **Rideau Canal** opens navigation between **Ottawa** and **Kingston** on Lake Ontario. The **Ottawa** and **Rideau Canals** thus connect **Montreal** and **Kingston** *via* Ottawa, and form an alternative route between Montreal and the Lakes. The St. Lawrence Rapids are, however, so gradual, that passenger vessels "shoot" them without danger ; freight vessels, however, descend and ascend by the canals. Another canal—the **Chambly Canal**—overcomes the rapids of the *Richelieu River*, and connects the *St. Lawrence* with *Lake Champlain*, which the Champlain Canal connects with the *River Hudson*, thus completing the waterway between **Montreal** and **New York**. By the canals which connect the lakes—the Canadian **Welland Canal**, between **Lake Ontario** and **Lake Erie**, and the Canadian and American *Sault Sainte Marie Canals*, between **Lake Huron** and **Lake Superior**—seagoing vessels of moderate tonnage can proceed to **Port Arthur** on Lake Superior, or to **Chicago** on Lake Michigan, without breaking bulk.

RAILWAYS : Canada has over 15,000 miles of railway open for traffic, and about 4,000 miles in course of construction or arranged for. The three principal systems are the **Canadian Pacific Railway** (5,186 miles), the **Grand Trunk Railway** (3,114 miles), and the **Inter-Colonial Railway** (894 miles).[1]

The **Grand Trunk Railway**, which forms a complete network of lines between the Great Lakes, the St. Lawrence, and the Atlantic Coast, starts from *Portland* (Maine) and runs through *Richmond* to *Montreal*, and thence through *Toronto* and *Hamilton* to *Detroit*, and from thence to *Chicago* by the Chicago and Grand Trunk Railway. A branch from Richmond runs to *Pointe Levis*, opposite *Quebec*, where it joins

The **Inter-Colonial Railway** (894 miles), which runs along the right bank of the St. Lawrence and through New Brunswick and Nova Scotia to *Halifax*, with branches and connecting lines to Cape Breton Island, St. John, and the Grand Trunk system.

The **Canadian Pacific Railway** is the shortest of the three great trans-continental lines of North America, the distance from *Montreal* to *Vancouver*—2,906 miles—being 600 miles shorter than from New York to San Francisco. By this great railway, which has been the means of opening up Manitoba, and the fertile wheat areas and rich pasture lands of the North-West, and of bring-

---

1. Or 1,227 miles including the Prince Edward Island Railway.

G

ing British Columbia into direct communication with Eastern Canada, thus consolidating the Dominion into a whole, one and indivisible, the distance from Liverpool to Japan and China has been shortened by about 1,000 miles, and it is thus of immense strategic importance, as it furnishes an alternative route to the East for troops and munitions of war, which could be conveyed from Great Britain to China and Japan quicker than by any other route, to Brisbane and Sydney as quickly as by the Suez Canal, and to India in a very few days more. Then again, British fleets command both the Atlantic and Pacific ends of the line, and there are large coalfields near, and graving docks both at *Halifax*—the station of the North Atlantic fleet—and at *Esquimalt*, the headquarters of the Pacific squadron.

The main line of the Canadian Pacific Railway starts from QUEBEC and runs through *Montreal, Ottawa, Carleton Junction, Sudbury, Port Arthur* (on Lake Superior), to WINNIPEG, the half-way house of the Dominion ; thence through the Fertile Belt, via *Brandon, Regina, Medicine Hat*, and *Calgary*, and across the Rocky Mountains by the Kicking Horse Pass, then descending the Thompson and Fraser Valleys to its terminal port of VANCOUVER, on Burrard Inlet. The C. P. R. connects at several points with the Grand Trunk and Inter-Colonial Railways, and also with the American lines which run to Chicago, New York, and Boston. The most important branches are from *Sudbury* to *Algoma*, and from *Winnipeg* to *Emerson* for St. Paul and Chicago. In Manitoba, several lines radiate from Winnipeg, and the line to *Port Nelson*, on Hudson Bay, is being constructed. The railway from *Regina* on the main line to *Prince Albert* on the Saskatchewan, 234 miles in length, is open and will have branches to the elbow of the North Saskatchewan and to *Battleford*. The Manitoba and North-Western line is open to *Yorkton* and will be continued to *Prince Albert*. The North-Western Railway from *Calgary* to *Edmonton*, 210 miles in length, is also open. The North-Western Coal Company's railway from *Dunmore* on the main line to *Lethbridge* has been extended to the frontier of Montana, thus making the coalfields of Alberta the nearest source of supply for the mining industries of Helena, Butte City, &c.

In Vancouver Island, a short line—the Esquimalt and Nanaimo Railway—connects the naval station of *Esquimalt* with VICTORIA, the capital of British Columbia, and with the *Nanaimo* and *Wellington* coalfields.

∴ **Canada** is said to have been discovered in 1497 by Sebastian Cabot, but its history dates only from 1534, when the French took possession of the country. The first settlement—Quebec—was founded by them in 1608. A series of wars, between the English settlers in the New England States and the French Canadians, culminated in 1759 in the capture of Quebec by Wolfe, and the whole territory became, subsequently, a British possession by the *Treaty of Paris* (1763). Nova Scotia was ceded in 1713 by the *Treaty of Utrecht*—the Provinces of New Brunswick and Prince Edward Island being afterwards formed out of it. British Columbia was formed into a Crown Colony in 1858, having previously been a part of the Hudson Bay Territory, and was united with Vancouver Island in 1866. By the British North America Act, passed by the British Parliament in 1867, the Provinces of Canada, Nova Scotia, and New Brunswick, were united under the title of the DOMINION OF CANADA, and provision was made in the Act for the admission, at any subsequent period, of the other Provinces and Territories of British North America. The Province of Manitoba was formed in 1870, and, with the remainder of the Hudson Bay Territory (now called the North-West Territory), was admitted into the Dominion. British Columbia joined the Confederation in 1871, and Prince Edward Island in 1873, thus uniting, under one Federal Government and Parliament, the whole of the British Possessions in North America, from the Atlantic to the Pacific, with the single exception of the island of Newfoundland, which still remains a separate colony.

GOVERNMENT : The Government of Canada is Federal. The *executive authority* of and over the Dominion is vested in the Queen, in whose name the **Governor-General**, aided by a Privy Council, carries on the government. The *legislative power* is vested in a **Parliament**, consisting of an Upper House, called the Senate, and a House of Commons. The political capital and seat of the **Federal Government** is OTTAWA.[1]

The ordinary public revenue amounts to about 8 millions, and the ordinary public expenditure to 7½ millions. The public debt, which amounts to nearly 49 millions sterling, has been chiefly incurred in the construction of railways, canals, and other public works.

With the exception of a garrison of 2,000 men at Halifax, there are no Imperial troops in the Dominion. The colonial forces comprise an active volunteer and marine militia of about 37,000 men. The total number of men, liable to be called on for active service, exceeds a million.

The members of the **Senate** of the Dominion Parliament are nominated for life by the Governor-General. There are 80 senators, namely, 24 from the Province of *Ontario*, 24 from *Quebec*, 10 from *Nova Scotia*, 10 from *New Brunswick*, 3 from *Manitoba*, 3 from *British Columbia*, 4 from *Prince Edward Island*, and 2 from the *Territories*.

The Canadian **House of Commons** consists of 215 members—one for every 20,000 of the population—elected every 5 years at longest : *Ontario* has 92 members, *Quebec* 65, *Nova Scotia* 21, *New Brunswick* 16, *British Columbia* 6, *Prince Edward Island* 6, *Manitoba* 5, and the *North-West Territories* 4.[2]

Each of the Provinces forming the Dominion have a **separate Parliament**, with a **Lieutenant-Governor**, appointed by the Governor-General, at the head of the Executive.

The Provincial Parliaments of Quebec, Nova Scotia, New Brunswick, and Prince Edward Island consist of two Chambers—a Legislative Council and a Legislative Assembly ; but Ontario, Manitoba, British Columbia, and the North-West have each only one Chamber—a Legislative Assembly. The Provincial Governments have full powers to regulate their own local affairs, but all matters affecting the Dominion as a whole are dealt with by the Central Government at Ottawa, which is supreme. Unlike the States in the Union to the south, the Provinces have no power to organize or maintain any military force, nor have they final legislation, as the Dominion Government can veto their acts.

∴ We now proceed to describe separately the various Provinces and Territories included in the Dominion of Canada, and these will be dealt with very briefly in the following order :—(1) The Maritime Provinces of Nova Scotia, New Brunswick, and Prince Edward Island ; (2) Quebec and Ontario ; (3) Manitoba and the North-West ; and (4) British Columbia.

---

1. Professor Selwyn points out that "under the United States' system the Central Government is the delegate of the sovereign States, whilst under the Canadian Confederation the Imperial Government is supreme, and has delegated to the Dominion and Provincial Parliaments a limited sovereignty, with control in the departments entrusted to them respectively."
2. These numbers indicate the relative importance of the several Provinces.

THE MARITIME PROVINCES.

The MARITIME PROVINCES of Canada include **Nova Scotia**, the most easterly Province of the Dominion, and the adjoining Provinces of **New Brunswick** and **Prince Edward Island.**

NOVA SCOTIA.

The Province of NOVA SCOTIA[1] includes the peninsula of **Nova Scotia** and the island of **Cape Breton,** which together have an *area* of over **20,000** square miles—one-fifth of which is covered with lakes and small rivers—and a *population* of about half a million.

Nova Scotia is united to New Brunswick by the narrow and fertile Isthmus of Chignecto,[2] and is divided from Prince Edward Island by Northumberland Strait, and from Cape Breton Island by the Gut of Canso. Cape Breton Island[3] is nearly bisected by a remarkable fiord, the **Bras d'Or.** The coasts of both divisions are indented by numerous inlets, some of which form magnificent harbours.

Both Nova Scotia and Cape Breton contain an abundance of valuable timber, but the Province is chiefly famous for its coal mines. Iron ore and gypsum are plentiful, and some gold is produced. About 5 million acres (out of a total acreage of 13 millions) are fit for tillage, and the soil in the western portion of the Province, "especially in the Annapolis Valley and in the famous Basin of Minas, is unsurpassed for fertility, owing to the rich marine deposits left on the shoreland by the tides of the Bay of Fundy." *Fruit-growing, dairy-farming,* and *stock-breeding* are gradually progressing, but the chief industries at present are mining, lumbering, and fishing. The climate is not so rigorous as that of Canada Proper, and is remarkably healthy. Halifax and the other ports on the eastern coast are open all the year round, while the St. Lawrence to the north is annually frozen over. The people are mainly of British or French descent, but there are a few thousand coloured people and some two thousand Indians.

The Provincial Government is administered by a Lieutenant-Governor, aided by an Executive and a Legislative Council and a Legislative Assembly.

The principal *towns* are Halifax and Sydney. **HALIFAX**[4] (42), the capital of the Province, is distinguished chiefly for its fine harbour and as the principal winter port of Canada, with all parts of which it is connected by rail. It is also the chief British naval station and the headquarters of the British Army in North America. **Sydney,** the chief town of Cape Breton Island, has a considerable trade, chiefly in coal, but fishing[5] is the main industry of the islanders.

Nova Scotia was the first settlement in America, and, with the neighbouring islands and New Brunswick, formed the famous colony of *Acadia* or *Acadie*, immortalised by Longfellow in his exquisite poem—"Evangeline." "He touched the Grand Pré and made every meadow and dyke beautiful with a new beauty.

---

1. *Nova Scotia* was discovered by Cabot in 1497, and was colonised by the French. It was taken or retaken four times by the English, and was finally ceded to England by the Treaty of Utrecht, 1713.
2. Across the Isthmus of Chignecto (17 miles in width), the *Marine Transport Railway*—and the first railway of the kind in the world—connects the Bay of Fundy with Northumberland Strait and the Gulf of St. Lawrence. Vessels up to 2,000 tons are raised 40 feet by hydraulic lifts to the level of the railway, carried along on trucks, and let down again by the same means into the water at the other end.

3. *Cape Breton Island* was also discovered by Cabot in 1477. The fortress of *Louisbourg*, on the south-east coast, was taken by the English in 1768, and the island was finally ceded to Great Britain by the Treaty of Paris, 1763.
4. Halifax is 2,463 miles from Liverpool, and letters are 8 to 10 days in transit.
5. The *fisheries* of Nova Scotia employ 27,000 hands, and the value of the catch is double that of New Brunswick, and three or four times as large as that of any other of the Provinces in the Dominion.

Every year tourists flock to visit Evangeline's country, and to see the descendants of the grand old settlers who raise their crops round the basin of Minas, and build ships from the 'forest primeval' on Cape Blomidon, and not only build them, but own and sail them on every sea." In *shipbuilding*, indeed, Nova Scotia has, in times past, eclipsed all other countries in proportion to population, and the Province has immense facilities for becoming a great manufacturing country, especially in the abundance of coal and iron and boundless water-power.

NEW BRUNSWICK.

NEW BRUNSWICK[1] borders on the south-western side of the Gulf of St. Lawrence, and is bounded on the *south* by the Bay of Fundy, on the *east* by the State of Maine, and on the *north* by the extreme south-eastern portion of the Province of Quebec. The Province has an *area* of 27,300 square miles, and a population of about 400,000.

The boundary between New Brunswick and Quebec is formed by the River Restigouche, which flows into Chaleur Bay, an inlet of the Gulf of St. Lawrence. On the west, the River St. John, a straight line from the Grand Falls of the St. John to the Chiputneticook Lakes, and thence the River St. Croix, which flows into Passamaquoddy Bay (an inlet of the Bay of Fundy), divide this Province from the State of Maine. On the south, the boundary is formed by the Bay of Fundy and Chignecto Bay, and a line drawn across the Isthmus of Chignecto, which unites New Brunswick and Nova Scotia.

The chief physical feature of New Brunswick is the River St. John (400 miles long), which is navigable for small vessels to FREDERICTON (85 miles inland), and for boats to the Grand Falls, 200 miles from the sea. The valley of the St. John forms a narrow and, on the whole, level plain, rising in the east into a plateau of considerable height, which extends to the level belt along the east coast. North of the uplands, the country is drained by the Miramichi (which enters Miramichi Bay) and the Restigouche.

Both the uplands and valleys are covered with magnificent forests[2] of pine and other woods, and the produce of the forest forms the chief export. Agriculture is also much attended to in the lower districts—the *intervale* lands along the rivers are extraordinarily fertile—but, next to the forests, the chief wealth of the Province lies in its valuable fisheries, in which over 10,000 men are employed. Shipbuilding is also an important industry.

The people are mainly of British descent, but there are many descendants of the old French settlers and a few Indians. The Government is administered by a Lieutenant-Governor, aided by an Executive Council, and a Legislative Council and Assembly.

The chief towns are Fredericton and St. John. FREDERICTON[3] (8), the capital of New Brunswick, stands on the River St. John, 80 miles above its mouth ; but the town of St. John (50, including Portland), at the outlet of the river into the Bay of Fundy, is much more populous and commands the chief share in the maritime trade of the Province. Both St. John and Fredericton are connected by rail with all parts of Canada and the New England States.

1. New Brunswick was colonised by the French in 1672, and formed part of the French colony of Acadia until 1713, when it was ceded to England. 2. Only about one-sixth part of the cultivable portion of New Brunswick is cleared, and the whole country north and east of Aroostook, with the exception of a few small settlements, is one huge *forest*, stretching right up into Quebec. Like the adjoining provinces, New Brunswick suffers frequently from *forest-fires*. The great fire, in 1825, devastated a district 100 miles in length and destroyed many towns. One large fire at Miramichi extended over an area 140 miles in length by 70 miles in breadth.

3. Fredericton is 2,748 miles from Liverpool *via* Cape Race, or 2,535 *via* the Strait of Belle Isle and Chatham, N.B.

PRINCE EDWARD ISLAND.

PRINCE EDWARD ISLAND[1] is within the southern portion of the Gulf of St. Lawrence, and lies opposite the shores of Nova Scotia and New Brunswick, from which it is divided by Northumberland Strait. The island, which is 130 miles long and 34 miles broad, is the smallest of the Canadian Provinces, being a little over 2,000 square miles in area with a population of 115,000.

The coasts of Prince Edward Island are so deeply indented that no part of it is more than 8 miles from the sea. The interior is, on the whole, level and is still largely covered with forests. The soil of the cleared districts is very fertile, and agriculture is the chief industry. Unlike the adjoining Provinces of Nova Scotia and New Brunswick, its mineral productions are unimportant. The fisheries, however, are valuable, and large numbers of *horses* and *cattle* are reared. The climate is extremely healthy, and it is no unusual thing on this favoured island to find people who have reached the age of a hundred years, without having known a day's illness.

The Provincial Government is similar to that of the other Provinces of Canada. A railway runs right through the island connecting all the chief places with Charlottetown, the capital and chief port.

QUEBEC.

The Province of QUEBEC[2] includes that portion of the St. Lawrence valley which is towards the mouth of the river and below the junction of the Ottawa. On the north, this Province is bounded by James Bay, the East Main River, and the Esquimaux River; and on the west by the Ottawa River, which divides it from the Province of Ontario. South of the St. Lawrence, the boundary between Quebec and the United States is marked partly by the 45th parallel, the Green Mountains, and the rivers St. John and St. Croix. In the extreme east, the River Restigouche and Chaleur Bay divide it from New Brunswick.

The *area* is over 250,000 square miles, and the perimeter of the whole Province is about 3,000 miles, 740 miles of which are sea-coast and 2,260 miles land frontier. The island of Anticosti—a large uncultivated island 145 miles long and 30 miles broad, in the Gulf of St. Lawrence, with the Magdalen Islands, a barren group also in the Gulf, 50 miles north of Prince Edward Island, the Isle of Orleans and Montreal Island in the St. Lawrence, Allumette and Calumet in the Ottawa, &c., belong to Quebec. The *population* of this vast Province is only about 1½ millions, an average of only 6 persons per square mile.

The great natural features of Quebec are the St. Lawrence and its tributaries, of which the principal are the Ottawa, St. Maurice, and Saguenay on the left bank, and the Richelieu, St. Francis, and Chaudière on the right. The northern affluents either rise in, or are connected with, a labyrinth of lakes, of which the most extensive is Lake St. John, drained by the Saguenay. Of the southern tributaries, the Richelieu rises in Lake Champlain, which is within

1. Prince Edward Island was discovered by Cabot in 1497; taken by the English, 1758; finally ceded to Great Britain at the peace of 1763; admitted into the Dominion, 1873. 2. The Province of Quebec was formerly called Canada East or Lower Canada.

the United States. The "Height of Land" forms the limit of the Lower St. Lawrence basin on the north.

South of the St. Lawrence, the country is for the most part level, fertile, and well-cultivated, and, except in the extreme east, well-peopled. North of the St. Lawrence, the settled and cultivated districts are confined to a narrow belt along the river between the mouths of the Ottawa and the Saguenay. North and east of the latter, the climate is so severe that cultivation is impossible. In the cultivated districts, large quantities of wheat, &c., are grown, but the chief wealth of the Province lies in its vast forests and productive fisheries.

More than three-fourths of the people of Quebec are descendants of the old French settlers, for Quebec was originally settled by, and long remained a valued possession of, France. Though they still adhere to their language and faith, the French inhabitants are intensely Canadian, and, since the union of the provinces, have lived in perfect harmony with their neighbours of British descent. The English portion of the Province is almost limited to the Eastern Townships, which lie close to Vermont and the United States frontier, and were originally settled by English Loyalists, who left the United States at the time of the War of Independence. The Government is vested in a Lieutenant-Governor, who is appointed by the Governor-General, and an Executive Council and two Legislative Chambers.

The principal towns are Quebec, Montreal, and Three Rivers. QUEBEC[1] (65), the capital of the Province and the former capital of all Canada, stands on the north bank of the River St. Lawrence, in a commanding position, and is the seat of a very large timber trade. MONTREAL[2] (200), further up the St. Lawrence, is situated on an island in the river, immediately below the junction of the Ottawa. It is by far the largest town in Canada, and has a very large trade and considerable manufactures. Three Rivers, at the confluence of the St. Maurice and the St. Lawrence, has a large lumber trade.

ONTARIO.

ONTARIO,[3] the most populous and wealthy Province in the Dominion, lies between Quebec on the *east*, the North-East Territory and James Bay on the *north*, and the Great Lakes on the *south*. It extends from east to west for nearly 1,100 miles, and from north to south for 700 miles, and has an area of 223,000 square miles[4] and a population of 2½ millions.

This Province is divided from Quebec by the Ottawa River, and from the North-East Territory by the Albany River, which flows into James Bay, the southern extension of Hudson Bay. Between Ontario and the United States are the Upper St. Lawrence, Lake Ontario, Niagara River, Lake Erie, the River Detroit, Lake St. Clair, the River St. Clair, Lake Huron, the River St. Mary, and Lake Superior. Of the rivers running north, besides the Albany, the longest are the Moose and Abittibi Rivers, the latter flowing from a lake of the same name. Of the numerous lakes in the interior, the chief are Lake Nipigon in the west, and Lake Nipissing in the east.

1. Quebec is frequently called the "Gibraltar of America." Its fortifications are considered impregnable. The victory gained by the gallant Wolfe in 1759, on the *Plains of Abraham*, immediately outside the town, ensured the transfer of Canada from French to English rule, and preserves to Quebec a conspicuous place in the page of history.

2. The Grand Trunk Railway crosses the St. Lawrence at Montreal by the famous *Victoria Bridge*, the longest tubular bridge in the world.

3. Formerly Canada West or Upper Canada. In 1794, the total population only amounted to 65,000.

4. Ontario is nearly 4 times as large as England and Wales.

The settled portion of Ontario is enclosed by the Ottawa and the St. Lawrence, and Lakes Ontario, Erie, and Huron. The rest of the country to the north and west is covered with **immense forests** of pine, beech, oak, &c. Formerly, the whole Province was forest-covered, and lumbering then formed the only occupation of the colonists. The extent of land under cultivation, however, has enormously increased within the last few years, and agriculture is now the chief industry in the southern counties.[1] The vast **mineral resources** of the country are being actively developed. *Iron, lead, copper, gold, silver, tin*, and other metals are found in the neighbourhood of the Great Lakes. The *petroleum* wells of the peninsular portion of the Province are extremely productive.

Ontario is rapidly becoming an important manufacturing country, and now makes all kinds of *agricultural implements* in iron and wood, *waggons, carriages, railway rolling stock* (including locomotives), *cotton* and *woollen fabrics, leather goods*, &c.

The population has rapidly increased within recent years, and the people are mostly of British descent, but there are several thousand German and Dutch settlers. The Provincial Government is similar to that of Quebec, and is administered by a Lieutenant-Governor, aided by an Executive and a Legislative Council. For administrative purposes, the Province is divided into 96 counties.

The principal towns of the Province are Ottawa, Toronto, and Kingston. **OTTAWA** (45), the Federal capital, lies on the River Ottawa, 90 miles above its junction with the St. Lawrence. It is also the centre of the Ontario *lumber-trade*, and its saw-mills are the largest in Canada. **TORONTO** (175), the Provincial capital, on the north-west shore of Lake Ontario, is, however, the largest city of Upper Canada. The "Queen City of the West," as Toronto is called, has great shipping interests on the lakes, and is the chief centre of the industries and trade of the Province. Kingston (16) is situated at the outlet of the St. Lawrence[2] from Lake Ontario. Of the smaller towns, the most important are Hamilton[3] (45), the "Birmingham" of Canada, on Burlington Bay (Lake Ontario), and London (27), on the River Thames, which flows into Lake St. Clair.

MANITOBA.

The Province of **Manitoba** is situated in the very centre of the continent, being midway between the Atlantic and Pacific Oceans on the east and west, and the Arctic Ocean and the Gulf of Mexico on the north and south. It has an area of over **60,000** square miles and a population of over **125,000**.

---

1. Ontario is highly favoured by its splendid and unique geographical position, especially its southern half, swept by the circle of the Great Lakes, which moderate the heat of summer and temper the cold of winter, and, along with the numerous rivers and canals, form a system of inland navigation unparalleled in the world. A fertile soil, capable of producing in abundance all the grains, fruits, and plants of the Temperate Zone, as well as many semi-tropical products, a healthy and invigorating climate, vast forests, great mineral wealth, and widely available water power, with, above all, hardy and energetic people, combine to make Ontario a veritable Land of Promise.

2. The portion of the St. Lawrence immediately below Kingston is known as the *Lake of the Thousand Isles.*

3. The country from Hamilton to Lake Huron on the west and Lake Erie on the south, is the best settled, the best farmed, and the most attractive part of Canada. Melons, grapes, peaches, and tomatoes, besides grain and the commoner kinds of fruits and vegetables, grow plentifully in the open air near Niagara Falls, also round Chatham, and in most other parts of this highly favoured district. The cultivation of the apple has been carried to great perfection in Ontario, many farmers devoting a large proportion of their farms to this crop. In a favourable year, Ontario produces about 1 million barrels of apples.

From its geographical position and its peculiar characteristics, Manitoba may be regarded as the keystone of that mighty arch of sister Provinces which spans the continent from the Atlantic to the Pacific. It was here that Canada, emerging from her woods and forests, first gazed upon her rolling prairies and vast North-West, and learnt, as by an unexpected revelation, that her historical territories of the Canadas, her eastern seaboards of New Brunswick, Labrador, and Nova Scotia, her Laurentian lakes and valleys, corn lands and pastures, though themselves more extensive than half-a-dozen European kingdoms, were but the vestibules and the anti-chambers to that, till then, undreamt of dominion, whose illimitable dimensions alike confound the arithmetic of the surveyor and the verification of the explorer.[1]

Manitoba, which only came into existence as a Province in 1870,[2] has no natural boundaries, its shape being that of a parallelogram, about 200 miles from east to west and 270 miles from north to south. It is bounded by Ontario and the North-East Territory on the east ; Keewatin and Saskatchewan bound it on the north ; Assiniboia, on the west ; and the United States (Minnesota and North Dakota), on the south.[3]

The climate of Manitoba, as the Marquis of Lorne remarked, has honest heat in summer and honest cold in winter ; but, in spite of the extreme temperatures, the summers are very pleasant and the winters most enjoyable, the dry cold air being bracing and invigorating. The soil is a deep, rich, vegetable mould—the product of centuries of crops of grass which have grown, seeded, and withered on the prairie—and produces the finest and heaviest wheat in the world. Other grains and vegetables grow equally well, and horses, cattle, and sheep thrive on the nutritious prairie grasses.

Manitoba had no railway communication with the outside world until 1878 ; it is now traversed by the Canadian Pacific Railway, which passes through WINNIPEG,[4] the capital of the Province, PORTAGE LA PRAIRIE, and BRANDON —the three largest of its towns. From Winnipeg, branches run southwards to *St. Paul* in the United States, and westwards to the Souris coalfield and Turtle Mountain. The Winnipeg and Hudson Bay Railway has also been commenced. An important line also runs north-west from the C. P. R. main line at Portage La Prairie, *via* Birtle and Yorkton to *Prince Albert* on the North Saskatchewan River. Winnipeg has now 30,000 inhabitants—in 1870 it had only 300. Its position at the junction of the Assiniboine and the Red River, and the extraordinary fertility of the surrounding prairie-lands, which, "acre for acre, could support a thicker population than any similar tract on the globe," mark it as the ' city of the future' in Central Canada. Not only is it connected by rail with Eastern Canada and British Columbia, but, during the open season, steamers run (*a*) on the Assiniboine River for 320 miles, (*b*) on the Red River, south to the United States and north into Lake Winnipeg, (*c*) up Lake Winnipeg to the mouth of the Saskatchewan, and (*d*) up the Saskatchewan to *Edmonton* (1,500 miles from Winnipeg) in Alberta. At the end of the summer, the rivers get very low and navigation is difficult, and in winter both rivers and lakes are frozen over.

1. Extract from speech by Lord Dufferin.
2. A rebellion of the French half-breeds under Riel in 1870 was suppressed by an expedition under Colonel (now Viscount) Wolseley. The North-West rebellion in 1885, also under Riel, was promptly suppressed by a colonial force.
3. The boundary between Manitoba and the United States is formed by the 49th parallel of North latitude.

4. Winnipeg is 1,423 miles west of Montreal, or about 900 miles further than Madeira from London, and the creation of the Province of Manitoba at so great a distance from the old provinces is another illustration of the immensity of our Canadian domain. Yet, Manitoba is only in the centre of the continent after all, and is as far from British Columbia on the west as from Quebec on the east.— Greswel's *Geography of Canada*.

THE NORTH-WEST TERRITORIES.

The NORTH-WEST TERRITORIES embrace a vast region, which stretches from Manitoba to the Rocky Mountains and the Arctic Ocean. It is bounded on the south by the 49th parallel of North latitude, which divides it from the United States.

The basin of the Saskatchewan is the most fertile and valuable portion of these territories. Extensive forests, alternating with tracts of prairie land, cover the southern division of this great region, which becomes colder with each succeeding parallel of latitude, until it passes, towards the extreme north, into a dreary and barren wilderness.

The fur-bearing animals, which have their homes in this extensive region, formerly supplied its sole produce of value, and the collection of their skins forms the object for which it is still frequented by the servants of the Hudson Bay Company.[1] But it has immense capabilities of another description. Large portions of it abound in mineral deposits; and there are extensive tracts well suited for the purposes of agricultural settlement, which have, within recent years, attracted a large population, while, along the Canadian Pacific Railway and on the banks of the chief rivers, towns and villages are springing up with wonderful rapidity.

PROVISIONAL DISTRICTS: Four Provisional Districts have been formed out of the North-West Territories, namely, **Saskatchewan, Assiniboia, Alberta, and Athabasca.**

These Districts are at present under the rule of a Lieutenant-Governor and Council. The capital and seat of Government is REGINA, on the Canadian Pacific Railway, in the District of Assiniboia.

The provisional District of **Keewatin** is under the jurisdiction of the Government of Manitoba.

ASSINIBOIA.

ASSINIBOIA, which lies between **Manitoba** and **Alberta** and adjoins the **United States** on the south, has an area of **95,000** square miles, or nearly twice that of England.

The **Qu'Appelle Valley** in this district is one of the most favoured parts of the North-West, and settlement in it is proceeding with surprising rapidity. Many towns and villages have sprung up along the line of the *Canadian Pacific Railway*, which traverses the district from east to west. Among these may be mentioned **Broadview, Indian Head, Qu'Appelle, REGINA** (the capital of the district), **Moose Jaw, Swift Current,** and **Medicine Hat.** From Regina a branch railway runs to *Prince Albert* in Saskatchewan, and from Medicine Hat another branch runs to *Lethbridge* in the Belly River coalfield, and beyond it to the borders of Montana. The Dominion Experimental Farm for the Territories is at Indian Head.

SASKATCHEWAN.

SASKATCHEWAN is an immense district, 114,000 square miles in extent, situated to the north of Assiniboia and Manitoba and traversed by the two branches of the great Saskatchewan River.

---

1. At various intervals throughout these territo- | posts or stations for the purpose of collecting the ries, the Hudson Bay Company maintains fortified | furs.

This vast district has immense resources, and now that **PRINCE ALBERT**, the capital, is connected by rail with Regina on the C. P. R. main line, the many fertile tracts will soon be occupied by thousands of industrious and prosperous farmers. Prince Albert, which is 500 miles west of Winnipeg, occupies the true centre of the great " Fertile Belt " of the Saskatchewan.

ALBERTA.

ALBERTA has an area of about 100,000 square miles, and is bounded on the *south* by the United States, on the *east* by the Districts of Assiniboia and Saskatchewan, on the *north* by the District of Athabasca, and on the *west* by British Columbia, from which it is separated by the main ridge of the Rocky Mountains.

The winter in Alberta is not so severe as in the districts further east. When the *Chinook wind* blows from the Pacific—and this is the prevailing wind—the weather becomes mild and the snow rapidly disappears. The region of the Chinook wind is the country of the great cattle and horse ranches, and here, in the latitude of Labrador, cattle and horses range, during both winter and summer, without shelter. The rich and luxuriant grasses which cover the foot-hills of the Rocky Mountains make Alberta a very paradise for cattle, and the district will be, in future, the pre-eminent dairy-region of America. But it is not only in agricultural resources that it is rich, for there are in it immense coalfields (worked to some extent at Lethbridge on the Belly River and on the main line of the Canadian Pacific Railway, near Banff, in the recently-formed National Park), while the Rocky Mountains and their foothills are known to contain rich deposits of *iron, gold, silver, galena,* and *copper,* which, now that the C. P. R. passes through this district, will not long remain without development. The railway passes through **CALGARY**, the chief town of Alberta, which is beautifully situated at the confluence of the Bow and Elbow Rivers.

ATHABASCA.

ATHABASCA comprises an area of 122,000 square miles, and is bounded on the *south* by Alberta, on the *west* by British Columbia, and on the *north* and *east* by the as yet unorganized territories of the North-West.

The eastern boundary of this vast district is formed by the Athabasca and the Slave Rivers—the parallel of 60° N. lat. limits it on the north and the meridian of 120° W. long. on the west. It includes the middle and lower Peace River Valley, which has been proved to be suitable for *wheat-growing,* but, owing to its northern position, it is as yet out of the range of immediate settlement. **DUNVEGAN,** on the Peace River, is the chief settlement, and there are numerous posts of the Hudson Bay Company and several mission stations in the district.

KEEWATIN.

The District of KEEWATIN extends along the western side of Hudson Bay to the north of Manitoba.

This district was formed out of the North-West Territories in 1876, and, with the portion of Manitoba added in 1883, it has an area of about 400,000 square miles. This otherwise unimportant region includes the only directly accessible seaboard to the Prairie Provinces of Central Canada, and a railway is being constructed from WINNIPEG to PORT NELSON, so that the grain and other products of Manitoba and the North-West may be shipped to Europe by the shortest and cheapest route, namely, *viâ* Hudson Bay.   Unfortunately, this route is ice-bound for more than half the year, but both the strait and bay are perfectly open during the summer months.

BRITISH COLUMBIA.

BRITISH COLUMBIA, the westernmost Province of the Dominion, is also the largest and yet the least populous.   Its area, including **Vancouver, Queen Charlotte,** and other islands along the coast, is about **357,000** square miles, or 6 times that of England and Wales, but the population does not exceed **100,000,** of whom only about **80,000** are whites, the rest being Indians and Chinese.

British Columbia is, in many respects, the most remarkable portion of the "country of magnificent distances," as Canada might well be called.   This Province, which is 760 miles in length and about 500 miles in breadth, is in itself larger than any other organized division of the Confederation, and has an area exceeding that of England and Wales, Scotland, Ireland, France, Holland, and Belgium taken together by about 8,000 square miles.   Between the Rocky Mountains and the meridian of 120° W. long., which divide it on the east from Alberta and Athabasca respectively, and the Pacific Ocean on the west, the United States frontier (49° N. lat.) on the south, and the 60th parallel on the north, all the great natural features of the other Provinces are reproduced on a magnified scale.   This "Sea of Mountains" has a greater variety of climate than all the other Provinces together, for the upper slopes of the Rockies are as cold as Labrador, while, on the southern coast, oranges and grapes ripen in the open air.   Its wonderful coast-line, its unrivalled fisheries, its magnificent forests, its incalculable wealth in those minerals which are the most valued and the most necessary to man, and its splendid geographical position on the Pacific Ocean—almost the counterpart of that of Great Britain on the Atlantic—all indicate a great future for the " England of the Pacific."

The natural features of British Columbia are extremely diversified.   A deeply-indented coast-line fringed with hundreds of islands, lofty mountains, numerous rivers and lakes, long, narrow, well grassed valleys, with dense forests of gigantic pines, combine to make this Province the most picturesque portion of the continent.   " New wonders," says Lord Lansdowne, "are revealed at every turn of the road.   Snow-capped pinnacles of vast height and fantastic shape, great glaciers, precipitous cliffs, raging torrents, and tranquil lakes, while there rise on all sides trees, the like of which I had dreamt of, but never seen."   Of the British Columbian coasts, the Earl of Dufferin says, " Such a spectacle as this coast-line presents is not to be paralleled anywhere by any area in the world. Day after day for a whole week, in a vessel of nearly 2,000 tons, we threaded an interminable labyrinth of watery lanes and reaches that wound endlessly in front of a network of islands, promontories, and peninsulas, for thousands of miles, unruffled by the slightest swell from the adjoining ocean, and presenting at every turn an ever-shifting combination of rock, verdure, forest, glacier, and snow-capped mountains of unrivalled grandeur and beauty.   When it is remembered that this wonderful system of navigation (*i.e.*, the channels between Vancouver Island and the mainland), equally well adapted to the largest line-of-battle ship and the frailest canoe, fringes the entire seaboard of the Province, and communicates at points, sometimes more than a hundred miles from the coast, with a multitude of valleys stretching eastwards into the interior, while at the same time it is furnished with innumerable harbours on either hand, one is lost in admiration of the facilities for intercommunication which are thus provided for the future inhabitants of this wonderful region."

Physically, British Columbia may be divided into four districts: —(1) the islands, (2) the mountains along the coast of the mainland, (3) the high interior plateau, and (4) the lofty mountain ranges that rise along the eastern border.

The vast range of the **Rocky Mountains** forms the eastern boundary of British Columbia from the International Boundary (49° N. lat.) to the Smoky River Pass (54° N. lat.) ; from thence, north to the 60th parallel of N. lat., the boundary is formed by the 120th meridian of W. longitude, and thus an extensive territory to the east of the Rockies, comprising the valleys of the **Upper Peace** and the **Liard Rivers**, are included in the Province. Between the main range of the Rocky Mountains and the Pacific Coast are a number of minor ranges, such as the **Purcell** and the **Selkirk Ranges**, within the "great bend" of the *Columbia River* ; the **Gold Range**, between the *Columbia River* and the *Thompson River* and *Shushwap Lake* ; the **Cariboo Mountains**, in the "great bend" of the *Fraser River* ; with the **Peak** and other mountains in the north, and the **Cascade Range** along the coast. The interior plateau of British Columbia, between the **Gold Range** and the Coast or Cascade Range, has an average width of 100 miles and an elevation of 3,500 feet. It is traversed by the Fraser and its tributaries, which, with the Upper Columbia and its affluents, the **Kootenay** and the **Okanagan**, are the chief rivers of the Province. The numerous ridges of the Coast Range extend inland from the coast for about 100 miles. They are extremely rugged and, as they receive on their western slopes abundant moisture from the sea, they have a rich vegetation. The main mass of Vancouver Island and of the Queen Charlotte Islands to the north-west, may be regarded as another partially submerged range, rising in **Mount Arrowsmith**, in Vancouver Island, to a height of 6,000 feet, and continued southward in the Olympian Mountains in the United States and northward in the islands and peninsular portion of Alaska.

Of the *resources* of British Columbia it may be said that its wealth of timber has not been appreciably touched. **The forests are magnificent**, and more than half the Province is covered with the Douglas or Oregon pine (which frequently grows to a height of over 300 feet, with a diameter of 8 or 9 feet), the white and red cedar, hemlock, maple, spruce, birch, and other valuable trees. [1]

In the rich valley of the Lower Fraser or New Westminster district, and on the south and east coasts of Vancouver Island, the **soil is exceedingly fertile**, and the climate is favourable to agriculture and fruit-growing. In the interior, also, the soil is, over very considerable areas (far exceeding in the aggregate the arable areas of the coast region), as fertile as the best on the coast, but the climate is so dry in summer that irrigation is necessary, except in a few favoured localities. As regards pasture, the interior as a whole is probably unequalled for horse and cattle ranches. About 5,000 or 6,000 square miles of the Peace River district of British Columbia are also of considerable agricultural value.

The fisheries are as rich as those of Eastern Canada, but they have yet to be developed. The whole of the seas, gulfs, bays, rivers, and lakes of the Province swarms with fine food fishes. The **salmon**[2] of British Columbia is famous the world over. Millions of them make their way up the rivers, and the

---

1. Cedars sometimes attain a diameter of 17 feet, and it is from the wood of this tree that the Indians make their celebrated canoes.
2. In 1889, the output of the canneries, most of them at the mouth of the Fraser, amounted to the enormous total of 420,000 cases, each case containing 4 dozen 1 lb. cans.

annual take from the Fraser River alone is over 10 million lbs. Sturgeon, sometimes exceeding 1,000 lbs. in weight, are numerous; halibut abound, especially
off the west coast of Queen Charlotte Island ; cod and seals are caught on the
north coast ; while the delicious oolachan¹ or candle fish enters the Fraser and
the Nasse rivers and other streams by the million, for several weeks. The lakes
and rivers in the interior are full of salmon, trout, perch, and other fish. The
whale, seal, and sea otter fisheries are important, and the coast abounds with
oysters, a very large and excellent crayfish, crabs, &c. Next to the salmon, the
most valuable sea-product is the fur-seal, which yields over 350,000 dollars a year.

MINERALS, however, form the chief wealth of the Province. As for gold,
there is scarcely a stream in which the colour of gold cannot be found, and
paying mines extend through a region of some 600 miles in length. The largest
mines are in the *Cariboo* district, whence 10 millions sterling have been obtained since 1858. Coal mines are worked at *Nanaimo*, *Wellington*, and *Comox* on
the east coast of Vancouver Island, and there are inexhaustible deposits of iron
ore on Texada Island and elsewhere. Copper, silver, and other metals are
widely distributed, but more labour and capital are wanted to develop the
rich mineral, timber, ranching, and fruit-farming resources of this immense
country, which, until 1887, had no railway communication with the outside
world. In that year, however, the Canadian Pacific Railway, which enters the
Province at STEPHEN in the *Kicking Horse Pass*, and, crossing the Columbia
River, runs along the valleys of the Thompson and the Fraser Rivers to the
seaboard, was completed.

The principal *towns* are VICTORIA, the provincial capital, which is picturesquely situated on a lovely harbour on the south-east coast of Vancouver
Island, and has about 24,000 inhabitants ; VANCOUVER (16), on the southern
side of Burrard Inlet, the terminal port of the Canadian Pacific Railway, connected by a magnificent line of steamers with Yokohama and Hong-Kong ;
and New Westminster (8), a growing river-port, very pleasantly situated on
the Fraser River about 8 miles above its mouth and 12 miles from Vancouver.
Yale is a small town at the head of navigation on the Fraser, and Lytton stands
at the confluence of the Thompson and the Fraser. There are several other
small places on the mainland, such as Kamloops on the Thompson, at the head
of Kamloops Lake, Lillooet, Hope, Alexandria, Quesnellemouth, and Fort
George on the Fraser. In Vancouver Island, the Island Railway runs from
Esquimalt, which has a magnificent harbour and graving dock and is the headquarters of the *North Pacific Squadron*, through Victoria and the forest
country beyond to Nanaimo (8), a thriving coal-mining town, and on to Wellington, another coal-mining centre, 7 miles north of Nanaimo. Coal mines
are also worked at Comox, 60 miles by steamer north of Nanaimo. Alberni is
a small settlement at the head of the long *Alberni Canal*, which opens into
Barclay Sound on the south-west coast of the island.

∴ Vancouver Island was discovered in 1792 by Captain Vancouver, who gave a glow-
ing description of "the serenity of the climate, the innumerable and the abundant fertility
that unassisted nature puts forth." It was secured to England by treaty in 1846, and 20
years later was united to British Columbia, which, until 1858, had formed part of the
Hudson Bay Company's Territory, when the rush of goldseekers forced the British Government to proclaim and govern it as a Crown Colony. In 1870, the united colony joined
the Dominion, and in 1887 the bond was completed by the opening of the Canadian Pacific
Railway. Since then, the establishment of a line of steamships to China and Japan has
still further increased the importance of British Columbia as a connecting link between
Europe and Asia.

---

1. This remarkable fish is smaller than a herring | oil or any other fish oil known, and it is a staple
and is so oily that, when dried, it will burn like a | article of food and barter among the Indian tribes,
candle. Its oil is considered superior to cod-liver | who catch them in immense numbers.

BRITISH HONDURAS.

The Crown Colony of BRITISH HONDURAS or *Belize*, the only
British possession in Central America, is bounded on the north by
the Mexican State of **Yucatan**, on the west and south by **Guatemala**,
and on the east by the **Caribbean Sea**. It has an area of **7,500**
square miles and a population of about **30,000**, the majority of whom
are Negroes and Indians.

Physically, British Honduras has a like configuration to that of the adjacent
Central American States—flat and swampy along the coast-line ; then pine and
cohoon ridges ; next, primeval forest-land, broken here and there by lofty hills
and sometimes considerable savannahs; and, finally, mountain ranges, which
run with more or less continuity along the western frontier.

The climate and soil along the coast are adapted for the luxuriant growth of
almost every tropical product, and the vast forests teem with an exhaustless
wealth of **mahogany, cedar, logwood, iron-wood, pine and india-rubber trees,**
with **sarsaparilla** and other useful shrubs and plants.   Sugar is still largely exported, and there are large **coffee** plantations, while bananas, plantains, cocoa-
nuts, pine-apples, oranges, mangoes and other fruits are extensively grown.
But the cutting of **mahogany** and **logwood** is the chief industry, and mahogany,
logwood, and bananas are the most important articles of export, principally to
Great Britain and the United States.   The total exports now amount to over
2 million dollars a year.   Capital and a larger supply of labour alone are wanted
to develop the great resources of this Colony and make it one of the most
valuable of the many " tropical gardens " of England.   The capital and centre
of trade is the neat and picturesque little town of BELIZE, at the mouth of
the **Belize River**.

THE BERMUDAS.

Far out in the wide Atlantic—600 miles from Cape Hatteras—
" the remote Bermudas ride in the ocean's bosom unespied." Of the
numerous islands in this isolated group, only about 20 are inhabited—these have a population of **16,000**, of whom 6,000 are
whites.

In these " Fortunate Islands," as old Andrew Marvell quaintly sings, " eter-
nal spring enamels everything," and here " hangs in shade the orange bright
like golden lamps in a green night, while the pomegranates disclose jewels
more rich than Ormuz shows."

The 300 islands and islets of Bermuda are all of coralline formation ; the
climate is so remarkably equable and salubrious that HAMILTON is a favourite
winter resort ; the soil produces arrowroot of the finest quality as well as an
abundant supply of fruits and vegetables ; while the forests yield a valuable
and durable cedar.

The geographical position of the group, a number of admirable harbours, a
royal dockyard and naval establishments so placed as to be unassailable, combine to make the Bermudas one of the most important of England's " sentry-
boxes " in the ocean.

BRITISH WEST INDIES.

The British portion of the West Indian Archipelago includes the large island of **Jamaica**, in the Greater Antilles, and **Trinidad, Barbados,** and other islands in the Lesser Antilles.

The **British West Indies** have an area of about **13,750** square miles, and a population of over 1¼ millions, most of whom are **Negroes** or **Mulattoes.** Europeans or people of European descent (**Creoles**) are comparatively few in number. Of the aboriginal inhabitants—the savage and warlike **Caribs**—a few families are still found in Dominica and St. Vincent, and thousands of the race are said to inhabit the north of Haiti, where, however, they are more or less tinged with Negro blood.

The **Administrative Divisions** of the British West India Islands are (1) Jamaica, (2) the Bahama Islands, (3) the Leeward Islands, (4) the Windward Islands, (5) Trinidad and Tobago, and (6) Barbados.

JAMAICA.

The island of JAMAICA,[1] the *Xaimaca,* or "land of wood and water," of the old Caribs, is by far the largest and most important of the British West India Islands. It is about **150** miles in length, and **50** miles in greatest breadth, and has an area of **4,200** square miles, and a population of over **600,000**, not more than 3 per cent. of whom are whites.

As when discovered by Columbus in 1494, so now, this island is "most charming, beautifully wooded, well watered, and abounding in picturesque mountains and fertile valleys." The fruitful soil produces sugar, coffee, spices, &c., in such abundance that the great mineral and forest wealth of the colony is almost neglected. Owing to its mountains and plateaux, there is a wonderful variety of climate—a few hours' ride enables one to exchange the tropical heat of the coast for the cool and salubrious climate of the uplands.

A glance at the map will show the physical character of the island. The grand central chain of the **Blue Mountains** rises in some peaks to 7,300 feet above the sea. Of the numerous rivers, only one, the **Black River,** is navigable, and that only for boats. There are excellent harbours ; the island is intersected by good roads, and there are about 90 miles of railway.

" Most of the staple products of tropical climates are raised. **Sugar** is the principal export in quantity ; Jamaica rum is still counted the best in the world; and the **coffee** grown in certain districts in the Blue Mountains fetches the highest price in the London market. There is an extensive trade in fruits with the United States. **Maize** and corn grow luxuriantly. The guinea grass, from 5 to 6 feet in height, grows wild, and is superior to any other for pasturage, while the woods furnish an abundance of rich dye stuffs, drugs, and spices, and the rarest of cabinet woods."

1. Jamaica was discovered by Columbus in 1494, and was first settled by the Spaniards in 1509. In 1655, it was taken by an English fleet sent out by Cromwell, and was formally ceded to Great Britain in 1670. In 1807, the planters' 'golden age' came to an end by the emancipation of the slaves, who, in 1864-5, broke out into an open rebellion, which was promptly and sternly repressed by Governor Eyre.

The trade of Jamaica is mostly with the United States (50 per cent.) and the United Kingdom (40 per cent.).   The chief articles of export, in order of value, are—dye-woods, fruits, coffee, sugar, and rum.   The centre of the external trade is KINGSTON, on the south coast, and its harbour—Port Royal—is the finest of the 30 good harbours of the island.   Montego Bay and Falmouth, on the north side of the island, are also important ports.

The Government is administered by a Governor, aided by a Privy Council and a Legislative Council—9 members of the latter are elected.   The island is divided into three counties—*Surrey* in the east, *Middlesex* in the centre, and *Cornwall* in the west.   KINGSTON in Middlesex is the seat of government ; SPANISH TOWN, the former capital, is about 11 miles west of Kingston.

TURK'S and CAICOS ISLANDS are under the Government of Jamaica, although they geographically form a part of the Bahama Archipelago.   Only 6 of the 30 'cays' included in the group are inhabited—the largest is Grand Caicos, but the most important is Grand Turk.   The only important industry is salt-making, but there is a small sponge fishery.

The Cayman Islands in the Caribbean Sea, to the west of Jamaica, are similarly attached to the Jamaican government.   There is a population of some 4,000, who rear cattle and export coco-nuts and turtles.   The Morant Cays and Pedro Cays are also attached to the Government of Jamaica.

THE BAHAMAS.

The BAHAMAS are the most northerly of the West Indian Islands. They lie to the north of Cuba and Haiti, and have an area of about 5,800 square miles, and a population of 50,000, nearly one-third of whom are whites.

The Bahama chain of islands and reefs is nearly 800 miles in length, rising in water 10,000 to 13,000 feet in depth.   Of over 500 islands and islets, not more than 20 are inhabited, and of these the most important are New Providence (containing the capital, NASSAU), San Salvador or Watling Island, Abaco, Grand Bahama, Long Island, Eleuthera, Great Inagua, and Andros, with the Turk's and Caicos Islands, which are politically attached to Jamaica.

Fruit-growing is the chief industry in these lovely islands, and enormous quantities of oranges, bananas, and pine apples are shipped, chiefly to the United States.   There are advantages here for the cultivation of oranges not known in Florida, as the islands are proof against frost, which often visits Florida.   The sponge fishery is important, but the fibre industry promises to become the mainstay of the colony.

∴ One of the Bahamas, Watling Island or San Salvador, is famous as the first land in the New World seen by Columbus.   On the morning of the 4th of October, 1492, it was that the simple natives hurried to the shore to see "the people from heaven," as they supposed the gorgeously dressed Spaniards to be.   Columbus was delighted with the loveliness of the island, and was enthusiastic in its praises.   "It seems to me," he wrote, "that I could never quit so enchanting a spot, as if a thousand tongues would fail to describe it, as if my hands, spell-bound, would never be able to write concerning it."   Some of the islands are, indeed, perfect paradises of beauty, with a wealth of tropical vegetation, and a pleasant salubrious climate, attracting hundreds of invalids from the States and elsewhere every winter.   Visitors to Nassau say they never tire of the lovely walks in the flower-covered woods, which are like immense gardens, or of the innumerable boating excursions among the bays and lagoons, the waters of which are marvellously clear and transparent.

THE LEEWARD ISLANDS.

The LEEWARD ISLANDS, extending from the Spanish island of Porto Rico to the French island of Martinique, belong to Britain, with the exception of St. Thomas, Santa Cruz, Guadeloupe, and a few other islands.

The British Colony of the Leeward Islands is a federation, formed in 1871, of the five presidencies of (1) Antigua, (2) St. Christopher and Nevis, (3) Dominica, (4) Montserrat, and (5) the Virgin Islands. Each Presidency is in charge of a Commissioner, and has its own Legislature—the whole being united under a Governor and a general Legislative Council for the whole colony.

ANTIGUA, the second largest of these charming islands, is the seat of the general government of the Leeward Islands, and its capital, ST. JOHN'S, is the residence of the Governor-in-Chief. The soil of this island is fertile, and sugar and cotton are grown ; still only a third of its area is cultivated. The climate is healthy, and all the elements of material wealth are here, but the Antiguan negro, though quiet and orderly, is no great believer in work, and makes no effort to better himself, with the result that plantations go out of cultivation from want of labour, while the planter also suffers from frequent droughts. Thirty miles north of Antigua is its dependency—Barbuda—a flat and fertile island, producing corn, cotton, pepper, and tobacco, but no sugar.

ST. CHRISTOPHER and NEVIS,[1] with ANGUILLA, form one presidency. St. Christopher, better known as St. Kitts, is in the interior lofty and rugged, and a semicircle of hills encloses a beautiful and fertile plain—in which stands the town of BASSETERRE. The climate is healthy, the soil exceedingly fertile, and agriculture is in a decidedly advanced state. Sugar is the principal crop in St. Kitts, as also in the neighbouring island of Nevis, which is simply one circular mountain mass, rising from the sea to a height of 3,200 feet. Half the island is cultivable, and might be made a very garden. The little island of Anguilla, a dependency of St. Kitts, exports much salt and phosphate of lime and some farm produce.

DOMINICA is a mountainous and picturesque island, abounding in rivers and streams, but, out of 186,000 acres, only some 20,000 are under cultivation—the rest are covered with virgin forest. It exports all the ordinary West Indian produce. The position of the island is unique—it lies between the two French islands of Guadeloupe and Martinique, and once belonged to France, and has remained French in speech ; and its history of alternate capture and cession is most eventful. The capital is CHARLOTTETOWN or Roseau.

MONTSERRAT, so widely and well-known for its healthful lime-fruit, is a small island, hilly, but fertile and healthy—the healthiest of the Lesser Antilles. Besides lime-juice, much sugar and some cotton and arrowroot are exported.

The VIRGIN ISLANDS, which form an archipelago, '' picturesque to view, but dangerous to navigation,'' were discovered by Columbus in 1493, and became partly British so early as 1666. The three inhabited British islands are Tortola (on which is ROADTOWN, the capital of the group), Anegada, and Virgin Gorda, all of which have suffered severely from hurricanes.[2]

---

1. Both these islands were discovered and named by Columbus—the one by his own Christian name, Christopher—the other from its cloud-covered peak.
2. Of that of 1867, Sir Arthur Rumbold, the then Governor, wrote—'' All was bright and verdant, the withering blast passed over it, and not a fruit or other tree remains. The works of the few remaining estates are all totally destroyed, two-thirds of the town blown down, and scarcely a hut or habitation is left standing.''

THE WINDWARD ISLANDS.

The WINDWARD ISLANDS extend from Martinique to Trinidad, and include St. Lucia, St. Vincent, the Grenadines, Grenada, Tobago, and Barbados. Of these, Tobago is politically attached to Trinidad, and Barbados forms a separate colony; the rest are included in the British Colony of the Windward Islands, which has a total area of over 500 square miles and a population of about 100,000, not one-twentieth of whom are whites.

The Windward Islands Colony has no general Legislative Council like the Leeward Islands; each of the three colonies—St. Lucia, St. Vincent, and Grenada—retains its own Council and Administrator, subordinate to the Governor-in-chief, who is also Governor of Grenada.

ST. LUCIA,[1] the most northerly of our Windward Islands, is hilly, wooded, and well watered, and is famous for its wild scenery, its volcanic *Soufrière* with crater basins in constant ebullition, and the adjoining picturesque *Pitons*. Sugar, molasses, cocoa—now extensively cultivated—and logwood are the chief exports. CASTRIES, the chief town, is fortunate in the possession of the best harbour in the Antilles, and is now the second naval station in the British West Indies. The port is being strongly fortified by the Imperial Government.

ST. VINCENT[2] lies between St. Lucia and Grenada and almost due west of Barbados, and the colony includes some of the Grenadines, a cluster of small islands between Grenada and St. Vincent. Though not so picturesque as St. Lucia, St. Vincent also boasts a remarkable *Soufrière*, which was in violent eruption in 1812. The undulating slopes of this fertile and comparatively healthy island are only partially utilized. Sugar is the chief crop, but some arrowroot, coffee, cocoa, and cotton are produced; nearly one-half the tillable land is, however, still unoccupied. The capital is KINGSTOWN, on the southwest coast.

GRENADA[3] is a mountainous and highly picturesque island, abounding in springs and streams, and fortunately out of the line of hurricanes, so that its bay of ST. GEORGE is the safest and snuggest of the ports in the Windward Islands, and the town itself seemed to Trollope more like a goodly English town than any other in the smaller islands.

The GRENADINES are a group of small islands between Grenada and St. Vincent, to the Governments of which they are attached.

---

1. St. Lucia has, perhaps, the most interesting history of all the smaller islands. "Some of the greatest names in England's naval and military annals earned their first lustre in operations connected with St. Lucia, among whom may be cited Sir John Moore, Sir Ralph Abercrombie, Lord St. Vincent, and Lord Rodney. The father of Her Gracious Majesty, the Duke of Kent, took, as a subaltern, a distinguished part in the storming of the stronghold of Morne Fortuné, on the 4th of April, 1794."—*Her Majesty's Colonies.*

2. Since its discovery by Columbus, in 1498, St. Vincent has passed through troublous times—alternately taken and re-taken by French and English fleets, and at various times visited by physical and commercial disasters.

3. Grenada was discovered by Columbus in 1498, and was first settled by the French in 1650. In the northern part of the island, a lofty cliff, overhanging the sea, still preserves the memory of the brave stand made by the old Caribs against the brutal French invaders. From this cliff, called the *Morne des Sauters*, the remnant of the natives, finding further resistance useless, leaped into the sea. Even more horrible than the massacre of the Caribs were the atrocities perpetrated by the French settlers during the insurrection of 1795.

TRINIDAD AND TOBAGO.

TRINIDAD,[1] the most southerly of the long chain of the Antilles, lies off the eastern coast of Venezuela—the Gulf of Paria separating it from the mainland—immediately north of the mouth of the Orinoco. The smaller island of Tobago, to the north, was annexed to the government of Trinidad in 1889.

Trinidad has an area of 1,754 square miles, and a population of about 200,000. It is an island of extraordinary resources, and its fertile soil is admirably adapted to the cultivation of almost every tropical product. Sugar and cocoa, coffee and tobacco, are at present the chief objects of culture, and the wonderful *Pitch Lake*, a bituminous deposit some 90 acres in extent, about 30 miles from Port of Spain, is a source of considerable wealth. The seaport town of PORT OF SPAIN is the capital, about 30 miles to the south is another seaport, SAN FERNANDO, with one of the finest harbours in the West Indies. No less than 26 steamers a month call at Trinidad, and the trade amounts to over 4½ millions sterling a year—imports 2 millions, exports 2½ millions. But there is ample room for development—only one-eleventh of the island is under cultivation, and the coal and pitch and the numerous and valuable plant-products only require labour and capital to double the trade of the island in a short time.

TOBAGO, a long and hilly island to the north of Trinidad, is administratively attached to it.

Tobago, which has an area of 114 square miles, and a population of about 20,000, is well-watered by numerous short streams, and produces tropical fruits and all the ordinary vegetables in abundance. Sugar, indigo, spices, coffee, and cocoa are produced, and the culture of cotton and tobacco has been introduced, but forests still cover two-thirds of the island, which is said to be one of the healthiest in the West Indies.

BARBADOS.

BARBADOS, the most easterly of the West Indian islands, is nearly a hundred miles distant from St. Vincent, the nearest of the Windward Islands.

Though only 21 miles long and 14 miles broad, with an area of not more than 166 square miles, Barbados has a population of nearly 200,000, or considerably more than a thousand to the square mile, and is, after Jamaica, the most important of the British West Indian islands. A favourable geographical position and a healthy climate, with the industry and prosperity of its inhabitants[2]—of whom about 50,000 are mulattoes and 15,000 whites—make this distant island one of the most valuable of our colonial possessions. There are numerous elementary schools, and secondary, and even university, education is obtainable at BRIDGETOWN, the neat, well-built capital of this productive little island, which has "the appearance of a well-kept garden." The Governor is aided by a nominated Legislative Council and an elected House of Assembly.

---

1. Though discovered by Columbus in 1498, and by him proclaimed a Spanish possession, Trinidad—so named by its discoverer from the *three* peaks he saw on it—made so little progress until 1783, that a French planter of Grenada obtained from the Court of Madrid a "cedula" or decree, offering unusual advantages to settlers from all nations, which resulted in an extensive immigra-

tion. The island was temporarily occupied by the French in 1676, and in 1797 the Spanish governor capitulated to Sir Ralph Abercrombie. The Peace of Amiens, in 1802, secured its possession to Great Britain.

2. The imports of Barbados are about 1½ millions sterling a year (half a million from Great Britain), and the exports over a million.

BRITISH GUIANA.[1]

BRITISH GUIANA,[2] the only British territory on the mainland of South America, extends along the north-eastern coast, from the mouth of the Orinoco to that of the Corentyn, and stretches inland for more than 400 miles to an as yet undetermined boundary. The colony has an area of 109,000 square miles and a population of 300,000, of whom about 20,000 are whites.[3]

With a total area of twice that of England, only 83,000 acres are as yet under cultivation in this the most favoured of the three Guianas. Settlement is, in fact, almost confined to the narrow but exceedingly productive coast region ; the mountain country in the interior is almost unknown, and is occupied only by a few scattered tribes of simple and inoffensive Indians. Numerous rivers cut their way through the hills by gorge and cataract, and the accumulated debris, swept down in the course of ages, has formed the rich alluvial coastlands, still in many parts so low as to necessitate the construction of vast dykes and dams along the river banks and the sea shore. Georgetown itself, at the mouth of the Demerara, needs the protection of its "sea wall," which does the double duty of resisting the sea and as a health-giving promenade.

The predominant industry is the cultivation of the sugar-cane, and sugar forms 70 per cent. of the export trade of the colony.[4] The growing of the much-prized Berbice coffee and cotton, once largely exported, might be made remunerative were the necessary capital forthcoming. In British Guiana, as in other colonies, there is "boundless land clamouring for population" and capital. Nature has been more than generous ; the climate, though hot, is not unhealthy, and, when the mountainous interior is opened up, there will be, as in Jamaica, an ample choice of climates, and both forest and field—along the banks of the Berbice, the Essequibo, and the Mazaruni—will yield their increase to prosperous and contented settlers. Several rich deposits of gold have been recently discovered and are now being actively worked.

The government of the colony consists of a Governor, appointed by the Crown, and a Court of Policy—a legislative council of 9 members, 5 of whom are elected. GEORGETOWN (50), on the Demerara, a short distance above its mouth, is the capital. The only other considerable town is NEW AMSTERDAM (8), on the Berbice.

THE FALKLAND ISLANDS.

The British Crown Colony of the FALKLAND ISLANDS is situated in the South Atlantic, about 300 miles east of Magellan Strait. They consist of two large islands—East Falkland and West Falkland—and about a hundred smaller islands. The total area of the colony is about 7,500 square miles and the population is about 2,000.

The staple industry on these treeless but well-grassed islands is pastoral—large numbers of sheep, cattle, and horses are reared, and wool, tallow, hides and skins, and sheep form the chief exports from STANLEY, a free port at the head of Port William on the coast of East Falkland, and which is not only the capital, but also the only important settlement in the colony.

The Falkland Islands were discovered by Davis in 1592, and were taken by the French in 1763, then fruitlessly held by Spain until 1771, and occupied by the Republic of Buenos Ayres in 1820. The Argentine settlement was destroyed by the Americans in 1831, and, two years later, the islands were taken possession of by the British Government, as a station for the protection of the whale fishery.

---

1. Guiana is so called from an Indian tribe, the *Guayanoes*.
2. Guiana was discovered by Columbus in 1498, and was first settled by the Dutch in 1580. The British portion of the country was finally ceded to England in 1814.
3. Although a British colony, most of the whites are Portuguese from Madeira and the Cape Verde Islands.
4. The imports are nearly 2 millions sterling, and the exports of produce and manufactures of the colony only over 2½ millions sterling. The exports to the United Kingdom, chiefly sugar, gold, rum, and molasses, are valued at 1½ millions, and the imports of British manufactures over 1 million.

BRITISH POSSESSIONS IN OCEANIA.

OCEANIA, the fifth grand division of the land surface of the globe, embraces the vast "world of islands" in the Pacific Ocean, and also includes the great island-continent of Australia and the smaller islands between the Pacific and the Indian Oceans.

Oceania thus includes the two great divisions of **Australasia** and **Polynesia**, and some geographers include in addition the **Malaysian**, or **East Indian, Archipelago.**

AUSTRALASIA.

AUSTRALASIA, that is, Austral or Southern Asia, is the general name given to the larger British Colonies and Possessions in Oceania. Australasia thus includes the great island-continent of **Australia**, the islands of **Tasmania** and **New Zealand**, together with the **Fiji Islands** and **British New Guinea.**

Australia, which is, strictly speaking, an island, but an island of such an immense size that it may well be regarded as a continent, is politically divided into **five distinct colonies**, at present independent of each other.

The **Five Colonies of Australia** are *New South Wales* (the Mother-Colony of Australia), *Victoria, Queensland, South Australia,* and *Western Australia ;* and their absolute and relative position may be readily grasped by regarding the continent as divided into three parts—Western, Central, and Eastern—by the 129th meridian east, and by another line formed by the 138th meridian in the north and 141st in the south. The *Western* part consists entirely of **Western Australia**, the *Central* section comprises South Australia and its Northern Territory, while the *Eastern* division includes the three colonies of **Queensland** in the north, **New South Wales** in the middle, and **Victoria** in the south.

The Colony of **Tasmania** is an island, nearly as large as Ireland, to the south of the Colony of Victoria, and separated from it by Bass Strait. It is much smaller in area than any of the five colonies of Australia.

The Colony of **New Zealand** consists of two large islands, and a number of smaller islands, in the South Pacific Ocean, about 1,200 miles to the south-east of Australia. The two large islands are known as the **North Island** and the **South Island** ; the smaller island to the south of the latter is called **Stewart Island**. The smaller outlying islands are collectively named the **Off Islands**.

The **Fiji Islands**, a British Crown Colony, are also situated in the Western Pacific, and about 1,000 miles north of New Zealand.

British New Guinea includes the southern and south-eastern part of the island of New Guinea, which lies off the north-eastern coast of Australia, at a distance of about 60 miles from it.

EXTENT : Some idea of the immense extent of Australasia may be gained by comparing the areas of the various colonies with that of Great Britain and other countries. Australia alone has an area of nearly 3,000,000 square miles, or 33 times that of Great Britain, while New Zealand, Tasmania, Fiji, and British New Guinea have together an area of 226,000 square miles, or nearly 4 times that of England and Wales.

Australasia has thus a total area of no less than 3,161,000 square miles, or 26 times the size of Great Britain and Ireland, 15½ times the size of France, and rather larger than that of the United States of North America, and only about one-sixth smaller than that of all the countries of Europe taken together.

The proportion in size of the Australian colonies to each other and to the whole continent may be readily seen by the following comparison :—If the continent were divided into 100 equal parts, Victoria would comprise 3 such parts ; New South Wales, 10 ; Queensland, 23 ; South Australia, 30 ; and Western Australia, 34.[1]

DISCOVERY : The "Great South Land" was probably first seen by a French navigator in 1503, but it was not practically made known to the world until 1770, when the famous Captain Cook explored the whole eastern coast, from Cape Howe to Cape York, and took formal possession of the country, to which he gave the name of New South Wales, from a real or fancied resemblance to the southern part of the Principality.

During the long period between its first discovery and the visit of Captain Cook, portions of the coast of New Holland, as it was then called, were sighted by Portuguese, Spanish, Dutch, French, and English navigators, and their discoveries may be traced by the names which they gave to various portions of the coast. Thus, **Dirk Hartog**, the captain of a Dutch vessel, gave his own name to the large island on the west side of Sharks Bay, and another captain of a Dutch ship discovered and named **Cape Leeuwin** (Lioness), the south-western extremity of Australia, after his vessel. The first Englishman who trod Australian soil was the bold buccaneer, **Dampier**, who sailed along the western coast as far as Cape Lévêque, in 1688, and again in 1699 explored the north-west coast of West Australia, leaving his name to be perpetuated in Dampier Archipelago and Dampier Land. **Torres**, a Spaniard, passed through the strait which bears his name in 1606, and **Tasman**, an enterprising and

1. The Australian Handbook (Gordon & Gotch).

skilful Dutch navigator, in 1642 discovered Tasmania, which he named Van Diemen's Land, after Van Diemen, the Governor-General of the Dutch East Indies, who had commissioned him to explore the "Great South Land." Tasman also was the first European to make known the existence of New Zealand: he sighted it in 1642, and gave it the name, first of all, of Staatenland, afterwards altering it to Nova Zeelanda. No European landed on the shores of New Zealand until 1769, when the great navigator, Cook, disembarked at *Poverty Bay*, on the east coast of the North Island, and subsequently took formal possession of both North and South Island in the name of His Britannic Majesty King George III. In April, 1770, Cook discovered and named Cape Howe on the south-eastern coast of Australia, and subsequently explored and named the chief features along the whole eastern coast. In 1791, **Captain Vancouver** explored the south coast, and in 1801 a French expedition, comprising the *Géographe* and the *Naturaliste*, under Baudin and Freycinet, explored and named a considerable portion of what is now the south-western coast of West Australia. In the following year, Flinders sailed round the island-continent and gave it the name of "Australia."

The Fijian Archipelago was discovered by Tasman, the discoverer of New Zealand and Tasmania, in 1643, but remained unvisited until Captain Cook touched at one of the eastern islands. As for New Guinea, it was sighted by a Portuguese navigator as early as 1526, and, two years later, another Portuguese explorer landed on its shores. In 1545, a Spanish mariner coasted along the northern shore, and gave it the name of Nueva Guinea, from some fancied resemblance it bore to the Guinea Coast on the west of Africa. The Spaniard, Torres, sailed through the strait named after him, in 1606. In 1643, Tasman explored part of the coast; and 56 years later, in 1699, Dampier, in the *Roebuck*, circumnavigated the island.

SETTLEMENT : The first settlement in Australia was formed in 1788 at **Sydney Cove**, the neighbourhood of **Botany Bay**, which had been, on Cook's recommendation, chosen as the site of the new settlement, being found utterly unsuitable.

For many years after the arrival of the "First Fleet" with convicts and soldiers, who were landed at **Botany Bay**, but shortly afterwards removed to **Sydney Cove** and settled on the site of the present city of Sydney, settlement was restricted to the coastlands, and it was not until 1813 that a passage across the wild and rugged ranges known as the Blue Mountains was found, and the fine western plains, on which the city of Bathurst now stands, discovered, and a road opened into the vast interior. Settlements were formed in Western Australia, on King George Sound and the Swan River, in 1829, in order to forestall the French, who were suspected of having an idea of forming settlements in that part of the island. Melbourne was founded, in what was then the Port Phillip District of the colony of New South Wales, in 1835, but **Victoria** was not formed into a separate colony until 1851. **South Australia** was colonized by British emigrants in 1836, and New Zealand in 1838, but a European settlement had been established at the Bay of Islands in 1814. Tasmania had been settled as a penal colony in 1803, and in the same year an attempt was made to establish a convict settlement at Port Phillip, but had to be abandoned. The settlement of Queensland dates from 1825, when the first batch of convicts was landed at Eagle Farm, near the site of the present city of Brisbane. But although all the Australian Colonies, with the exception of South Australia and Victoria, were founded as penal settlements and for many years received the rejected elements of society from the mother country, yet an increasing number of immigrants, attracted by the opening up of the rich pastoral lands in the interior, and, above all, by the discovery of gold, and a succession of intrepid explorers and wise and resolute governors, in time overcame all difficulties, and at length the colonies embarked on a career of steady progress, until Australia can now claim a foremost rank in the Empire in respect of efficient administration of the law, eminent security of life and property, and the most favourable social, moral, and material surroundings.

EXPLORATION : Since the discovery and occupation of Australia, a succession of dauntless explorers have crossed and recrossed the continent, in order to ascertain its character and capabilities for settlement.

The discovery and the opening up of the interior of Australia is associated with a crowd of famous names, such as Lawson, who first crossed the Blue Mountains in 1813 ; Evans and Oxley, who explored the Lachlan and the Macquarie in 1817-18—Oxley, some years later, in 1823, discovering the Brisbane River ; Hume and Hovell, who, in 1824, crossed the Murray (then called the Hume), and traversed what is now the Colony of Victoria ; the well-known botanist, Allan Cunningham, who, in 1827, discovered the rich pastoral and agricultural country now known as the Darling Downs ; Captain Sturt, who, in 1828, traced the Macquarie and the Darling, and, in the following year, the Murrumbidgee to its confluence with the Murray, the "Queen of Australian rivers," which he followed down to Lake Alexandrina, but was unable to find its outlet into the sea ; Batman and Fawkner, famous, not as explorers, but as the founders of Melbourne ; Count Strzlecki, the explorer of the Australian Alps and the discoverer of Mount Kosciusko (1840) ; Edward John Eyre, who, in 1840, made known the vast salt lakes to the north of Spencer Gulf ; Ludwig Leichhardt, a dauntless explorer, who crossed the continent from Moreton Bay to Port Essington, and perished, in 1848, in another attempt to cross the continent from the Eastern to the Western Sea ; the three brothers Gregory, who gave 15 years to the exploration of Western Australia ; Sir Thomas Mitchell, who discovered several rivers in what is now Central Queensland ; Baron Von Mueller, engaged from 1847 to 1862 in various important explorations (the Australian Alps, the Kimberley District, &c.) ; John McDouall Stuart, the most celebrated of Australian explorers, who determined to cross the continent from south to north, and camped in the centre of Australia in 1860, but was compelled to return by the hostility of the natives ; he again tried and failed in 1861, but, nothing daunted, started the third time, and on July 25th, 1862, planted the British flag on the shores of the Indian Ocean ; the intrepid but unfortunate Burke and Wills, who successfully crossed the continent from Melbourne to the Gulf of Carpentaria, but perished miserably of starvation at Cooper's Creek, on the return journey, only one of the party—King, who was found barely alive at a native camp—survived ; Major Warburton, who with a troop of camels crossed the solitudes of Western Australia, between the Trans-continental Telegraph line and the De Grey River ; John and Alexander Forrest and Ernest Giles, the explorers of Western and Central Australia. Other surveyors and explorers have since made known large areas of the interior, and in time the whole country will be mapped out.

"Unexplored Australia of the present day," says Mr. Ernest Favenc, a recent explorer, "is almost, or quite, confined to two colonies—Western Australia, and a large portion of the Northern Territory of South Australia. To arrive at the

extent of unknown territory that we possess, we may assume :—New South
Wales and Victoria have an area of over 400,000 square miles, almost every
mile of which is fairly known ; Queensland may still have a comparatively small
extent of unknown land in the far northern peninsula ; South Australia has at
least 250,000 square miles unexplored or but little known ; and the huge Colony
of Western Australia can claim more than half a million of square miles, just
crossed at intervals by the tracks of Giles, Forrest, and Warburton.   The extent
of unknown country on the Australian continent, therefore, is still more than
double that of the well-known portion, and this after years of continued toil and
the advancement of settlement.   The continent of Australia is almost fairly
bisected by the overland telegraph line, which may be considered a line of demarcation between the explored and the unexplored portions."[1]

GOVERNMENT : All the Australasian Colonies, except Fiji
and British New Guinea, which are Crown Colonies, possess responsible government.

The form of government is similar to that of the United Kingdom.   At the
head of the executive in each colony is a Governor, representing the Queen, and
appointed by the Crown.   The legislative power is vested in a Parliament of
two Houses—the Legislative Council, nominated or elected, corresponding to
the British House of Lords, and the Legislative Assembly, elected by the people,
exercising similar powers to the British House of Commons.

The first step towards the Federation of the Australasian Colonies was taken
in 1836, when a Federal Council, at which representatives from Victoria, Queensland, Tasmania, Western Australia, and Fiji, were present, met at Hobart for
the first time.   The Council met subsequently in 1888 and 1889.   In 1889,
South Australia took part in the Conference, but New South Wales and New
Zealand were not represented.   In 1890, a Conference of Representatives of all
the Australasian Colonies met at Melbourne, and resolved to take steps towards
the holding of a " National Australasian Convention," to consider and report
upon a scheme of Federal Government.

At the National Australasian Convention, which held its sittings in Sydney,
in 1891, the Constitution drafted for the proposed " COMMONWEALTH OF
AUSTRALIA " was submitted to the delegates representing the whole of the
Australasian Colonies.

The Bill, as finally adopted by the Convention, provides for the appointment of a
Governor-General, with powers similar to those of the Governor-General of Canada, as
the Queen's representative, and, as such, to be also the Commander-in-Chief of all the
military and naval forces.

The several 'States,' as such, are to be represented in a 'Senate,' composed of eight
members from each State, 'directly chosen by the Houses of Parliament of the several
States for a term of six years.'

The people of the Commonwealth are to be represented by one member for every 30,000,
in a " House of Representatives," chosen every three years.   The powers of the Federal
Government are to be exclusive with regard to customs' duties and excise, posts and telegraphs, military and naval defence, ocean beacons and lighthouses, and quarantine.

1. Mr. Favenc is of opinion that good habitable country and enormous areas of well-watered pasture land may still be found in what is generally supposed to be, on the authority of the explorers who have crossed it—Giles, Forrest, and Warburton—the Central Desert, unfit to sustain human life, and impracticable for settlement.   According to the *Australian Handbook*, several successful boring operations were carried on in several parts of Queensland, revealing the fact that, in many of the driest tracts, unlimited water supply is to be found at a few hundred feet below the surface.   Where there is a permanent stream, otherwise worthless land can be cultivated by means of irrigation, and successful experiments in this direction have been made on the great Irrigation Colonies, established by Messrs Chaffey Brothers at Mildura in Victoria and at Renmark in South Australia.

The **Executive Government** is vested in the Governor-General, who can summon and dismiss at his pleasure the officers who, as the "Queen's Ministers of State for the Commonwealth," are to administer the great departments of State, and are to form the Executive Council, which is to aid and advise him in the government.

As to **Finance** and **Trade**, uniform duties of customs are to be imposed, but, until the uniform tariff is enacted and absolutely free trade between the States in the Union is established, the present duties will remain in force.

As regards the Federal Judiciary, a "**Supreme Court of Australia**" will be the final court of appeal in all cases referred to it, but subject, in some cases, to an appeal to the Queen in Council.

The powers, privileges, and territorial rights of the several existing Colonies, shall remain intact, except in respect to the powers exclusively vested in the Federal Government. But no State law can be passed antagonistic to, or inconsistent with, a federal law ; although the Colonial Governors are to be appointed by the Queen as before, all communications with the Home Government must pass through the Governor-General. The boundaries of existing States may be altered and new States formed with the consent of the Colonies affected.

The proposed Constitution must, of course, be accepted by the legislatures of the several Colonies, before the long-expected Federation of Australasia is accomplished, but the marvellous progress made under a decentralized government, and the general satisfaction of the people with matters as they are, may retard for some time to come the formation of the "**Commonwealth of Australia**."[1]

AUSTRALIA.

AUSTRALIA, the "Southern Land," is, strictly speaking, an island, but it is of such an immense size that it may well be regarded as a continent.

As an island, Australia is by far the largest in the world. Greenland is the next in size, and New Guinea, the third largest island, is only one-tenth the size of Australia.

As a continent, Australia is the smallest of the six great land-masses on the surface of the globe. But it is not much inferior in extent to Europe, which is only about one-fifth larger.

The greater part of South America, and a considerable portion of Africa, are within the Southern Hemisphere, but Australia is the only one of the continents wholly to the south of the Equator, its extreme northern point being in nearly 11° South lat., or 770 miles from it.[2]

BOUNDARIES : Australia is bounded by the ocean on every side—by the Pacific on the east, the Indian Ocean on the west, and the Southern Ocean on the south, while Torres Strait, the Arafura Sea, and the Timor Sea, separate it from the Melanesian and East Indian Archipelagoes on the north, and Bass Strait from the island of Tasmania on the south-east.

1. A consummation no doubt ardently desired by advocates of Imperial Federation, and indeed, if South Africa and Australasia would but follow the example of Canada, some definite scheme of Federation might be formulated on the bases of a *Zollverein* and a *Kriegsverein*, or a union for trade and for defence. The Zollverein would certainly be of incalculable benefit to the empire, while the Kriegs- verein appears to be absolutely essential to the maintenance of its integrity. See further the article on "The Commonwealth of Australia," by G. H. Reid, M.P. for Sydney, and Mr. Boulton's paper— Sir John Macdonald on Imperial Federation—in the *Nineteenth Century* for July, 1891.

2. The southernmost point of Europe is 35° N. lat., or 2,502 miles north of the Equator.

EXTENT : The greatest length, from Steep Point on the west to Cape Byron on the east, is about 2,400 miles. The greatest breadth, from Wilson Promontory on the south to Cape York on the north, is nearly 2,000 miles. The **total area** of Australia is nearly **3,000,000** square miles.

The actual area of Australia is computed at 2,944,628 square miles, or 26 times that of Great Britain and Ireland. The largest of its five divisions, **Western Australia**, has an area of over **1,000,000** square miles, or more than 8 times the size of the United Kingdom ; the smallest colony, **Victoria**, with an area of 87,884 square miles, is nearly as large as Great Britain. **New South Wales** comprises an area of **309,175** square miles, or 2½ times the size of Great Britain and Ireland. **Queensland**, with an area of **668,224** square miles, is 5½ times as large as the British Isles, while **South Australia**, with an area of over **903,425** square miles, is nearly 8 times the size of the mother country, or 15 times as large as England and Wales alone. The five Colonies together are very nearly equal in area to the United States, exclusive of Alaska.

COASTS : Australia is much more **solid** and **unbroken** in shape or external contour than Europe—more so, indeed, than any of the other continents, except Africa and South America. The **total length of coast-line** is estimated at **10,000** miles—an average of 1 mile of coast to every 300 square miles of area.

Of the few large indentations which penetrate the coast-line of Australia, but do not materially affect the general solidity of the whole mass of land, the great bight known as the **Gulf of Carpentaria**, on the north, and the corresponding incurve of the **Great Australian Bight**, on the south, are by far the most extensive.

Of the smaller inlets, the most noteworthy are **Port Phillip, Encounter Bay, the Gulf of St. Vincent, Spencer Gulf**, and **King George Sound**, on the south coast ; **Géographe Bay, Shark Bay, Exmouth Gulf**, and **King Sound**, on the west coast ; **Cambridge Gulf, Van Diemen Gulf**, and **Arnhem Bay**, on the north coast ; and **Princess Charlotte Bay, Halifax Bay, Broad Sound, Hervey Bay, Moreton Bay, Broken Bay, Port Jackson**, and **Botany Bay**, on the east coast.

A large portion of the Australian coast is "ironbound," and much of it is absolutely deficient in any inlets which might afford shelter to shipping, besides being most uninviting in appearance. The shores of the **Great Australian Bight**, especially, for 600 miles, present nothing but a wearisome stretch of sandy beach, backed by barren cliffs, varying in height from 400 to 600 feet, and absolutely unbroken by a single stream or creek. The west coast, though richer in bights and inlets, is also low and, on the whole, monotonous ; but the Pacific coast, and parts of the northern and south-eastern coasts, are comparatively picturesque, with here and there inlets, such as **Port Jackson**, of surpassing beauty.

The north-eastern coast is skirted by the wonderful **Great Barrier Reef**, a chain of coral reefs extending southwards from Torres Strait for about 1,200 miles, at a distance of from 10 to 150 miles from the coast, and presenting, throughout its entire extent, only one absolutely safe opening for ships, though broken here and there by many deep channels, through which vessels may pass into or out of the open ocean. Steamers generally take the sheltered channel between the reefs and the coast, but this route requires careful navigation, and sailing vessels prefer the open ocean route outside the Reef. The navigation of **Torres Strait** is also obstructed by coral reefs and sandbanks ; here the most frequented channel is the sound in which **Thursday Island** is situated. This island, which is about the smallest and most central of the Prince of Wales group, is in the direct track of all vessels reaching Australia *via* Torres Strait, and has an excellent harbour—Port Kennedy—one of the finest in Australia, vessels of the largest tonnage being able to enter and anchor at any state of the tide and in all weathers. **Bass Strait** in the south, between the Victorian coast and Tasmania, is much frequented by shipping, but its navigation is dangerous.

CAPES : The principal capes are **Cape York**, the most northerly point ; **Cape Byron**, the most easterly ; **Wilson Promontory**, the most southerly ; and **Steep Point**, the most westerly point.

Other notable points on the Australian coasts are **Cape Arnhem**, on the north ; **Cape Lévèque** and **North West Cape**, on the west ; **Cape Naturaliste**, and **Cape Leeuwin**, on the south-west ; **Cape Catastrophe**, **Cape Spencer**, **Cape Jervis**, and **Cape Otway**, on the south ; and **Cape Howe**, **Sugar Loaf Point**, **Point Danger**, **Sandy Cape**, and **Cape Flattery**, on the east.

ISLANDS : With the single exception of **Tasmania**, there are no large islands off the coasts of Australia.

The principal are **Kangaroo Island**, off the coast of South Australia ; **Stradbroke**, **Moreton**, **Great Sandy**, and **Hinchinbrook Islands**, off the east coast of Queensland ; **Wellesley Islands** and **Groote Eylandt** (Great Island), in the Gulf of Carpentaria ; **Melville** and **Bathurst Islands**, off the coast of the Northern Territory ; and **Dirk Hartog Island**, at the entrance to Sharks Bay, off the coast of Western Australia.

NATURAL FEATURES: The surface of Australia is distinguished from that of all the other continents in that there are no mountains at all comparable to those of Eurasia, America, or even Africa, while there is only one really large **river**, and its volume, though swollen by enormous floods during the rains, is generally much inferior to that of streams of equal length in other parts of the world. There are also but few permanent **lakes**, the largest of them are very shallow, even in the rainy season, and become a series of mere pools in dry weather.

The physical structure of the continent is thus described by the late Rev. J. Tenison-Woods :—"Australia is an immense plateau, with a narrow tract of land sometimes intervening between the edge of this elevated area and the sea. The *east* side is the highest, averaging about 2,000 feet above the ocean. The *west* side is not more than 1,000 feet above the same. The *north* is a little higher. The *south* side is either level with the ocean, or abuts in cliffs upon the sea, ranging from 300 to 600 feet in height.

"The general character of all the seaward side of the table-land is precipitous, but on the south-east angle of the continent the tabular form disappears, and there is a true cluster of mountains—the Australian Alps—whose highest elevation is a little over 7,000 feet. This group is near the sea (Bass Strait), and to the southward there is another group of almost equally high mountains which forms the island of Tasmania.

" The inland portion of the table-land slopes by a very gradual incline towards a central depression, which is south and east of the true centre of the continent. Thus the incline is greater and shorter for the east side of Australia, and it is on this side alone that there is what can be properly termed a river-system.

" The elevation of the west side of Australia being only half that of the east, or even less, and the distance of the central depression being twice as great, there is no drainage towards the interior at all, and whatever water falls from the clouds collects in marshes, which are generally salt.

" The soil on that side of the continent consists generally of disintegrated granite rocks, and is therefore sterile and dry, forming little better than a sandy desert. All the table-land is more or less interrupted by ranges of mountains which do not run for any distance, and are not sufficiently high to give rise to any river-system. Their general direction is north and south, or east and west, and they seem to be quite independent of each other and of the general axis of the continent. The most conspicuous of them are the Flinders Range, which commences at Cape Jervis on the south coast, and continues without interruption for five or six hundred miles into the salt lake area, where it abruptly terminates."

". As to the character of the country according to its physical structure, the same writer remarks[1] that it may be stated generally that the narrow strip which lies between the table-land and the sea is well watered by mountain streams, and that the alluvial land in the neighbourhood of these channels is rich and fertile. On the table-land, where the mountains are not too rocky and rugged, the soil includes some fertile areas ; but this is generally on the volcanic strata, which are fortunately of wide extent. The lands of the interior are, as a rule, poor, except in the river valleys, and towards the central basins of the continent they are in all respects like the Sahara or the table-lands and prairie lands of North America.

The Colony of Victoria is better suited with regard to its lands than any other. It is well watered, and has a large share of the fertile basaltic areas between the table-land and the sea, while the portions of the table-land itself which fall to its inheritance are rich in volcanic tracts.

The Colony of South Australia may be said to be, as far as the richness of its lands is concerned, confined to the valleys and slopes of the Flinders Range, and as this is about 500 miles long and of gentle elevation, the tracts available for agriculture are considerable. Towards the north of a line parallel with the head of St. Vincent's Gulf, the rainfall is small and uncertain, which renders both agricultural and pastoral enterprises subject to great losses from drought.

New South Wales and Queensland are in the same position relatively to the table-land. The capitals of these colonies are built on the slopes between the table-land and the sea. Portions of the upper part of the high lands included in both colonies are volcanic areas of some richness, but the lower lands are poor and sterile, except in the river valleys, and these are very numerous in both colonies.

The interior of Australia is not, however, doomed to perpetual sterility, inasmuch as "the actual amount of the rainfall on the interior slopes must be largely in excess of the drainage by the rivers, and a great portion therefore soaks into the ground and drains along the incline towards the interior. On this account, and the structure of the rocks, the central basin must be especially favourable for the formation of artesian wells." Many such wells have been already sunk with the most gratifying results, and in this way, as well as by irrigation from running streams, large areas, otherwise barren and unfit for settlement, have been rendered productive, and, in some cases, more so than equally good soil even with an ample rainfall.

---

1. The teacher should carefully study Mr. Wood's | logy of the continent in the *Australian Handbook*
lucid exposition of the physical structure and geo- | for 1895, pp. 116-119.

MOUNTAINS: In Australia, as in Southern Africa, the higher grounds run from south to north, at no great distance from the eastern coast. These elevations on the eastern side of Australia form a continuous, though most irregular, cordillera or chain of heights, extending from Cape Howe to Cape York, and known by the general name of the **Great Dividing Range.**

Various names are applied to the different portions of this long range. The southern portion bears the name of the **Australian Alps** ; further north, the range forms the well-known **Blue Mountains**, and, still further north, it is known as the **Liverpool Range**, &c. The Australian Alps are the loftiest part of the chain, and contain the highest of all the Australian Mountains—*Mount Kosciusko*, which is 7,336 feet in height. *Mueller's Peak*, the second highest summit in the same group, has a height of 7,268 feet. The picturesque Blue Mountains are much lower, and seldom exceed 3,000 feet, but their highest summit, *Mount Beemarang*, reaches 4,100 feet. In the Liverpool Range, *Oxley's Peak* stands conspicuous at a height of 4,500 feet, while still further north, *Ben Lomond*, in the New England Range, reaches 5,000 feet, and *Wooroonooran*, in the Bellenden Ker Range, 5,400 feet, and *Mount Dalrymple*, 4,250 feet, in the coast range of Queensland. All these mountains are below the limit of perpetual snow, but on the south-eastern slopes of the Snowy Ranges, and on the summits of the high peaks of the Australian Alps, snow sometimes lingers throughout the year.

In Victoria, the high lands to the west of the Great Dividing Range culminate in two distinct ranges running north and south, and known as the **Grampians** and the **Pyrenees.**

In South Australia, the principal range runs along the eastern side of the Gulf of St. Vincent and Spencer Gulf. It is known in the south as the **Mount Lofty Range**, and in the north as the **Flinders Range**. In Eyre's Peninsula, on the western side of Spencer Gulf, is a rugged chain of hills called the **Gawler Ranges**. North-west of Lake Torrens is **Stuart Range**, and, further north in the interior, there are other mountains, or rather hills, all of much less elevation than the coast ranges.

The western coasts are also backed by high grounds of moderate elevation, the principal portion of which, known as the **Darling Range**, runs parallel to the coast at a distance of from 10 to 25 miles. Further north, the coast range bears the names of the **Herschel Range** and the **Victoria Range**. Similar elevations adjoin the north-western and northern coasts, but they are insignificant in height, and nowhere exhibit the character of a true mountain-chain.

The **Mountains of Tasmania**, which are divided into two sections by the valleys of the Tamar and the Derwent, may be regarded as outliers of the great Australian Cordillera. Several peaks rise over 4,000 feet—the culminating point, *Cradle Mountain*, attaining an elevation of 5,069 feet.

RIVERS: Although a large number of rivers are met with on the coast of Australia, there is but one river—the **Murray**—which at all approaches the larger streams of other continents.

The chief characteristic of the rivers of Australia is their liability to sudden and violent floods, and too many of them are, unfortunately, mere surface-torrents, supplied by the rains, which are, over the greater part of the interior, both scanty and irregular. During seasons of drought, they are speedily dried up under the intense heat of an Australian sun, or converted into a chain of ponds. With the recurrence of the rainy season, vast floods of water are poured through their beds, and huge trunks of trees, masses of rock, and other *débris*, carried down by the stream, bear witness to the violence of the torrent. The **Murray** and its chief tributaries are perennial streams, but, although their volume of water undergoes great variation according to the season of drought or rain, they are navigable by river steamers, and the perennial streams are likely to play an important part in the settlement of the interior, now that public atten- tion has been directed to the feasibility of cultivating the soil by means of irri- gation. The gradual slope of the plains over which these rivers flow renders it a comparatively easy matter to irrigate large tracts of otherwise uncultivable soil, but which only require a regular supply of water to be rendered capable of grow- ing almost all the temperate and sub-tropical plants. "Where this can be done, the value of the land, so situated that it can be supplied with water in this manner, is always enormously increased. For irrigation has many advan- tages. It not only makes the supply of moisture certain, but it increases the yield of crops grown on a certain area of land, especially where the water run on to the land still contains a good deal of the sediment derived from the high grounds; it enables more valuable crops to be grown than those which can be grown without irrigation; and, in a climate like that of Australia, which is warm enough for vegetation all the year round, it enables crops to be grown almost without in- terruption, one crop following another in close succession; and for all these reasons, in regions where the climate is so dry as to make the growing of grain crops uncertain, land that can be irrigated will support a population ten or twenty times as large as that which can be supported without vegetation."[1]

Strictly speaking, Australia is surrounded by the Pacific and the Indian Oceans, which meet at Torres Strait on the north and at Bass Strait and along the meridian of South-West Cape (146° E.) in Tasmania on the south, and its rivers, with the exception of those in the interior which have no outlet to the sea, belong to one or other of these great ocean basins.

The **Australian Section of the Pacific River-System** presents the same characteristic features as the South American—in both a con- tinuous cordillera limits the area for river development to a narrow tract between the mountains and the sea. The Australian Alps and their northerly continuation are, of course, insignificant compared with the vast chain of the Andes; still, the streams that enter the Pacific from the latter are shorter and less navigable than those which descend from the former.

Of the coast streams of Queensland, the principal are the **Burdekin**, the **Fitzroy** (formed by the junction of the Mackenzie and the Dawson, with their subsidiary creeks), **Burnett**, and the **Brisbane**, all of which are navigable for steamers of considerable tonnage for some distance inland. In New South Wales, the coast plain is watered by many noble streams, the largest of which are the **Clarence**, **MacLeay**, **Manning**, **Hunter**, **Hawkesbury**, and **Shoalhaven**. The volume of all these rivers varies greatly, being liable in winter to sudden and violent floods, but they are not altogether worthless for navigation. The Clarence is half-a- mile wide at its mouth, and is navigable by sea-going steamers for 70 miles; the Hunter and other rivers are also regularly traversed by colonial trading

---

1. Chisholm.

vessels. In Victoria, the **Snowy River**, the **Mitchell**, and other smaller streams fall into the Pacific; the other rivers of the colony, and, indeed, of the whole continent, with the exception of those already named, and a few "continental"[1] streams in the interior, belong to the basin of the Indian Ocean.

The Australian Section of the River-System of the Indian Ocean comprises the *Murray*, which enters Encounter Bay through Lake Alexandrina ; the *Swan, Murchison, Gascoyne, Ashburton, De Grey, Fitzroy*, and other rivers of Western Australia ; the *Victoria* and *Daly*, in the Northern Territory ; and the *Roper, Flinders, Mitchell*, and other streams which fall into the Gulf of Carpentaria.

The **MURRAY**, which drains a large portion of Queensland, the whole of the interior of New South Wales, the northern half of Victoria, and a part of South Australia, rises on the western slopes of the *Australian Alps*, about 15 miles south of *Mount Kosciusko*, and becomes navigable at Albury, about 150 miles from its source. Throughout its upper and middle course, the Murray forms the boundary between the colonies of Victoria and New South Wales ; its lower course, below its junction with the Darling, is within South Australia. From Victoria, the Murray receives the **Mitta Mitta, Ovens, Goulburn, Campaspe**, and **Loddon**, the valleys of some of which are highly auriferous. On the north, its two great tributaries—the Murrumbidgee and the **Darling**—with their subsidiary creeks, drain the whole of New South Wales and a part of Queensland west of the Great Dividing Range. The Murrumbidgee rises near the Murray in the *Muniong Range*, and is joined, about 50 miles above its junction with the Murray, by the River **Lachlan**. The Darling is formed by numerous periodical streams, of which the largest are the **Macquarie** and the **Condamine**. The length of the Murray is 1,300 miles, the average width of the main stream is about 240 feet, and its depth about 16 feet, but it undergoes great variation according to the season of drought or rain.[2]

The Murray is regularly navigated, at certain seasons of the year, to ALBURY, 1,000 miles above its mouth ; small steamers also ascend the Murrumbidgee for 500 miles to WAGGA WAGGA, and the Darling as far as BOURKE, a distance of 600 miles. The Murray and the Murrumbidgee alone are perennial streams, the rest are, for the most part, mere surface torrents supplied by the rains, and are consequently liable to sudden and violent floods.

Of the numerous streams that traverse the more settled portion of Western Australia, the Swan River alone is navigable to any extent. The Murchison, Gascoyne, **Ashburton, De Grey, Fitzroy**, and other rivers drain the northern half of Western Australia.

In the Northern Territory of South Australia, the **Victoria** has long since been proved to be navigable for a considerable distance ; while the **Roper**, which flows into the Gulf of Carpentaria, is known to be a magnificent stream, easily ascended by large steamers and sea-going vessels for 100 miles from its mouth.

The **Gregory, Flinders, Norman**, and **Mitchell** are the largest of the many streams that converge into the south-eastern portion of the Gulf of Carpentaria.

---

1. By a "continental" stream is meant, of course, not a stream continental in magnitude, but one which has no direct or indirect outlet to the ocean.
2. The level lands along the banks of the Murray seem particularly well adapted for irrigation, and especially, as "in ages now long past, the river, in-

stead of being a sickly dried-up stream, was a kind of Southern Nile, bearing upon its wide waters the same annual gift of fertilizing mud; and hence on its banks are now stored up the elements of an un-exhaustible fertility."

The fine country through which they flow is being rapidly occupied and utilised for both pastoral and mining purposes.

The largest "continental" rivers of Australia are the **Diamentina** and **Cooper's Creek** or the **Barcoo River**.

Both these streams enter South Australia from the south-west of Queensland, but while Cooper's Creek, in the rainy season, when it swells to a breadth of two miles and rises to a depth of 20 feet, carries a vast amount of water into Lake Eyre, the Diamentina dries up and disappears in the stony desert to the north of the Delta of Cooper's Creek. Further north, the **Finke River**, which rises in the *MacDonnell Range* in the centre of the continent, also disappears some distance north of the Macumba River, another "continental" stream, but which, when flowing, enters Lake Eyre.

LAKES: Lake Alexandrina, through which the Murray passes immediately above its mouth, is the largest fresh water lake in Australia. It is a shallow expanse of water, difficult to navigate. Most of the other lakes that are marked on the maps of Australia are only salt marshes, or mere surface ponds, with dry beds during the larger portion of the year. Of these, the most extensive are Lakes **Eyre**, **Torrens**, and **Gairdner**, to the north of Spencer Gulf, and Lake **Amadeus**, in the interior, 270 miles north-west of Lake Eyre.

CLIMATE: Generally speaking, the climate of Australia may be said to be uniformly **warm** and intensely **dry**, but exceptionally **healthy,**[1] and well suited to Europeans.

So vast a continent necessarily exhibits great differences in climate, which, in fact, ranges from the tropical heat of the north to the cooler and more enjoyable climate of the south. The most densely-peopled districts of Victoria, New South Wales, South Australia, and Western Australia, have a climate resembling, in the main, that of the countries of Southern Europe—genial and delicious in autumn, winter, and spring, and disagreeable only in summer, during the prevalence of the hot winds which now and then blow from the interior, fortunately only for very brief periods. These hot winds, while they last, are almost as unbearable as the simoom of the African deserts. "Sometimes they are succeeded by a cold south wind of extreme violence (called the *Southerly Burster*), the thermometer falling 60 or 70 degrees in a few hours. In the desert interior, these hot winds, nearer to their source, are still more severe. On one occasion, Captain Sturt hung a thermometer on a tree, shaded both from the sun and the wind. It was graduated to 127° F., yet the mercury rose till it burst the tube! The heat of the air must have been at least 128° F., probably the highest temperature recorded in any part of the world, and one which, if long continued, would certainly destroy life. The constant heat and drought in the interior, for months together, are often excessive. For three months, Captain Sturt found the mean temperature to be over 101° F. in the shade; and the drought during this period was such that every screw came out of their boxes, the horn handles of instruments and combs split up into fine laminæ, the lead dropped out of pencils, their hair ceased to grow, and their finger nails became brittle as glass."

All the Australian colonies suffer more or less from periodical droughts, but the total rainfall is, on the whole, greater than in England. The rains fall with

1. The mean death-rate in the Australasian colonies | 14.11 in Victoria, to 15.12 in Western Australia. In 1891, varies from 10.2 per thousand in New Zea- | In Europe, the average rate is about 25 per thou-land, 13.44 in South Australia, 13.75 in New South | sand, varying from about 20 in the United Kingdom Wales, 13.47 in Tasmania, 13.34 in Queensland, and | to 25 in Germany, 30 in Austria, and 34 in Hungary.

great violence at particular seasons, more especially during the winter of the Southern Hemisphere, that is, from May to August.[1]   During nine months of the year there is often little or no rain, and the plains in the far interior are sometimes without rain for two or three years consecutively.

But although the climate of Australia is, on the whole, so warm, snow is not unknown. Many of the higher mountains are covered with snow all the winter, and, although the highest peaks fall short of the limit of perpetual snow, patches of snow remain unmelted all the summer in the higher valleys and ravines. The higher sections of the railway between Sydney and Bathurst have been covered with snow, and at Kiandra, a gold mining village in the Australian Alps, at an elevation of 4,640 feet above the sea, the thermometer sometimes registers 1° F. below zero, while the mean annual temperature in the shade is only 46° F., and snow falls from May to November, sometimes for a month together.

PRODUCTIONS: The native flora of Australia is peculiar and unique, and its characteristic animals present no analogy to those of other continents. The mineral wealth of the continent is enormous, and apparently inexhaustible.

The native vegetation of Australia is altogether different from that of other parts of the globe. Australia is, in fact, the only one of the continents that has no characteristic food-plants of its own—neither grains, fruits, nor esculent roots of any value—almost all the plants that are indigenous to the soil are valueless as human food.   But the soil is capable of producing every variety of the grains, fruits, and vegetables of temperate and even tropical regions in abundance, wherever there is sufficient moisture.

The native animals of Australia are also unique, and, like the indigenous plants, are of no obvious service to man, but all the domestic animals of Europe have been successfully introduced, and millions of *sheep, cattle*, and *horses* are now reared on the vast pasture lands, and some of the imported animals, such as the *rabbit* and the *sparrow*, have taken so kindly to their Australian home, and have multiplied so fast, that they have become veritable pests.

As regards the mineral wealth of the continent, the *gold mines* of Victoria have yielded more of the precious metal than those of any other country ; the *coal-fields* of New South Wales are among the most extensive in the world ; the *copper mines* of South Australia laid the foundation of the commercial prosperity of that colony ; the *tin mines* of Queensland, and the *silver mines* of New South Wales, are apparently inexhaustible, while there are rich deposits of *iron ore* in most of the colonies.

VEGETATION : Australian vegetation is of a strange and peculiar character, and is not less noticeable for the large number of distinct species, than for their dissimilarity to those of other countries.

The characteristic trees of Australia are the eucalypti or gum trees, and the acacias or wattles. There are altogether about 150 different kinds of gum trees, most of them found in Australia alone, and many of them of great value for their timber.   One of them, the blue gum, is noted all over the world on account of its power of dispelling the miasmatic influences of swampy land and marshy

---

1. The seasons in Australia are, of course, the reverse of those in the Northern Hemisphere. In Australia, Spring commences on September 23rd ; | Summer, on December 21st (the longest day) ; Autumn, on March 21st ; and Winter, on June 21st (the shortest day).

districts, and rendering them dry and healthy. Certain varieties of the eucalyptus attain colossal heights in the mountain ravines, and some of the peppermint trees of Victoria overtop even the famous mammoth trees of California. In the Dandenong Range, about 40 miles east of Melbourne, the ravines contain numerous trees over 400 feet high, and one fallen tree was discovered of the enormous length of 480 feet—undoubtedly the highest tree in the world. Several of the acacias are also magnificent woods, and the bark of the black wattle is valuable for tanning. Ferns are numerous, particularly in the mountain gullies —one variety has fronds six feet in length. Timber trees of great value are found in the brush forests along the coast—here the magnificent red cedar often attains a height of 150 feet, with a girth of over 30 feet. Extensive forests of the leafless casuarina or shea oak are found in the south and west, and even in the barren wastes of the interior. The Australian "bush" is generally an open forest country, easily traversed, and with large areas of good pasture for sheep and cattle. But the desolate "scrub" country and the "Mulga scrub" are the dread of the explorer, while the spinifex grass regions are most difficult and often impossible to penetrate. The Mallee scrub is a dense growth of stunted eucalypti, growing so close together as to be almost impenetrable, and extending, in some parts, in an unbroken expanse thousands of square miles in extent. The appearance of the mallee scrub regions is gloomy and monotonous in the extreme ; nothing can be seen but a dark-brown mass of low bushes, which seem like "a heaving ocean of dark waves, out of which here and there a tree starts up above the brushwood, making a mournful and lonely landmark." The mallee scrub is, however, less dreaded than the Mulga scrub, which consists of acacia bushes armed with strong and sharp spines, and, where matted with other shrubs, are absolutely impenetrable. But "the most terrible production of the Australian interior is the spinifex or porcupine grass, which extends for hundreds of miles over sandy plains, and probably covers a greater amount of surface than any other Australian plant." Fortunately, this hated shrub is mostly confined to the deserts of the west and the centre of the continent.

Except in the luxuriantly wooded ravines and humid valleys of the Pacific slope, Australian forests are not attractive, and the prevalence of the gum trees, with their dull greyish-blue foliage and vertical leaves, gives them a rather monotonous aspect. On the other hand, the flowering plants, of which there are no less than 10,000 distinct species—a greater number than is found in the whole of Europe—display a wealth and vividness of colour that produce the most beautiful and striking effects. The "flame trees" of New South Wales, with their gorgeous bunches of red flowers, render the Illawarra Mountains conspicuous for miles out at sea, and the "fire tree" of Western Australia, with its orange-coloured blossoms, can really be compared to nothing but a tree on fire. The giant rock lily, 30 feet in height, is crowned with a mass of flowers several feet in circumference, and other plants, or rather trees, are, when in bloom, simply masses of crimson flowers, gaudier, but not more beautiful, than the white stars which crown the rugged stems of the grass trees in winter. In fact, scarcely any other part of the world affords a greater variety of aromatic plants and odoriferous flowers of the most graceful forms and brilliant colours than Australia.

The numerous native grasses are also highly nutritious, and are capable of resisting great extremes of heat and cold. The kangaroo grass, especially, can withstand a long period of drought, while the invaluable salt bush thrives on a dry and saline soil, and affords good pasture to sheep in districts otherwise utterly useless for stock.

CULTIVATED CROPS : All the grains, fruits, and vegetables, whether European or tropical, planted in the Australian soil, yield abundant crops, wherever there are sufficient moisture and suitable temperature.

Although the Australian Colonies, compared with European countries, have barely emerged from the pastoral stage, the agricultural produce is of considerable value and now amounts to about three-fourths of the animal products.

Wheat is the principal crop in South Australia and Victoria, and although the average yield is less than in most other countries, yet the quantity produced suffices, in most years, to supply the home demand and leave a large balance for export, while the quality is such that it invariably realizes high prices in the London market. Maize is the chief crop in Queensland, and is also an important product in New South Wales, but it is not grown to any great extent in any of the other colonies. Oats and barley are most largely grown in Victoria, and of potatoes, which everywhere yield abundantly, Victoria and New South Wales are the largest growers. The vine is extensively cultivated, especially in Victoria, New South Wales, and South Australia, and the climate and soil of these colonies are peculiarly adapted to its successful culture on the largest scale. The growth of the sugar-cane and the manufacture of sugar are important industries in Queensland and New South Wales, and tobacco is grown in many of the warmer districts in the Eastern Colonies. The olive is systematically cultivated in South Australia, and all kinds of fruit are grown in all the colonies. "There is nowhere else on earth such a climate for fruit," and the oranges of New South Wales, the pine-apples of Queensland, the peaches and apricots of Victoria, the raisins and currants of South Australia, and the apples and strawberries of Western Australia, rival the productions of the most favoured of other fruit-growing countries.

ANIMAL LIFE : The native animals of Australia are even more peculiar and anomalous than the plants, and are of no obvious service to man.[1]

The characteristic mammals of other continents are entirely wanting in Australia, where there are "no apes or monkeys; no oxen, antelopes, or deer; no elephants, rhinoceros, or pigs; no cats, wolves, or bears; none even of the smaller civets or weasels; no hedgehogs or shrews; no hares, squirrels, porcupines, or dormice." A number of small rats and mice, the great fruit-eating bat or flying fox, and the native dog or dingo, are the only ones which resemble those of other regions; while of the peculiarly Australian mammals—the *marsupialia*,[2] or pouch-bearing animals—the only representatives elsewhere are the opossoms of America. These animals are distinguished not only by their strange forms and motions, but also by the bag or pouch in which they carry their young for a considerable time after they are born.[3]

The *Marsupialia* are the characteristic mammals of Australia. The most remarkable marsupials are the kangaroos, the largest of which are about 5 feet

1. Some of the indigenous animals yield commercial products, but they are of small value compared with the damage they do to the pastures, &c. The leather made from the skin of the kangaroo is soft and durable, that of the smaller species resembling kid. The opossum is also ruthlessly pursued for the sake of its skin, of which handsome and comfortable traveling rugs are made. The skin of the rabbit, the most obnoxious of all, has also a certain value, but the amount derived from the sale of the skins of all these animals is absolutely insignificant in comparison with the loss they inflict upon the squatter and the settler.

2. From the Latin, *marsupium*, a pouch.

3. "The adaptation of the peculiar structure of these creatures to the country in which they live, where long periods of drought are not uncommon, where there are extensive tracts without water, and where there is always more or less difficulty in obtaining that necessary element of life, is worthy of notice."

high and weigh some 200 lbs.   The smaller wallabies, hare kangaroos, and
rat kangaroos, are much more numerous.   The Australian opossums are marsupial animals of arboreal and nocturnal habits, but quite distinct from the true
opossums of America.   The koala or native bear is a kind of sloth, not much
larger than the opossum, and of similar habits.   The thick-limbed and clumsy
wombat, also a marsupial, lives on roots and burrows underground.   The
bandicoot is a small rat-like kangaroo.   But the most remarkable, perhaps, of
all the animals found in Australia are the duck-billed platypus *(Ornithorhyncus
paradoxicus)* or water mole, which is a mammal, but has a bill like a duck, and
lays eggs ; and the echidna or ant-eating porcupine, which somewhat resembles
the English hedgehog.   The elegant native cat, a carnivorous marsupial, is
fierce and intractable, but the dingo or native dog is much more formidable.
The ferocious and untameable pouched hyena, and the native devil,[1] now only
found in Tasmania, were formerly also found in Australia.[2]

The abundance and variety of bird life in Australia are remarkable, and of over
600 distinct species, less than one-twentieth are found elsewhere.   Parrots,
cockatoos, and paroquets, are more numerous and beautiful than in any other
part of the world ; eagles, falcons, hawks, and owls abound, but vultures and
woodpeckers, which are found almost everywhere else, are quite unknown in
Australia.   Birds that, like the humming-bird of tropical America, feed on
flowers, are very numerous, and these honey-suckers are often most gorgeously
clothed ; the lyre-bird, the bower-bird, with the numerous species of pigeons
and doves, are all famed for the beauty of their plumage ; while the laughing
jackass, or great kingfisher, and the mocking bird, arrest the attention of the
traveller by their extraordinary cries.   Among the larger kinds of birds are the
black swan, the brush turkeys or mound makers, the native companion, a water
bird somewhat like a gigantic crane in appearance, and the emu, a kind of
ostrich, the largest of all Australian birds.

The *reptiles* of Australia include numerous varieties of snakes and lizards.
Several species of snakes, particularly the death-adder, the black snake, and the
tiger snake, are venomous, and some of the lizards, which are very common,
attain a large size.   Several species of alligators infest the rivers of Queensland
and Northern Australia.

There are over a hundred different species of edible sea-fish, of which the
schnapper, which sometimes attains a weight of 30 lbs., is the most valuable
and abundant.   The Australian salmon is a fish of great size and beauty, but
is inferior to the true salmon.   The large sea mullet is unrivalled in richness
and delicacy of flavour, and mackerel, whiting, herring, and numerous other
varieties of food fishes are found in Australian seas, which are, however, also
infested by sharks and other destructive fishes.   The king of river fish is the
Murray cod, which is sometimes caught weighing as much as 100 lbs.   Oysters
are plentiful and excellent.   But the most remarkable of all the Australian
fishes are the frog fish, which has fins adapted for walking on the ground
rather than for swimming ; the hopping fish, with fins developed into legs, so
that it can hop along the mud flats which it frequents ; the sea-horse, so named
from a resemblance in the shape of the head and the fore part of the body to

1. Fossilized remains of both these 'devils' have been discovered in the caves of New South Wales.
2. "Geological research has brought to light the remains of numerous extinct species of gigantic mammalia. The largest of these pre-historic animals was the *Diprotodon*, a marsupial quadruped, which, according to Professor Owen, was as large as the rhinoceros or hippopotamus. Fossil remains of another large marsupial, the *Nototherium*, somewhat resembling a tapir, and of a marsupial lion *(Thylacoleo)* have also been discovered. Bones of gigantic kangaroos *(Macropus titan)* have been frequently met with, as well as those of a species of wombat larger than that actually in existence."—*Coghlan.*

that of the horse; the still more singular phyllopteryx, which looks like the ghost of a sea-horse, with its winding-sheet all in ribbons around it; and the dugong, a warm-blooded graminivorous ruminant, with a stomach exactly like that of an ox, and therefore incorrectly described as a fish.[1]

In the insect world, Australia occupies a foremost position, whether as regards number, peculiarity, or activity, the latter quality being unpleasantly conspicuous in the mosquito.

Introduced Animals : All the domestic animals of Europe have been introduced into Australia, and immense flocks of sheep and countless herds of cattle, besides a very large number of horses, are now reared on the vast pastures which, not so long ago, only 'carried' kangaroos and wallabies.

It is interesting to compare the stock of animals landed by Governor Phillip with the first expedition with the stock depastured in the colonies a hundred years later. Governor Phillip, in 1788, landed 1 bull, 4 cows, 1 calf, 1 stallion, 3 mares, 3 colts, and a few sheep, goats, and swine. In 1888[2] there were, in the five colonies of Australia, no less than 80 million sheep, 8 million cattle, and over 1¼ million horses. Including Tasmania and New Zealand, the number amounted to 96½ million sheep, 9¼ million cattle, and 1½ million horses.[3]

Camels have also been successfully introduced into South Australia, but instead of loading them as in Arabia and Africa, they are harnessed to light but strong waggons, and eight of them will draw four tons and travel 15 miles a day. "They have been known to travel, heavily laden, for nine successive days without a drink of water, and with no other food than that which they could pick up for themselves from the scanty bushes."[4]

Noxious Animals : The dingo or native dog is not the only noxious animal; kangaroos, wallabies, and rabbits, which consume the pasturage, are even greater pests. The rabbits are the greatest pests in many parts of Australia, and over a million sterling has been spent in their destruction. At one time, over 100 million acres were infested by them in New South Wales, and more than 25 millions were destroyed in one year. Large sums have also been paid for the destruction of kangaroos, wallabies, kangaroo rats, hares, and wild pigs, and an immense number of these animals is destroyed every year.

MINERAL WEALTH : Australia abounds in mineral wealth, and its marvellous progress and prosperity are largely due to the enormously rich gold mines of Victoria, Queensland, and New South Wales, the productive copper mines of South Australia, the valuable coal fields and rich silver mines of New South Wales, the famous tin mines of Tasmania, and two extensive deposits of other useful and valuable metals and minerals.

Gold may be said to be the "creator" of Australia, for the discovery of fabulously rich goldfields attracted a large and energetic population, and ad-

1. "The dugong, which is now seldom seen south of the Brisbane River, frequents the flats and shallows along the margin of the shore, and feeds upon the grass which is found thereon. It attains a large size, and sometimes measures 14 feet in length and 10 feet in girth."—Coghlan's Wealth and Progress of New South Wales, p. 139.

2. A year in which the whole of Australia suffered from one of the most disastrous droughts on record.
3. The latest returns give the following number for Australia, with Tasmania and New Zealand:—sheep, 118 millions; cattle, 12 millions; horses, nearly 2 millions; and pigs 1 million.
4. Chisholm.

vanced the progress of the country hundreds of years at a bound.   The richest gold fields are those of Victoria, but the gold mines of New South Wales and Queensland have produced a vast amount of the precious metal, which is also found in South Australia, Tasmania, and Western Australia.   In 1823, the assistant-surveyor of New South Wales found particles of gold in the sands of the Fish River, about 15 miles from Bathurst, and, in 1839, Count Strzlecki found gold in the Vale of Clwyd, and, two years later, the noble metal was also discovered by the Rev. W. B. Clarke in the Macquarie Valley, but it was not until 1851 that payable deposits were proved to exist in New South Wales and Victoria.   Up to this time, the Government had done all in its power to prevent men searching for gold and to withhold the knowledge of its presence in the soil from becoming public.   But the finds, coming so rapidly one after another, created a gold fever, and all the obstacles the Government could present were swept away as a sand barrier would be by the rising tide.   Finding that it was totally in vain to stem the flood, the Government gave way—the first licenses to mine for gold being issued on the 1st of September, 1851.   No sooner had gold-digging been declared a recognized pursuit, than the entire population became, as has been aptly said, "drunk with gold."   Settlers left their homesteads, merchants their desks, professional men their offices, tradesmen their avocations, seamen their ships, and all engaged in the hazardous and laborious search for the precious metal.   For a time there was an excitement that nothing could allay ; and it was not until the hardships, privations, and dangers of a digger's life, to say nothing of its uncertain results, began to prove that none but the strong, hardy, and experienced could hope to succeed, that things began to resume anything like their normal course.   Though there is now no such excitement as in the early days, gold-mining is still a most important industry, affording constant employment to over 45,000 miners, considerably more than half of these being in Victoria and the rest chiefly in Queensland and New South Wales.   The gold occurs embedded in rock, generally quartz reefs, or in alluvium.   Quartz-mining is now mostly carried on by wealthy companies, equipped with costly machinery, and able to command the best skilled labour. The alluvial diggings are easier and cheaper to work, but they are soon exhausted, while the quartz mines are apparently inexhaustible, some being worked at a depth of from 2,000 to 2,600 feet with no appreciable diminution in the yield of gold.   Some of the claims at Ballarat have yielded fabulous amounts, and veins of gold of extraordinary richness have been worked in New South Wales, but the **Mount Morgan Mine** in Queensland, which has already paid over a million in dividends, is the most wonderful auriferous deposit in the world.   Mount Morgan is a huge mound, 1,225 feet in height, of extraordinarily rich gold ore, a veritable "mountain of gold."[1]

Metals other than gold are also found.   The silver-lead mines of the Barrier Ranges and Broken Hill districts of New South Wales, have, since their discovery in 1883, produced 16½ million pounds' worth of silver and silver lead ore, and the output in 1893 was over 3 million oz.   Much silver and ore are also exported from Queensland and Western Australia.   Lead is found in all the colonies, but it is only worked when combined with silver in paying proportions. **Copper** also exists in all the colonies, and has been mined extensively in South Australia, New South Wales, and Queensland.   The celebrated Burra Burra Copper Mine in South Australia was for many years the richest in the world,

1. Large 'nuggets,' or masses of pure gold, have at various times been unearthed in Victoria and New South Wales.   The "Welcome Stranger," found in 1869 in Victoria, weighed 190 lbs. ; the "Welcome" nugget, found in 1858, weighed 184 lbs. 9 oz. 16 dwts., and sold for £10,500.   In New South Wales, a mass of gold weighing 106 lbs. was found on the Turon in 1851.

while the Kapunda, Wallaroo, and Moonta Mines in the same colony, the Great Cobar Mine in New South Wales, and the Cloncurry Mines in Queensland, are also very rich, and rich lodes have been found in Western Australia and in the Northern Territory of South Australia. The richest deposits of tin are in Tasmania—Mount Bischoff in the north-west of the island is a mass of tin ore, yielding 80 per cent. of the pure metal. But nearly twice as much tin is produced in New South Wales, and considerably more in Queensland than in Tasmania. There are large deposits of excellent iron ore in almost all the colonies, but none of them are worked, except in New South Wales.

Mineral fuels are abundantly distributed throughout Australia, and the coalfields of New South Wales are among the most extensive in the world. Coal is also found in Queensland, Tasmania, and Victoria, but the great development of the industry in New South Wales, and the facilities for its distribution thence have hitherto prevented the working of the coal areas in other colonies on anything like so large a scale. Kerosene shale, which yields petroleum oil and other products, is found in several parts of New South Wales, and a large number of diamonds has been found in the same colony.

The total value of the metals and minerals produced in Australia, between 1850 and 1890, amounted to about 450 millions sterling, of which 300 millions represent the value of gold, and of this amount considerably more than two-thirds were contributed by Victoria. About 30 million pounds' worth of copper, and 20 million pounds' worth of tin, and not much less than 25 million pounds' worth of coal, with 4 million pounds' worth of silver and silver lead ore, have been won in the Australian colonies during the last 40 years, in addition to the gold, an average yearly production of over 11 millions sterling. The gold mines are not now nearly so productive as they were, and the gold-mining population has decreased, but the number of men engaged in mining for other minerals has so largely increased, that probably the mining population is now as large as at any time during the prevalence of the "gold fever."[1]

INHABITANTS : The people of Australia are mainly settlers from the British Isles, or their descendants, and now number over 3 millions. There are also about 40,000 *Chinese*, 10,000 *Polynesians*, 1,000 *Malays*, and perhaps 100,000 *Aborigines*.

The Australian Aborigines are among the most degraded members of the human race. A few roots and berries, with shell-fish, insects, grubs, and other repulsive objects, form the food resources of the Australian savage, who will eat almost anything—lizards, snakes, and frogs being especially esteemed, while most of the wilder tribes are also cannibals, not from necessity, but from choice. "They have no fixed habitations ; in the summer time, they live almost entirely in the open air, and, in the more inclement weather, in bark huts of the simplest construction. Their implements are of wood, stone, or the bones of animals or fish. Their religious and intellectual condition is apparently of the lowest kind." Some of them are occasionally employed as shepherds by the colonists, but they dislike continuous work, and soon return to the bush. Others, especially in Queensland, are employed as a kind of native police, and, as such, are found very useful in dispersing dangerous tribes, but they are so cruel and bloodthirsty that they can scarcely be restrained from the wholesale slaughter of those against whom they are led. The natives of Northern Queensland are

---

1. See further the "Mineral Wealth of Australia," in Mr. Coghlan's *Statistical Account* | *of the Seven Colonies of Australasia* (Sydney, 1890).

stronger and far more fierce and intractable than those of the south, but their intelligence and habits are much the same. Altogether, the Australian natives occupy the lowest position of the whole human family, and are decidedly inferior to most other savage races. They have made no advance in civilization, and are rapidly disappearing in the settled districts. There are still a few hundreds in Victoria, and they are rather more numerous in New South Wales, South Australia, and Western Australia, but most of them are confined to the unsettled parts of Queensland, and even there, as elsewhere, they are gradually dying out, although, from the nature of the country which they inhabit, it will probably be a long time before they are entirely extinct. In Tasmania, the last of the aboriginal race died in 1876.

The Chinese,[1] who number about 50,000, are most numerous on the goldfields, principally in Northern Queensland, where the Polynesians and Malays are also mostly found, having been introduced as labourers on the sugar or cotton plantations.

There is a considerable German element in Australia, especially in South Australia and Queensland, and there are several thousand Scandinavian, American, and French settlers in the various colonies.

With these insignificant exceptions, the Australian people are British or of British origin. The native-born Australians now largely outnumber the settlers from the United Kingdom and all other countries, and the increase of the Australian-born section of the population is much greater than the increase due to immigration, except in Queensland and, perhaps, in Western Australia.

The birth rate in the Australian colonies generally is much higher, and the death rate considerably lower, than in most European countries, and the increase of population due to the excess of births over deaths is proportionately higher than in any part of the world. Owing to this, and to an active immigration, which is still very large — the total excess of arrivals over departure from 1871 to 1890 amounted to no less than 800,000 people — the population of Australia has increased with a rapidity hardly rivalled either in the United States or in Central Canada. From 1,030 whites in 1788, 60,000 in 1833, a quarter of a million in 1841, and less than half a million at the time of the discovery of gold, the subsequent "rush" increased the population to such an extent that, in 1861, it amounted to over a million. Twenty years ago, the continent of Australia contained nearly 2¼ million inhabitants, and now the population amounts to over 3½ millions — British in blood, language, and tradition. These Britons of the South are thus united to the Britons of the North by the strongest of all ties, and among them an Englishman is at home, "almost as much at home as between the hedgerows of an English lane. Everywhere he finds the same language, the same institutions, the same beliefs, the same books. The English society is everywhere, and nowhere can he feel himself an exile."[2]

---

1. Masters of vessels are now forbidden, under a heavy penalty, to bring more than one Chinese to every 300 tons, and a poll-tax of £100 is charged on landing, with the result that the Chinese immigration has almost entirely ceased.

2. Professor Seeley.

NEW SOUTH WALES.

NEW SOUTH WALES, the "mother colony" of Australia, extends along the eastern coast of the continent from **Point Danger** on the north to **Cape Howe** on the south, a distance of 700 miles, and stretches inland for a distance of from 500 to 850 miles.

New South Wales originally included the whole of Australia east of the 135th meridian, but, in 1836, its south-western section was included in the new colony of South Australia, while, in 1851, the southern or Port Phillip district was separated from it and formed into the colony of Victoria, and, in 1859, the immense northern division was erected into the colony of Queensland. Still, the mother colony possesses an ample territory, its total area amounting to 310,700 square miles, or 2½ times that of Great Britain and Ireland, and, as regards population, it has almost outstripped Victoria, although the number of inhabitants—nearly 1¼ millions—is very small compared to the extent of the country they occupy, there being on an average scarcely 4 persons to the square mile, or less than 125th part of the density in England and Wales.

BOUNDARIES: New South Wales is bounded on the *north* by Queensland, on the *west* by South Australia, on the *south* by Victoria, and on the *east* by the **Pacific Ocean**.

The boundaries are partly natural and partly artificial. The 141st meridian forms the whole western frontier; the northern boundary is formed, for the most part, by the 29th parallel, and then by the Macintyre and Dumaresq rivers and the Macpherson Range to the coast at Point Danger; while the southern boundary is formed by the Murray, from its intersection with the 141st meridian—the converging point of the three colonies of New South Wales, South Australia, and Victoria—to its source, and thence by an imaginary line to Cape Howe.

COASTS: Though there are no extensive indentations such as those on the coast of South Australia, the bold and, on the whole, regular coast-line of New South Wales is broken up by numerous bays and inlets.

Many of these bays and inlets along the coast afford ample shelter and safe anchorage, and some of them, such as **Port Jackson**, on which Sydney is situated, **Broken Bay**, the estuary of the Hawkesbury River, 20 miles north of Sydney Heads, and **Port Stephens**, still further north, are among the finest natural harbours in the world. Others, such as **Port Hunter**, on whose shores stands the great coal-port of Newcastle, have been rendered secure and commodious by breakwaters and training walls.

The scenery along some parts of the rock-bound coast of New South Wales is very fine—bold headlands alternating with sandy beaches, the line of coast broken here and there by irregular inlets or skirted by coastal lakes. Some of these openings are as picturesque from a scenic, as they are valuable from a commercial, point of view. The magnificent expanse of Port Jackson, especially, with its lake-like scenery, stretching some miles inland, is unrivalled as a harbour, both for beauty and convenience. The high and rocky coast of the Pacific is suddenly broken, and the cliffs form the portal to an estuary of sufficient capacity to shelter all the navies of the world. So completely is the harbour shut in that, until an entrance is fairly effected, its capacity and safety cannot be conjectured. A vessel, making the port, sails, in a few moments, out of the long swell of the ocean into calm, deep water, protected on every side by high lands. The elevated shore is broken into innumerable bays and inlets, some of which form of themselves capacious harbours with a depth of water sufficient to float the largest vessels. Anthony Trollope describes it as "so inexpressibly lovely that it makes a man ask himself whether it would not be worth his while to move his household gods to the eastern coast of Australia, in order that he might look on it as long as he can look at anything."

RELIEF : Physically, the colony presents considerable diversity. A comparatively narrow **coast plain** extends inland to the coast ranges, that generally form the edge of an **elevated tableland** (upon which lie the irregular ranges of the Great Dividing Chain), and which slopes gradually into the **great plains** of the interior.

New South Wales is thus physically divided into three great belts or sections —(1) the Coastlands, (2) the Tablelands, and (3) the Great Plains in the interior.

The **Coastlands** form a charming, diversified, well-watered, but comparatively narrow region, extending along the sea coast from the borders of Victoria to the Queensland frontier, often not more than 35 miles in width and sometimes less, and in other parts widening out to a broad expanse of 120 miles. Generally speaking, these coastlands have a fine fertile soil, and considerable portions of them cannot be surpassed for agricultural purposes by the best farm lands in any country in the world. The coast region has, moreover, an abundant rainfall, and is well-watered by many fine rivers, most of which periodically overflow their banks, but "although these floods undoubtedly do much mischief to farmers, they cover the flat and bottom lands with a fine, rich, alluvial deposit that will produce any kind of crop in abundance, and can be cropped for years without fallowing or manure."

The **Tablelands**, or Mountain Region of New South Wales, rise steeply from the coast plain, frequently presenting almost perpendicular escarpments, and extending westwards to about the 141st meridian, then sloping gently to the vast plains of the interior.

The **Great Plains** of the interior are not to be compared in productivity to the coast plain—there every European cereal and fruit comes to perfection in a rich soil, abundantly watered—here seemingly boundless "downs" are covered during the greater part of the year with rich natural grasses, on which millions of sheep are depastured. East of the mountains, the rainfall is heavier than in England. In these western plains, the rains are too often scanty and irregular, and the rivers, except a few of the larger streams, flow only during the rainy season.

MOUNTAINS : A series of mountain-chains, known as the Great Dividing Range, runs parallel to the coast at a varying distance of from 30 to 100 miles. East of it are smaller mountain-chains—the Coast Ranges. There are also several hill ranges and isolated hills in the interior.

The portion of the Great Dividing Range within New South Wales consists of seven main ranges and numerous smaller hills. These are, from south to north, (1) the Muniong Range in the Australian Alps, the highest points in which—*Mount Kosciusko*, 7,336 feet, and *Mueller's Peak*, 7,268 feet, above the sea—are only about 600 feet below the limit of perpetual snow in that latitude ; (2) the Monaro Range, in which the head of the *Kybean River* has an elevation of 4,010 feet ; (3) the Gourock Range, with *Jindulian*, 4,300 feet ; (4) the Cullarin Range, with *Mount Mundoonen*, 3,000 feet ; (5) the famous Blue Mountains, the highest point of which is *Mount Beemarang*, 4,100 feet; (6) the Liverpool Range, with *Oxley's Peak*, 4,500 feet; and (7) the New England Range, which is traversed by the Northern Railway near its highest point, *Ben Lomond*, 5,000 feet above sea level.

The Coast Ranges extend between the Great Dividing Range and the sea, and mark the seaward edge of the plateau or upland region. The average elevation is inconsiderable, but *Mount Sea View* in the Northern Coast Range attains a height of 6,000 feet. The Illawarra Mountains, though picturesque, are mere hills in elevation, and the Southern Coast Ranges are also low.

The Interior Ranges include the Grey Range, highest point, *Mount Arrow-smith*, 2,000 feet, and the now famous Barrier Ranges, which rise in *Mount Lyell* to about the same height.

There are several isolated mountains and hill groups in various parts of the colony. The loftiest is *Mount Imlay* or Baloon, 2,900 feet.

Scenery : The mountains of New South Wales, though their elevation is not great—the loftiest summits fall far short of the limit of perpetual snow, and the general elevation is inconsiderable compared to the lofty mountains of Eurasia or America—yet cover a wide extent of country and are very irregular and frequently cleft by precipitous valleys, while the scenery in some parts is extremely picturesque. The apparently impassable Blue Mountains afford some of the grandest panoramic views in the world—in many places cliffs, 2,000 feet in height, rise perpendicularly from the valley beneath. The Illa-warra Mountains, on the coast to the south of Sydney, are also famed for their charming scenery, while the deep gorges, luxuriantly wooded valleys, the running streams, and fine waterfalls of the National Park, on the coast about 15 miles south of Sydney, render it a favourite resort.

The caves of New South Wales are among the most striking features of the Australian wonderland. The Jenolan Caves, about 36 miles south-east of Bathurst, are of vast extent and singular beauty. Other caves are famous for their fossil remains—the walls of one of the Wellington Caves being literally studded with the bones of the carnivorous *Thylacinus*, or pouched hyena, long extinct in Australia, but still existing in Tasmania.[1]

RIVERS : The Great Dividing Range forms the main watershed of the colony, and contains the sources of almost all the rivers. The Hawkesbury and other rivers drain the comparatively short eastern slope into the Pacific ; the Darling, the Lachlan, the Murrumbidgee, and other streams drain the long western slope into the Murray, which itself flows along the southern frontier of the colony for over a thousand miles.

The rivers of the eastern slope have comparatively short courses, and are subject to sudden and violent floods. The lower courses of several of them are navigable for steamers of light draught, but their navigation is impeded by the sand bars at the entrances and by the fluctuations in depth. The Hawkesbury or Nepean, 330 miles in length, is the longest.[2] The Hunter is 300 miles long ; the Shoalhaven, 260 miles ; the Clarence, 240 miles ; the Macleay, 190 miles ; the Richmond, 120 miles ; and the Manning, 100 miles. The Hastings, the Clyde, and numerous other streams are under 100 miles in length. All these rivers have a perennial flow, but they vary greatly in volume in summer and

1. In January, 1887, remains of the long extinct Australian lion were also found in these caves, "consisting of several complete jawbones, with the teeth in an excellent state of preservation. Professor Owen is of opinion that the animal was a marsupial lion, fully equal in size to the now existing African species."—*Australian Handbook*.
2. The Hawkesbury, which at different points bears different names, flows through very beautiful scenery, and is justly called the "Australian Rhine."

winter, and the winter floods are sometimes very heavy, and cause enormous damage. The Hawkesbury at Windsor has at times risen 80 feet above its usual level, while the Richmond, on one occasion, rose to the enormous height of 93 feet.

Compared with the principal rivers of England or France, these eastern rivers appear considerable, but compared with the streams that drain the western slope of the colony, they are insignificant in length, if not in volume.

The rivers of the western slope, with the exception of a few 'continental' creeks in the extreme north-west, belong to the basin of the **MURRAY**, which, from its source in the Australian Alps to the border of South Australia, flows between New South Wales and Victoria, receiving from the latter several considerable affluents, but none equalling the Darling, the Murrumbidgee, and other great rivers of the western plains of New South Wales. The most westerly of these, the Darling, joins the Murray at Wentworth after a Nile-like course of over a thousand miles, "feeding the thirsty plains of the south with water falling many hundreds of miles distant on the downs of Queensland." The Lachlan, 700 miles in length, is the chief tributary of the Murrumbidgee, the second great affluent of the Murray. The Murray and the Murrumbidgee rise not far from each other in the Muniong Range of the Australian Alps : the Murray, before it enters Lake Alexandrina, attains a length of 1,300 miles ; the Murrumbidgee has a course of 1,350 miles before it falls into the Murray.

The facilities for water carriage, afforded by the Murray and its tributaries, are utilized at certain seasons of the year—small steamers then ascend the Murray as far as *Albury*, the Murrumbidgee to *Wagga Wagga*, and the Darling to *Bourke*, and, in time of freshets, to considerably above these places. The navigable portions of the Murray and its tributaries in the western division of New South Wales alone measure no less than 4,200 miles, but the principal value of these streams in the future will be for irrigation. The water in all of them is highly charged with sediment, and the soil along their banks is on the whole fairly, and in some places extremely, fertile, requiring moisture alone to bring forth abundant crops of grain and fruit.

The eastern rivers, although their volume fluctuates with the seasons, are perennial, but the western rivers, with the exception of the Murray and the Murrumbidgee, in dry weather shrink up into chains of water-holes. Even when not so completely dried up, there is then scarcely any current, owing to the "very gentle fall, the long course which is traversed, and the rapid evaporation in that hungry soil and under that ardent sun." In winter, however, all the streams of the great western plains resume their flow, and are not unfrequently very heavily flooded.

LAKES : The mountain and coastal lakes of New South Wales are few in number and small in size. The western lakes, which lie about 25 miles south-west of Goulburn, are, in times of great drought, absolutely dry.

The largest mountain lakes are **Lake George,** which varies in size, being on an average about 25 miles long and 8 miles broad, and **Lake Bathurst,** about 10 miles east of Lake George. As they have no outlets, the waters of both are saline.

The largest coastal lakes are **Lake Macquarie,** about 12 miles south of Newcastle, 20 miles long and 3 miles wide ; **Lake Illawarra,** about 50 miles south of

Sydney, 9 miles long and 3 miles broad; and **Brisbane Water,** an inlet of Broken Bay. These and other coastal lakes are formed by and connected with the sea, and are, strictly speaking, inlets rather than lakes.

The western lakes are shallow depressions filled by the overflow from the rivers when in flood, and emptied again when the floods subside. They vary, therefore, in size, and are not unfrequently absolutely dry.[1]

CLIMATE : The climate of New South Wales is, on the whole, warm and dry, and everywhere extremely healthy.

But the country is so large, and, physically, so diversified, that nearly every variety of climate, from the coldest in the British Isles to the warmest and most genial in the south of Europe, is experienced in different parts of it, and there is, in fact, an ample choice of climates, dry or moist, hot or cold. Sydney has a mean temperature of 63° F. in the shade, and there the temperature is rarely below 40°, but in the dry inland plains the heat sometimes rises to 130° in the shade, and is for the greater part of the summer over 100°, while at Kiandra, in the Australian Alps, the mean annual temperature is only 46°, and frost, snow, and hail prevail during the winter months.[2]

For eight months of the year, says Dr. Lang, namely, from March to November, the Australian climate is particularly delightful. The sky is seldom clouded, and day after day for weeks together the sun looks down in unveiled beauty from the heavens. The great heat of summer along the coast is moderated by the sea breeze, which blows regularly from nine in the morning until the evening, when the land breeze from the mountains sets in, while the winter is remarkably mild. The intense heat of summer in the interior is rendered bearable by the dryness of the air. Winter on the plains is refreshing and enjoyable.

The rainfall is ample and sometimes excessive on the coast plain, moderate on the high lands, and scanty on the great western plains. It averages from 50 inches at Sydney to 20 inches on the uplands and less on the plains. Much stock has been lost through want of water, but, even in the interior, the increasing number of reservoirs and wells makes the dry season and droughts less feared than they were.

PRODUCTIONS : The *indigenous trees, shrubs,* and *plants,*[3] are of less value than the cultivated cereals, fruits, and vegetables ; the *native animals* are absolutely valueless compared with the domestic animals now reared in millions on the rich pastures ; while the mineral resources of the colony are practically inexhaustible.

The flora of New South Wales, as of the adjoining colonies, is distinguished by the remarkable number of flowering plants, indigenous trees of great economic value, and native grasses, which form excellent pastures. No country has been favoured by nature with a greater variety and abundance of trees, yielding strong, beautiful, and durable timbers, and the rare luxuriance of the flowering plants is a convincing proof of the innate fertility of the soil. "Open forests" extend over the greater part of the colony, except in the basins of the Monaro, the Lachlan, and the Murrumbidgee, where the wide treeless plains are covered

1. The mountain and coastal lakes of the colony are much admired on account of their picturesque surroundings. Lake Illawarra and Brisbane Water are most charming expanses, while on either side of Lake George are gigantic towering mountains, rising in grassy slopes from the water's edge.

2. Slight frosts are experienced in the winter nights even at Sydney, but there has been no fall of snow since 1836.

3. Except, of course, the native grasses and shrubs, such as the salt bush, which form the natural pasturage for so many millions of sheep.

with salt bush, scrub, or natural grasses. Various species of *eucalypti*, or gum trees, and *acacias* or wattles, predominate in the open forests, and the red and white *cedar*, the *native beech* and *pine*, *tree-ferns*, *palms*, and *fig trees* abound in the "brush forests" along the coast. The "scrub forests" of the plains include some trees of great beauty, but the close-set stunted timber of the "Mallee" districts is said to impress the traveller even more unfavourably than would a barren waste.

Among the native animals, the most noteworthy are the *kangaroo*, the *wallaby*, *bandicoot*, *wombat*, *opossum*, *native bear*, *native cat*, the curious *native hedgehog* (echidna), and the extraordinarily singular *duck-billed platypus*. Bats of all sizes, from the large flying-fox to the tiny flying-mouse, abound. *Lizards* and *snakes* are numerous, and many of the latter are dangerous and their bite is sometimes fatal.

The birds are far more numerous than the animals, and some of them are equally as strange and interesting. They include over 60 species of *parrots* all with extremely brilliant plumage, and numerous *eagles* and *owls*, while the strange hoot of the *laughing jackass*, or great kingfisher, and the musical cry of the *magpie* and the *mocking bird* resound at early morning through the bush. Vast numbers of beautiful *pigeons* and exquisite little *doves*, gaudy *parrots*, *cockatoos*, and *love-birds*, fill the great primeval forests of the coast with life, but the *emu*, or Australian ostrich, is now found only in the solitudes of the Western Plains. Among aquatic birds are the crane-like *native companion*, the graceful *black swan*, and an enormous number of all kinds of water-fowl. The troublesome *mosquito* and the noisy *locust* and other insects are incredibly numerous, while fish swarm in the rivers and seas of the colony. *Edible fish*, both salt and fresh water, are always obtainable, and there are large *oyster* beds in the Clarence and other rivers.

The mineral wealth of the colony is very large, and includes rich deposits of the precious as well as of the more useful metals, and practically inexhaustible supplies of the most useful and essential of all minerals—coal. The auriferous area is very large; the silver-lead deposits are very rich, while the coalfields are the richest, most accessible and extensive in the Southern Hemisphere, and must ultimately make New South Wales the wealthiest and most important of all the Australian colonies.

INDUSTRIES : More than *one-fourth* of the adult male population of New South Wales are engaged in agriculture, *one-fourteenth* in pastoral pursuits, *one-eighth* in manufactures, *one-thirteenth* in mining, and *one-twelfth* in trade and commerce; the rest are chiefly engaged in professional pursuits, in transport by land and sea, or as skilled or unskilled labourers in building or construction.

THE PASTORAL INDUSTRY, the first in order of time, is still the first in order of value, and, though the number of workers directly employed in this industry is comparatively small, no less than 75 per cent. of the domestic exports of the colony consists of pastoral products, principally of wool—the annual clip amounting to about 270 million lbs. In fact, the staple product of the colony is wool, and scarcely any other country in the world has such resources for the development of this industry, or can show such progress in pastoral enterprise as New South Wales. Natural pastures exist in all parts, but especially in the Western Plains, where, in spite of the extreme dryness of the climate—which, however, so long as the water supply does not fail, is beneficial to stock rather than otherwise—many varieties of the best fattening grasses, herbage, and salt bush flourish. Many men have risen to wealth and affluence from the position of shepherds, and squatters have not infrequently become landed proprietors on a large scale, with " more than a hundred thousand sheep depastured on their own freehold estates." There are at present about 57 million sheep in the colony —a far larger number than in any other Australasian colony, and equal, in fact, to one-half the total number of sheep depastured in the whole of Australasia— and the number is increasing, though not so fast as formerly. In 1860, there were 6 million sheep ; in 1870, 16 millions ; in 1880, 35 millions ; and in 1890, 56 millions. The losses from drought and failure of grass are, in some years, very heavy—5½ millions were thus lost in 1877, and in 1884 it reached the enormous total of 8 millions. But though the number of sheep is still increasing, the number of cattle has decreased from 3 millions, in 1875, to 2 millions in 1894. The number of horses, also, although not actually decreasing, shows a very small increase, and does not now exceed half-a-million.[1]

AGRICULTURE : Although the variety of soils and climate is so great that almost any crop—" whether specially the produce of temperate and even cold climates, or of sub-tropical regions "—may be grown, only about one-half per cent. of the area of the colony is under cultivation.

Only one-fortieth part of the colony is absolutely unfit for occupation—the rest is suitable for settlement and capable of cultivation to some extent, and to a very large extent, if the rainfall were more regular and abundant.

The principal crops grown at present are wheat, maize, oats, barley, potatoes, sugar-cane, tobacco, grapes, oranges, and other fruits.[2] But *arrowroot, cocoa, coffee, raisins* and *currants, hops, mustard, olives, rice, spices, tea, vanilla,* and other articles in constant demand, could be successfully grown in various parts of the colony.

Some of the richest soil in the world is to be found in the periodically inundated river valleys along the coast. Maize and sugar are the chief crops on the northern rivers ; in the southern coast districts, the land is devoted chiefly to dairy-farming. The soil and climate are almost everywhere suitable to the growth of fruit-trees, both European and semi-tropical. The Parramatta Valley is specially famed for its oranges, and excellent grapes for table use and suitable for wine-making are grown in all parts of the colony, but principally along the coast and in the Riverine district of the Murray in the interior—the wines of the Albury district being specially famous throughout Australia.

There is much good agricultural land, well adapted for the cultivation of wheat, barley, oats, and other products of cold and temperate regions, in the central division of the colony, which embraces the high lands on either side of the Great Dividing Range.

In the western division, the alluvial deposits along the course of the Murray and its tributaries are extremely fertile, and, were the rainfall more regular and abundant or artificial irrigation more general, immense quantities of wheat might be grown, and the soil and climate are especially favourable for the growth of the vine. As it is, the cultivation of wheat is making great progress

---

1. "*Pastoral property and stock* form the largest factor in the wealth, not only of New South Wales, but also of all the other principal colonies of Australasia, and the return derived therefrom is the largest source of the income of its inhabitants. The total capital value of pastoral property, including land, improvements, and plant, as well as | stock, was estimated at the beginning of 1890 at £417,000,000, and of this large sum £167,000,000, or 40 per cent. belongs to New South Wales."—Coghlan's *Wealth and Progress of New South Wales.*

2. In 1893-4, nearly 15 million bushels of wheat, maize, and other grains were produced.

in the valleys of the Murray, Murrumbidgee, and Lachlan, but most of the western districts are devoted to pastoral pursuits. New South Wales is still essentially a pastoral country, and the cultivation of the soil has not yet become so important an industry as sheep-farming.[1]

MINING : Mining, especially for coal, silver, and gold, is an important industry, and about one-ninth of the population is dependent upon the yield of the mines.

Gold-mining was only for a brief period the leading industry ; the discovery of gold in 1851, however, gave an impetus to the colony by attracting a large population, but the yield of gold decreased from 2½ millions in 1852 to 1½ millions in 1872, and is now about half-a-million a year.

Silver-mining has, since the discovery of the rich silver fields of the *Barrier Ranges* and the *Broken Hill* districts in 1883, become of great importance, and the annual export of silver and silver-lead ore amounts to nearly 3 millions sterling.[2] BROKEN HILL and SILVERTON, the centres of the silver region, are connected by rail with Port Pirie and Adelaide.

Tin and copper are also largely mined, but the most important of all the mineral products of the colony is coal, of which the annual production is now over 3½ million tons. For hundreds of miles the coast districts of New South Wales may be said to be one vast coalfield, and in some places the seams are found associated with immense quantities of iron ore, and thus the colony will naturally become the great seat of the iron industry of Australia. Diamonds and other precious stones have been found in many parts of the country, and marble, building stones, fire-clays, and slates are abundantly distributed.

MANUFACTURES of various kinds employ about 50,000 people, and support about one-sixth of the whole population of the colony. Most of the factories are in Sydney and its suburbs. The most important are wool-washing establishments, metal-works, printing-works, &c.

COMMERCE : The foreign trade of New South Wales is larger than that of any of the other colonies of Australasia, both in absolute amount and in proportion to the population. It is, in fact, exceeded in value by that of no other British colony or dependency, India and Canada alone excepted. About 88 per cent. of the total trade is carried on with Great Britain and the other Australasian Colonies and the British Possessions in other parts of the world— the rest is chiefly with the *United States, Belgium, France, Germany,* and *China* (Hong-Kong).

The trade of New South Wales has increased from less than half-a-million in 1825, and 20 millions in 1865, to 41 millions in 1893. The advent of the French *Messageries Maritimes* and the *North German Lloyd's* steamships has increased the direct trade with France and Germany, but 88 per cent. of the shipping is still under the British flag, and while the tonnage of British vessels —considerably more than half of which is owned in Australia—trading to New South Wales amounts to 4½ million tons, that of all the foreign vessels entered and cleared is only about half-a-million tons.

---

1. "It is in the valleys of the Murray, the Murrumbidgee, and the Lachlan that the struggle between the squatter and selector has been fier est, and there the most remarkable increase of agricultural settlement, both as regards wheat-growing and agriculture generally, is now exhibited."— *Coghlan.*   2. Total value of the precious metals up to 1893: Gold, £19,853,952; silver and silver-lead ore, £16,692,435.

IMPORTS : The imports consist principally of *articles of food and drink, clothing, textile fabrics, iron and metal goods*, and other *manufactured articles, sugar, tea, &c.* Annual value, 18 millions sterling.

EXPORTS : The main articles of export are *wool, gold* (in coin), *coal, live stock, tin, silver, copper, skins*, and *tallow.* Annual value, 23 millions sterling.

WOOL is the staple export of the colony, and amounts in value to about one-half the value of the total exports, and with other animal products—live stock, meat, tallow, &c.—forms nearly three-fourths of the exports of home produce. Several million lbs. of wool exported from New South Wales ports are not the produce of the colony, but are sent through chiefly from Queensland ; and, similarly, much of the wool and other produce of New South Wales is exported, not through Sydney, but through Melbourne and South Australian ports.

Coal is also one of the staple exports. About one-third of the total quantity raised—over 3½ million tons—is consumed in the colony, the rest is sent principally to Victoria, the United States, New Zealand, Chili, South Australia, Hong-Kong, Tasmania, India, Java, Singapore, &c.

The export of silver and silver-lead ore has become important since 1884, and now amounts in value to 2½ millions sterling a year. The exports of copper and tin are now much less than formerly, but, since the opening of the mines, over 6 million pounds' worth of tin and 3½ million pounds' worth of copper have been produced and exported from the colony. There is also a very large export of gold bullion and specie, but the bulk of this is the produce of other colonies, sent to Sydney to be minted.

The re-export trade by sea is very large ; most of the articles re-exported are sent from New Zealand, Queensland, and the South Seas, for transhipment to Europe.

The overland trade with the adjoining colonies is also very large, amounting to between 9 and 10 millions a year. Nearly three-fourths of this is with Victoria. Much of the trade with South Australia depends on the state of the rivers, but the exports to both these colonies are about double the value of the imports from them.

Nearly nine-tenths of the foreign trade of the colony is carried on with *Great Britain* and the *British Possessions*, and the direct trade with Great Britain alone is even larger than the intercolonial trade.

Nominally, the British trade is less than the inter-colonial trade—the imports from Great Britain, in 1893, amounting to nearly 5 millions, and the exports to 9¼ millions ; while the intercolonial trade, in the same year, amounted to 19 millions—imports, 9 millions, exports, 10 millions. But large quantities of British goods, destined for New South Wales, pass through the neighbouring colonies, and *vice versa*, so that the actual value of the British trade is larger than the above amounts indicate.

Of the intercolonial trade of New South Wales, more than half of that with *Victoria* is overland and the rest by sea ; with *Queensland*, it is mainly gold (bars and dust) and stock ; with *South Australia*, the principal imports are flour and copper, and the chief exports, silver-lead, and wool—the greater part of this trade is overland ; to *Tasmania,* the imports are mainly potatoes, tin, and green fruit, while coal is the chief export ; to *New Zealand,* wheat and flour are the principal imports, and gold coin and coal the largest exports.

The trade with other British Possessions is not considerable, except with *Fiji, Hong-Kong,* and *India.*

The trade with countries outside the empire has increased in recent years, and now amounts to over 5½ millions sterling, that with the *United States* being by far the largest.   The rest is mainly carried on with *Belgium, France* and *New Caledonia, Germany,* and *China.*

TARIFF : The tariff in force in New South Wales is more simple in character than that of any other important country except Great Britain.

The **Customs Revenue**, which amounted, in 1893, to 2¼ millions sterling, is derived chiefly from fermented and spirituous liquors, tobacco, tea, and sugar.

PORTS : The chief ports are **Sydney, Newcastle, Wollongong,** and **Eden** on Twofold Bay.

The bulk of the foreign trade of the colony passes through SYDNEY, which ranks next to Melbourne among Australian ports, and is, in absolute tonnage, surpassed by five only of the English ports—London, Liverpool, Cardiff, Newcastle, and Hull—while, in point of value, its trade exceeds that of any port in Great Britain, except London, Liverpool, and Hull.   NEWCASTLE, the great coal port, ranks second among New South Wales ports, and for many years previous to 1880 its trade exceeded that of Sydney.   WOLLONGONG has also a considerable coal trade ; and EDEN, on Twofold Bay, formerly an important whaling port, does a large trade in agricultural produce and live stock.

COMMUNICATIONS : Internal communication is facilitated by a network of over 24,000 miles of **roads** and over 3,000 miles of **railways,**[1] together with several thousand miles of **navigable waterways.**

The roads, 7,600 miles of which wind their way through the forests of the interior, chiefly along the lines marked out by the cart-wheels of the earliest settlers, still remain the sole means of communication throughout a large part of the interior, although in rainy weather many of them are almost impassable. As arteries of trade, however, the great Northern, Western, Southern, and South Coast Roads have been superseded by **railways,** which, commenced by a private company, were taken in hand by the colonial government in 1855. But for a long time after the first settlement of the colony, the precipitous cliffs of the Blue Mountains barred all access into the interior, and the men who first attempted to force their way across the mountains had to creep through dense forests and scale tremendous precipices, before they reached the vast plains lying west of the great eastern cordillera.   Shortly after, the "Great Western Road" was completed to Bathurst, and the subsequent discovery of gold gave an impetus not only to road-making, but railways were extended across the mountains[2] and rendered the interior plains accessible, north, south, and west.

1. 2,500 miles open for traffic and 1,000 miles in course of construction.
2. " The remarkable triumphs of engineering skill, known as the *Zigzag* and the *Great Zigzag,* by which the lofty heights of the Blue Mountains are ascended and descended, form two of the leading sights of the colony."

With the exception of two private lines,[1] the railways of New South Wales are the property of the State, and over 38½ millions sterling have been spent on their construction.

The railways of the Colony are divided into the *Southern, Western,* and *Northern* systems, all three radiating from Sydney. The Southern Railway extends from SYDNEY to ALBURY on the Murray, where it connects with the Victorian line to *Melbourne* (387 miles). The Western Railway runs from SYDNEY across the Blue Mountains, and by BATHURST and WELLINGTON to BOURKE on the Darling, the chief town of Central Australia. The Northern Railway now runs from SYDNEY to NEWCASTLE, and thence *via* TENTERFIELD across the frontier to *Wallangarra*, where it connects with the railway to *Brisbane*. The North-Western Railway branches from the Northern line at WERRIS CREEK, and runs across the Liverpool Plains to NARRABRI. There are several other minor lines or branches, such as (1) from SYDNEY through the Illawarra district, (2) from GOULBURN on the Southern Railway to COOMA, (3) from JUNEE JUNCTION on the same railway to HAY on the Murrumbidgee, in the Riverina district ; &c.

"The Southern main line ' is the most important of the railway lines in the colony, as it passes through the richest and most populated districts, and, with the connecting lines, places the four great capitals of Australia—Brisbane, Sydney, Melbourne, and Adelaide—in direct communication with each other."

The postal and telegraphic services are most complete and effective. Over 77 million letters and nearly 3 million telegrams are annually conveyed. The telephonic system is rapidly extending in Sydney and the leading county towns.

External Communication : New South Wales is in easy and constant communication with all parts of the world. Four great lines of ocean steamships maintain regular communication between Sydney and other Australian ports and Europe. These are (1) the Peninsular and Oriental Steam Navigation Company, (2) the Orient Line, (3) the French Messageries Maritimes, and (4) the North German Lloyd's. By the Orient and P. & O. lines the passage from England to Sydney occupies less than six weeks. Other steamers and sailing vessels are constantly entering or clearing New South Wales ports for all parts of the world, but principally from or for Great Britain and the other colonies of Australasia.

GOVERNMENT : The government of New South Wales is vested in a Governor, who represents the Queen, and a Parliament of two Houses—the Legislative Council and the Legislative Assembly.

"Responsible government" was conceded in 1855—the executive power being entrusted to the Governor and a Cabinet or Council of Ministers, while the legislative power was vested in a nominated Legislative Council and an elected Legislative Assembly.

The Revenue, in 1893, amounted to 9¼ millions sterling, and the Expenditure to 10¼ millions, while the Public Debt was nearly 58 millions, or £47

---

1. The two private lines are (1) from *Deniliquin* in the Riverina to *Moama* on the Murray, opposite Echuca in Victoria, a distance of 45 miles, and (2) from *Broken Hill* and *Silverton*, the centres of the Barrier Ranges silver mining district, to the *South Australian border*, where it connects with the lines to Port Pirie and Adelaide.

2. This line was not actually open for traffic until 1883. The traffic is very heavy, and the express trains traverse the distance between Melbourne and Sydney (576 miles) in 18 hours, and ordinary passenger trains in 24½ hours.

millions, or £47 per head of the population.[1]   The Revenue is mainly derived from *taxation, crown lands, railways,* and other services.   40 per cent. of the expenditure is on account of public works, and 20 per cent. on account of the Public Debt.   For the defence of the colony there is a land force of 9,000 men, 7,500 of whom are volunteers and 600 the regular military force.   There is also a naval force of over 600 men.   The British ships of war on the Australian station are to be strengthened by the addition of extra vessels, and the cost of maintenance and 5 per cent. on the cost of construction of the new vessels, are to be borne, *pro rata,* by the Australian colonies.   Strong fortifications have been erected at various points on the coast.   Port Jackson is practically impregnable, and the great coal port of Newcastle is also strongly fortified.

PUBLIC WEALTH : The public and private wealth of the colony is estimated at over 503 millions sterling, or £485 per head of the population.   The public and municipal estate is valued at about 198 millions, or £154 per head, while the private property in the colony amounts to 404 millions, or £330 per head.

EDUCATION : Education is under State control, and is compulsory, and free for poor children.   There are not only primary schools, but also high schools for both boys and girls, and a large number of private schools and colleges.

The University of Sydney, founded in 1858, is well endowed, and its degrees in arts, law, and medicine are recognised as equal to those granted by British universities.   A State system of technical education is in operation throughout the colony.   A central Technical College was opened at Ultimo in 1892, and there are branch schools and classes at the principal centres.[2]

DIVISIONS and TOWNS : The colony is divided into 13 Pastoral Districts, and also into 141 Counties.   The " Old Counties " near the coast are more clearly defined and more densely peopled than the " New Counties " in the interior.   The chief towns are *Sydney* and *Newcastle,* on the coast ; *Bathurst,* on the tableland ; and *Broken Hill, Bourke,* and *Albury,* in the interior.

SYDNEY (400), the capital of New South Wales and the oldest city in all Australia, is most picturesquely situated on the southern shore of Port Jackson —a magnificent natural harbour, absolutely unrivalled for convenience of entrance, depth of water, and facilities for shipping.   The strongly fortified entrance to Port Jackson from the Pacific lies between perpendicular cliffs of sandstone several hundred feet high, and is only a mile in width.   The water area of the harbour is 15 square miles, but it is so broken by numerous promontories, that the shore line measures no less than 165 miles in circuit.   The depth is so great that the largest vessels can lie close inshore, and the water frontage of Sydney Cove[3] and the eastern side of Darling Harbour are skirted by an almost unbroken line of wharves and quays.   The foreign commerce of Sydney is very extensive, and is surpassed among Australasian ports only by that of Melbourne, while, in the value of its trade, it is exceeded only by London, Liverpool, and Hull among British ports.   Regular communication with Europe is

1. *Cf.* With the debt per head in Queensland, £71; in South Australia, £63; in New Zealand, nearly £57; in Tasmania, £52; in Victoria, £39; in Western Australia, £30; an average for all Australasia of £52 per head.
2. *The Australian Hand-Book,* 1895.

3. Circular Quay, at the head of Sydney Cove, has a length of 1,300 feet available for the largest vessels.   Here the P. & O. and other ocean steamers dis charge ; the smaller intercolonial and trading steamers generally berth at the Darling Harbour wharves.

maintained by the magnificent steamships of the *Peninsular & Oriental Co.*, the *Orient Line*, the French *Messageries Maritimes*, and the *North German Lloyd's*.

Sydney is less regular and more picturesque than most other new cities in Australia. The streets occasionally converge and bend and wind about so as to give intricacy and variety. The presence of the sea in numerous bays and coves, the jutting promontories, and the beautiful gardens, add further to its variety and beauty. The walks immediately around the city are unsurpassed for picturesqueness, while the public gardens probably excel any in the world, owing to their combination of sea and land, hill and valley, rock and wood, and grassy slope, with a climate that permits all the beautiful forms of vegetation, both of tropical and temperate zones, to luxuriate side by side. About 14 miles west of Sydney is Parramatta (12), on the Parramatta River, really an extension of Port Jackson. Parramatta, now famous for its orangeries and fruit gardens, was founded in November. 1788, and is therefore, next to Sydney, the oldest town in the colony.

The chief towns on the coast, besides Sydney, are *Newcastle*, the great port of the northern coalfields, *Wollongong*, the port of the southern coalfield, *Ballina*, at the mouth of the Richmond River, and *Eden*, on Twofold Bay.

Newcastle (53), at the mouth of the Hunter, is the second largest town in the colony, but, as the chief emporium of the coal trade and a shipping port for the wool of the northern districts, its tonnage sometimes rises above that of Sydney. The harbour is protected by a breakwater.[1] Wollongong (8), the third seaport of the colony and the principal port on the south coast, lies about 50 miles south of Sydney. Much coal and dairy produce are shipped here. Ballina (2), at the mouth of the Richmond, is the chief timber port of the colony. Eden, on the splendid natural harbour of Twofold Bay, which is second only to Port Jackson in size and security, is the outlet of the rich Monaro district.

The chief towns on the tableland are *Bathurst*, on the Macquarie River, *Tamworth*, *Mudgee*, and *Goulburn*.

Bathurst, the centre of the principal wheat-growing district in the colony, is, with the exception of Broken Hill, the largest town in the colony to the west of the Blue Mountains. There are several gold-fields, silver mines, and slate quarries in the neighbourhood. Tamworth (4) is the principal place on the Liverpool Plains, in the midst of a fine pastoral, agricultural, and mining district. Mudgee (3) is an important gold-mining centre, to the north of Bathurst, and is also noted for its fine wool. Goulburn (12) lies to the east of the mountains, and is the principal centre of the southern inland trade. Much wheat and other valuable agricultural products are grown in the district.

The largest towns on the plains are *Albury*, on the Murray, *Wagga Wagga* and *Hay*, on the Murrumbidgee, *Deniliquin*, in the Riverine district of the Murray, *Bourke* and *Wentworth*, on the Darling, and *Broken Hill* and *Silverton*, in the silver-mining region to the west of the Darling.

---

1. On the same river, 20 miles above Newcastle, is Maitland, the centre of the "Granary of New South Wales." The town and district are subject to, and have been frequently damaged by, great floods, but the alluvial deposits on either side of the river are so productive that they are largely cultivated in spite of the risk. East Maitland is so far protected by flood gates, and West Maitland by embankments.

Albury (3), famous for its wines and as the frontier town where the railway from Sydney is connected with the main line from Melbourne,[1] is situated on the Murray, at the head of navigation, about 150 miles from its source, and about 1,000 miles from the outlet of this great river in Lake Alexandrina. The Great Southern Railway, between Albury and Sydney, also passes through Wagga Wagga (5), a pastoral and agricultural centre, at the head of navigation on the Murrumbidgee. Hay (3), the present terminus of the South-Western Railway, is also on the Murrumbidgee, about 50 miles above its confluence with the Lachlan. Hay is the nearest railway station to Wentworth (2), an important pastoral centre and river-port on the Darling, near its confluence with the Murray. This town, which lies about 500 miles west of Sydney, 400 miles north of Melbourne, and 200 miles north-east of Adelaide, occupies a peculiarly advantageous position, as it will doubtless be the meeting point of the intercolonial railways, and it has also been proposed as the capital of Federal Australia. Bourke (5), on the Darling, 500 miles north-west of Sydney, is the centre of an extensive and wealthy pastoral and copper-mining district, and is at present the terminus of the Great Western Railway. During the rainy season the Darling is navigable to this town. The want of rain is often severely felt, but in 1890 the whole country was inundated, the railway line for miles carried away, and the town flooded. Between the Darling and the South Australian border lies the rich silver-mining district of the Barrier Ranges. Here Broken Hill (25), 800 miles west of Sydney, and 300 miles north-east of Adelaide, and Silverton, 17 miles north-west of Broken Hill, are the chief centres.[2] Both are connected by rail with Port Pirie and Adelaide. The nearest New South Wales railways are at Bourke, 300 miles to the north-east, and Hay, 280 miles to the south-east. Deniliquin (2), on a branch of the Murray, is the principal place in the Riverine district of the Murray.[3] At Kiandra, in the Australian Alps, at an elevation of 4,600 feet above the sea, snow often lies for two months in the winter, and the thermometer sometimes registers 1° F. below zero.

∴ All the Australian colonies may be said to be even yet new countries, and their history is as of yesterday. The earlier history of New South Wales—the pioneer State of Australasia—is virtually that of the whole island, but from the 20th January, 1788, the day when Captain Phillip landed at Botany Bay with a troop of convicts, New South Wales has three distinct eras in its history. The first era is the *penal* age, when the colony was altogether dependent for existence upon England, and during which the rejected elements of society in the mother-country were poured into the settlement; the second, or *pastoral* age, begins with the opening of the colony to emigrants, by whose aid it soon became self-supporting; and the third is its *golden* age, dating from the impetus given to its prosperity by the discovery of gold, and rendered permanent by the subsequent discovery and development of the more useful mineral—coal. We have no space here even to note a few of the landmarks in the history of the colony. Its early annals are by no means pleasant reading. Disputes with the blacks, lawlessness of freed and escaped convicts, bloodshed and discord, food supplies often at starvation-point, and the disinclination of English emigrants to settle among felons, produced a state of things from which it then seemed barely possible for it ever to escape. But an increasing number of immigrants, a few intrepid explorers, and a succession of wise and resolute governors, in time overcame all difficulties, and the colony at last embarked on a career of steady progress, interrupted occasionally by severe droughts, but never checked, so that, at the present time, New South Wales may fairly claim a high place among the most enlightened and progressive countries in the world.

---

1. The Victorian and New South Wales lines have a different gauge, on which account the Victorian trains run right through to Albury, which is the changing station both on the up and down journey.

2. One mine in the Broken Hill district is the largest silver mine in the world, not excepting even the celebrated Comstock Mine in Nevada, U. S. A.

This mine is worked day and night in three shifts, with its smelters always going, and employing over 2,000 hands.

3. Deniliquin is connected with the Victorian railways at Echuca by a line, 45 miles in length, constructed by a private company.

VICTORIA.

The "Gold Colony" of VICTORIA, called by the early explorers Australia Felix, from its beauty and fertility, is the smallest in area, but the foremost in wealth and enterprise, of the Australian Colonies. It forms the south-eastern portion of the continent, of which it occupies one thirty-fourth part.

BOUNDARIES: Victoria is bounded on the *north* and *north-east* by New South Wales ; on the *west* by South Australia ; on the *south* by the Southern Ocean and Bass Strait ; and on the *south-east* by the Pacific Ocean.

Victoria is separated from New South Wales (of which, from 1835 to 1851, it formed a part under the name of Port Phillip) by the River Murray and an imaginary line running in a south-easterly direction from its source in Forest Hill in the Australian Alps to Cape Howe, while the 141° East longitude divides it from South Australia, and Bass Strait from Tasmania.

EXTENT : The greatest length of the colony, from east to west, is about 420 miles, the greatest breadth about 250 miles, while its area of nearly 88,000 square miles is slightly less than that of Great Britain.

The exact area of Victoria is 87,884 square miles, and is therefore only a little more than *one-third* the size of New South Wales, *one-seventh* of Queensland, *one-tenth* of South Australia, and *one-eleventh* of Western Australia. Next to Tasmania and Fiji, it is the smallest in area of all the Australasian colonies.

But although Victoria is the smallest in area, it is the most densely peopled of all the Australian colonies. The actual populations of New South Wales and Victoria are nearly the same—both being over a million—but the density of population in the smaller colony is, of course, much greater than in the larger one, being, in Victoria, about 13, and, in New South Wales, about 4 to the square mile.

The people of Victoria are, with the exception of a few thousand Chinese and foreigners, and a few hundred aboriginal 'blacks' and half-castes, entirely British or of British origin. Considerably more than half the people were born in the colony—nearly all the rest are immigrants from England, Ireland, and Scotland.

COASTS : The coast line of Victoria is about 700 miles in length, and is indented by several inlets, the largest and most important of which is Port Phillip Bay.

The principal headlands are Cape Howe, the easternmost point of Victoria ; Wilson Promontory, the southernmost point of Victoria as well as of the whole continent ; Cape Schanck and Point Nepean to the *east* of Port Phillip Bay ; and Point Lonsdale, Cape Otway, and Cape Bridgewater to the *west* of Port Phillip Bay. Point Lonsdale and Point Nepean form the "heads"[1] at the entrance to Port Phillip Bay. Both are strongly fortified and stand about 2 miles apart, but the navigable channel between them is less than a mile in width.

---

1. The entrance to Port Phillip Bay is defended by strong batteries on the "heads," and a fort is being built between Point Nepean and Queenscliff. where the men of the Victorian Permanent Artillery are stationed

Port Phillip Bay is about 40 miles in length and the same in width.   Hobson's Bay, at its head, forms the harbour of Melbourne; Corio Bay, on its western side, is the harbour of Geelong.   Port Phillip Bay is easily navigable by, and affords safe anchorage to, the largest vessels.   It has an area of no less than 800 square miles, but its entrance is narrow—the distance between the "heads" of Point Lonsdale on the west and Point Nepean on the east being about 4,000 yards, but the navigable channel, called *the Rip*, through which the tide rushes with great velocity, is only about 1,600 yards wide.

Other considerable inlets are *Western Port*, *Waratah Bay*, and *Corner Inlet*, to the east of Port Phillip Bay; and *Portland Bay* and *Discovery Bay* to the west.

The largest islands are *French* and *Phillip Islands*, in Western Port Bay on the south coast, and *Snake Island*, off Corner Inlet on the south-east coast.

MOUNTAINS: A chain of mountains, the southern part of the Great Dividing Range of Eastern Australia, extends across the colony from east to west, at a distance of from 60 to 70 miles from the coast.

The eastern part of the Cordillera is known as the **Australian Alps**, and the range terminates to the west in the **Pyrenees** and the **Grampians**.

In these ranges, about 30 peaks rise over 4,000 feet, one-half of them over 5,000, and at least six over 6,000 feet—the loftiest, *Mount Bogong*, is 6,508 feet above the sea[1]—but the average elevation of the Victorian mountains is only about 3,000 feet, and some of the lowest gaps are little over 1,000 feet above sea level.

These long and irregular ranges divide Victoria into two, or rather three, well marked regions—(1) the **coast region**, from Cape Howe on the east to the mouth of the Glenelg near the western boundary, an agreeably diversified, comparatively well-watered, and certainly the most fertile and attractive, portion of the colony; (2) the **broken, undulating country** between the mountains and the Murray River; and (3) the wide, level **Wimmera district** in the north-west, the most sterile portion of the colony, covering an area of some 25,000 square miles, almost exclusively pastoral, scantily watered, and liable to severe droughts.   But much of the soil even here, and especially along the banks of the Murray, is extremely fertile, only requiring sufficient moisture to be rendered most productive.

RIVERS: The Australian Alps and the western ranges form the watershed between the **Murray river system** on the north and the basins of the numerous **coastal streams** on the south.

Owing to the nearness of the mountains to the sea, there is no space for the development of such long rivers as on the coast plain of New South Wales, still, the Snowy River is 300 miles long (only 120 miles of which, however, are within Victoria), the Latrobe 135 miles, the Yarra Yarra 150 miles, the Hopkins 155 miles, and the Glenelg 280 miles.

Into the **Murray**, which forms the northern frontier of the colony for 980 miles, flow the Goulburn, 345 miles in length, the Loddon 225 miles, the Campaspe 150 miles, the Ovens 140 miles, and the Mitta Mitta 175 miles.   The waters of the Avoca, 163 miles long, the Wimmera, 228 miles, in the dry and

1. Other lofty peaks are *Mount Feathertop*, 6,303 | 6,025 feet; and *Mount Pilot*, 6,000 feet. feet; *Mount Hotham*, 6,100 feet; *Mount Cobbras*, |

sandy Wimmera district, rarely reach the Murray, generally terminating in salt lakes or marshes. The Gippsland District, in the south-east, is exceptionally favoured with running streams ; and, with the exception of some of these, and the Yarra Yarra, the Goulburn, and the Murray, none of the Victorian rivers can be said to be navigable.

In winter time, the Victorian rivers, like other Australian streams, are "frequently swollen by the heavy rains into angry torrents, which carry all before them in their resistless course, and, overflowing their banks, devastate the country. Many of the smaller streams, colonially called 'creeks,' and indeed some of the larger rivers in the west, dwindle down into mere threads of water and occasional pools or water-holes, during the summer heat, and sometimes dry up altogether, a circumstance which, when it occurs, is productive of terrible loss to stockowners, cattle and sheep dying in thousands from thirst."[1]

LAKES : There are numerous salt and fresh water lakes, but most of them are shallow, and many are dry during the summer months.

The largest is **Lake Corangamite**, which lies about 50 miles to the west of Geelong and covers nearly 90 square miles, but, as it has no outlet, its waters are salt. **Lake Colac** (10 square miles), a few miles distant, is quite fresh. The Gippsland Lakes—*Wellington* (50 square miles), *Victoria* (45 square miles), *King*, and *Reeves*—are only separated from the sea by a belt of sand, through which there is a narrow entrance, usually navigable by small steamers.

". "The scenery of Victoria is diversified and pleasing. The hills and mountains are mostly clothed with dense forests, and the ranges of the Australian Alps offer much grand mountain scenery. Again, in the west, the Pyrenees and Grampians are very picturesque, and some of the rivers are broken by waterfalls of great beauty. The whole country, from Melbourne westward, is exceedingly rich in soil and varied and beautiful in scenery. Here there is an additional charm in the numerous extinct volcanoes which occur in extraordinary numbers. In many instances the craters are perfectly defined, leaving not the slightest doubt as to their former character. In general they appear as isolated cones, such as Mounts Elephant, Eles, Napier, and others, standing out conspicuously from the surrounding level ; in others, as the Warrior Hills, between the Lakes Colac and Corangamite, they assume the form of a small chain comprising about a dozen volcanic hills. Within and around the craters are strewed rocks of pumice and lava ; and the lower part is often occupied by a small lake, sometimes of fresh water, at others of salt, or nauseous to the taste from the presence of sulphuretted hydrogen. This fine country is also variegated with salt lakes and lagoons, some of which, by their circular form, their peculiar mineral water, and a sort of escarpment round them, have the appearance of craters, although not in the customary form of cones. Luxuriance of vegetation everywhere accompanies the volcanic deposits."—*A. R. Wallace.*

CLIMATE : The climate of Victoria is, on the whole, warm, dry, and distinctly healthy.

Victoria is not so hot as New South Wales or Queensland, still the temperature sometimes rises in January above 100° F. in the shade, and rarely falls in July, the coldest month, below the freezing-point. The average temperature at Melbourne is about 58°, or 8° above that of London ; the rainfall averages 25 or 26 inches, or a little higher than that of London, but the air is so dry and the soil such, that moisture is absorbed much more quickly than in England.

Even with the drawback of occasional hot north winds during the summer, the climate of "Australia Felix" is so genial and healthy, that the death-rate is much lower than in England,[2] and native Victorians consider it to be "the

1. *The Australian Hand-Book.* 1895, p. 220. | 14.11 per 1,000, while the birth-rate in the colony
2. The death-rate in Victoria, in 1893, was only | was 31.23 per 1,000.

finest climate in the world." " Though the summer," one writer remarks, " is invariably marked by a few days of great heat, yet even in that season there are many days when the weather is pleasant and cool, and nothing can exceed the climate experienced in the colony in autumn, winter, and spring. A cloudless sky, a bright sun, and a refreshing breeze are characteristic of the greater number of days in each of those seasons; and while the salubrity of the climate is shown by the absence of those diseases which yearly sweep off so many of the inhabitants of England, it is yet equally favourable to the growth of fruits and vegetables of the colder countries."

PRODUCTIONS : A rich soil, a warm and genial climate, and, in most parts, sufficient moisture, combine to give Victoria a high position in respect of her **vegetable productions** ; the Victorian merino **wool** is unsurpassed in length of staple, softness, and lustre ; while the colony is justly famed for its immense mineral **wealth**.

The indigenous plants and animals are similar to those of New South Wales, and the characteristic gum-trees and kangaroos, emus and black swans, parrots of beautiful plumage, snakes (many of them venomous and too numerous to be at all pleasant), and mosquitoes, are distinctly Australian. But numerous well-known English animals and birds have been introduced, and have taken very kindly to their new home; some of them, such as the rabbit and the sparrow, have multiplied so enormously as to become serious pests to the sheep farmer and fruit grower.

Pastoral Pursuits : Although mining (principally for gold), agriculture, manufactures, and commerce have been developed to a greater extent in Victoria than in any other of the Australian colonies, the **pastoral industry** stands second to none in importance, and **wool** and other animal products and live stock amount to considerably more than half the total value of the exports of the colony.

The genial climate, dry air, and rich natural pastures naturally gave an impulse to pastoral industry, and, soon after its settlement, the country became famous for the excellence of its flocks and herds. All the ordinary domestic animals have been brought to the highest perfection, but the greatest success has been achieved in the rearing of sheep and in the production of wool of the highest quality. The merino wool of Victoria is absolutely unsurpassed, and invariably commands the highest prices in the European markets. " Beside it, all other wools look mean and dull. In its brilliancy and softness it seems to reflect the sunny skies under which it grows. It early became the favourite with European manufacturers, and has ever since maintained its place as the most valuable merino wool in the world."[1]

At the present time over 13 million sheep are depastured in the colony, not, as formerly, over great " runs," each larger than many an English county, but in enclosed paddocks; in fact, almost the whole colony is now subdivided and fenced in, and provision is generally made for feeding and watering the sheep during the seasons of drought, when rivers and streams are either dry or mere chains of pools, and there is scarcely any grass.

Victoria is pre-eminently the " Land of the Golden Fleece." The old Spaniards used to say that " sheep have golden feet—wherever the print of their footstep

---

[1] " The Merino Sheep in Australia," by G. A. Brown.

is seen the land is turned to gold," and this is true of Victoria ; her golden fleeces have produced more wealth than even her fabulously rich gold mines.

Cattle-rearing is by no means so important an industry as sheep-farming, still there are nearly 2 million head of cattle in the colony, a vast increase from 155 head in 1836.

The number of horses is comparatively large—nearly half a million—and Victorians generally are quite at home in the saddle, while horse-racing is extremely popular with all classes, and almost every town, village, and hamlet has its race-course.

AGRICULTURE has made much progress of late years, and is becoming an important industry, not only along the coast, but also in the region north of the Dividing Range, where the rainfall is so scanty and precarious that it has to be supplemented by irrigation.

A large proportion of the soil in Victoria is **exceedingly fertile**, and, where the rainfall is ample or irrigation possible, it yields splendid crops. Nearly everything grown in England grows equally well in Victoria, and very many things that the cold, uncertain climate of England will not allow to come to maturity, thrive in the rich soil and warm climate of Victoria. Wheat of the finest quality, with **oats, barley, maize, root crops, hay,** and **English grasses,** are extensively cultivated, while both soil and climate are peculiarly adapted for the perfect growth of the **vine** and the **olive,** and wine and oil of the highest quality can be produced on a large scale. All the **fruits** of temperate regions, and many distinctively sub-tropical, come to perfection, and **vegetables** of all kinds can be grown all the year round. The productive area of the colony will be very largely increased when the **National Irrigation Works** in the Goulburn, Loddon, Campaspe, Wimmera, and other districts have been completed : these and private enterprises, such as the **Mildura Irrigation Colony,**[1] will transform some of the most arid districts into the most productive portions of the colony. In a warm and dry climate like that of Victoria, land that can be irrigated will support a population ten or twenty times as large as that which it can bear without irrigation.

MINING for **gold** has been by far the most important industry, and it is to its enormously **rich goldfields** that Victoria owes its extraordinarily rapid progress.

Victoria is *par excellence* the **Gold Colony,** and fully one-third of its area is occupied by gold-bearing rocks. The precious metal occurs in **quartz** and **alluvium.** Several of the quartz mines are now worked to a depth of over 2,000 feet—Lansell's, at Bendigo, is 2,640 feet deep—and even alluvial mining requires expensive machinery and a large amount of capital. No fortunes can now be made by individual diggers as in the early days, when gold was indeed "more plentiful than blackberries," and when large nuggets were not infrequently unearthed.[2] All that is passed : corn and wool have long been of more value than gold, and "gold-mining has become no more exciting than coal-mining, and the gold-fields are as quiet and orderly as the sheep runs."

The principal goldfields are in the BALLARAT district, west of Melbourne ; the BEECHWORTH district, in the valley of the Ovens ; the CASTLEMAINE dis-

1. Up to 1895, about £1,000,000 have been expended on this flourishing settlement, and 25,000 acres are now under irrigation.

2. The *Welcome Nugget*, found in 1858, weighed 184 lbs. 9 oz. 10 dwts., and was sold for £10,500.

Some claims at Ballarat, 8 feet square and 8 feet deep, yielded from £10,000 to £12,000, and a miner often made as much as £10,000 to £20,000 in a few months.

trict, around Mount Alexander, in the valley of the Campaspe; the SAND-HURST or Bendigo district, north of Mount Alexander; the MARYBOROUGH district, between the Avoca and the Loddon; the ARARAT district, around Mount Ararat in the Pyrenees, to the west of Ballarat; and the GIPPSLAND district.

Since their discovery in 1851, the goldfields of Victoria have yielded nearly 59 million ozs. of the precious metal, valued at 235 millions sterling, or two-thirds of the gold raised in all Australasia. In 1853, the gold produced reached the enormous amount of 3,150,000 ozs., but in recent years the output has averaged 600,000 to 700,000 ozs. In 1893, it was only 671,126 ozs. The gold-mining population numbers about 26,000, of whom 2,500 are Chinese.

Minerals other than gold are also found—silver, tin, copper, iron, zinc, and some coal, but the total output of these is inconsiderable, and scarcely 500 miners are employed in their extraction.

Manufacturing industry has made much progress in Victoria, but the products are almost entirely for home use.

Victorian manufactures are protected by heavy import duties, and there are now about 3,000 factories and works, employing nearly 45,000 hands.

COMMERCE : Nearly half the trade is with **England**—the rest is principally with the neighbouring colonies. The imports, in spite of heavy duties on most of the important articles, greatly exceed the exports, four-fifths of which consist of *wool, gold, wheat and flour*, and *live animals*. Annual value, 26½ millions sterling— imports and exports each amounting to 13¼ millions.

The principal exports, in order of value, are *wool, gold, live stock, leather, bread stuffs*, and *tallow*.

The chief imports are *wool, live stock*, and *coal*, from the neighbouring colonies; *cotton and woollen goods, metals and metal goods*, principally from the United Kingdom; *tea, sugar, timber*, &c.

The direct trade with the United Kingdom is very large, the imports therefrom amounting to about 6 millions sterling, and the exports thereto to between 3 and 4 millions.

The British imports into Victoria consist chiefly of *iron, machinery, cotton and woollen goods, hardware and cutlery*, and *paper*.

The staple articles of export from Victoria to the United Kingdom are *wool* and *gold*.

PORTS : Melbourne, Geelong, and Portland are the chief ports.

Five-sixths of the imports are received through, and eleven-twelfths of the exports are shipped from, Melbourne. About 5 million pounds' worth of wool are shipped from Melbourne—the greatest wool port in the world—every year, but less than half of this is produced in Victoria, the rest being sent to Melbourne from the other colonies for shipment. Geelong is also an important *wool port*.

COMMUNICATIONS : All the railways in Victoria are the property of the State. Over 3,000 miles are now open for traffic.

The policy of the Victorian Government has been to open up the interior by means of railways, so that railway communication should keep pace with settlement, be the latter ever so rapid. The consequence is that railways extend to the most remote parts of the colony.

The Victorian railways include four systems, all radiating from Melbourne, (1) the Northern System, extending from MELBOURNE to ECHUCA, a distance of 156 miles; [1] (2) the North-Eastern System, from MELBOURNE to WODONGA (187 miles), which is connected by a bridge across the Murray with *Albury*, the terminus of the New South Wales railway to Sydney ; (3) the Eastern System, from MELBOURNE *via* SALE to BAIRNSDALE ; and (4) the Western System, from Melbourne *via* GEELONG, BALLARAT, ARARAT, and STAWELL to SERVICETON, where a connection is made with the South Australian railway to Adelaide. The main lines have each several branches, and there are also Suburban Lines from Melbourne to *Williamstown, Port Melbourne, Sandringham, St. Kilda,* &c. [2]

GOVERNMENT : Victoria possesses responsible government. The Governor is appointed by the Crown, and there are two Houses of Parliament.

The Victorian Parliament consists of a *Legislative Council* of 48 members, and a *Legislative Assembly* of 95 members. The latter are elected by universal suffrage. The Revenue in 1892-3 amounted to 6¾ millions sterling, the Expenditure to 7¼ millions. The Public Debt is over 46¼ millions, nearly the whole of which has been incurred in the construction of railways, water-works, school buildings, and other public works.

For the defence of the colony there is a small but efficient navy, and a considerable military land force.

Education is amply provided for in the numerous State primary schools, and by private schools, technical colleges, and the three colleges affiliated to the University of Melbourne. The instruction given in the State schools is strictly secular, and so far compulsory and free. Practically all the children of school age are under instruction. There are public libraries or institutes in all the larger towns. There is no State Church in Victoria.

DIVISIONS : Victoria is divided into **4 districts** and **37 counties**.

The districts are Gippsland, in the south-east ; the Murray, in the north-east ; Wimmera, in the north-west ; and Loddon, in the north-central part of the colony.

TOWNS : More than one-half of the people of Victoria live in towns. [3] Of the 60 cities, towns, and boroughs, the largest are Melbourne, Ballarat, Bendigo, and Geelong.

MELBOURNE, the capital of Victoria, and the most important commercial centre in the Southern Hemisphere, is situated on the Yarra, not far from the shores of Port Phillip Bay, on which stand its lovely suburbs of *St. Kilda* and *Brighton*, with *Port Melbourne* (formerly called Sandridge), its port, and *Williamstown*, its outport. Melbourne is a well-built and stately city, with splendid

---

1. There is a double line as far as Bendigo.
2. Over 40 million passengers and a million tons of goods are carried yearly on the Victorian railways.

3. "Any aggregation of 300 houses can be constituted a borough. A borough with a revenue of over £10,000 per annum becomes a town, and with a revenue of over £20,000 becomes a city.

public buildings, beautiful parks, and charming public gardens. Including its suburbs, Melbourne now has a population of nearly half a million. Melbourne is a striking example of what enterprise and wealth can accomplish in a short time. Sixty years ago, no white man had trodden the ground on which this vast city now stands, and little did the settlers of 1835—Batman and Fawkner —think, when they built their mud huts on the then solitary banks of the Yarra, and surveyed the immense and desolate meadows around them, that in half a lifetime a colossal city would cover them ; that the dreary spot, which they bought from the blackfellows for two blankets and a bottle of spirits, would be the site of the ninth largest city in the British Empire, and, as its citizens would fain assert, second only, in commercial importance, to the great imperial capital itself.

Other noteworthy towns, besides Melbourne, are **Geelong, Ballarat, Ararat, Stawell,** and **Serviceton,** on the *Western Railway;* **Castlemaine, Bendigo,** and **Echuca,** on the *Northern Railway;* **Wodonga,** the terminus, and **Beechworth** on a branch, of the *North-Eastern Railway;* and **Sale** and **Bairnsdale,** on the *Eastern Railway.* On the coast, to the west of Port Phillip, are **Warrnambool, Port Fairy** or Belfast, and **Portland.**

Geelong (20), one of the oldest towns in Victoria and noted as a wool port and for its "tweed" manufacture, stands at the head of Corio Bay, the western arm of Port Phillip Bay. **Ballarat** (37), one of the most famous gold-mining towns in the world, stands on the south side of the Dividing Range, about 74 miles west of Melbourne. The formerly rich alluvial deposits have been worked out, and the gold is now mainly obtained from the quartz mines, some of which are as deep as some of the coal-pits in England, and have thus to be worked by expensive machinery. **Ararat** (4) and **Stawell** are mining towns, also on the Western Railway, which connects with the Eastern Railway of South Australia at **Serviceton,** a border town, 300 miles from Melbourne and about 200 miles from Adelaide. From Ararat a branch line runs to **Portland** (2), a seaport on Portland Bay and the oldest town in the colony, having been founded in 1834. Forty miles east of Portland is another rising seaport, **Port Fairy** (2), formerly called Belfast, and now connected by rail with Melbourne. Still further east is the port of **Warrnambool**—the outlet of a rich agricultural and pastoral district.

Castlemaine (7) is an important gold-mining town, and also the centre of a productive agricultural district. It is connected by a branch line of railway with **Maryborough** (5), another mining and agricultural centre, and by that route, with Ballarat and the western districts. **Bendigo** or Sandhurst (38), a gold-mining town, famous in the early annals of the colony, is still, and likely to remain, an important mining centre—its rich quartz reefs (worked in the Lansell mine at a depth of 2,640 feet) being practically inexhaustible. **Echuca** (5), the "Chicago" of Australia, is a border town, situated on a peninsula formed by the junction of the Murray and the Campaspe rivers. It is the terminus of the *Northern Railway* from Melbourne (156 miles distant), and is connected by a bridge across the Murray with Moama, and thence by rail with Deniliquin, in New South Wales.

Beechworth (3), the principal town in the north-eastern part of the colony, is the centre of the celebrated Ovens River gold-mining district. The town is connected by a branch line with the main line of railway between Melbourne and

Sydney; the Victorian terminus is Wodonga, on an arm of the Murray River, but the trains run right through to the New South Wales terminus, *Albury*, on the opposite side of the river. Beechworth, on one of the spurs of the Australian Alps, lies at an elevation of 1,800 feet, but Sale (5), the principal town in Gippsland, is only about 30 feet above the sea-level. Sale is at the head of navigation of the Gippsland lakes, and carries on a considerable trade with Melbourne by sea and by the Eastern Railway, which now terminates at Bairnsdale (3), another lake-port at the mouth of the Mitchell River.

QUEENSLAND.

QUEENSLAND is another "New England," growing daily in wealth and population under the bright sun and cloudless sky of Australia. This immense colony occupies the whole of the north-eastern portion of the continent, and is bounded on the east by the Pacific Ocean, on the north by Torres Strait and the Gulf of Carpentaria, on the west by South Australia, and on the south by New South Wales.

The boundary of Queensland on the south is formed by a line from Point Danger running westward along the Macpherson and Dividing Ranges, and the Dumaresq and Macintyre rivers, and thence along the 29th parallel to the 141st meridian East longitude. The western boundary is formed by the 141st meridian from the 29th to the 26th parallel, and thence by the 138th meridian to the shores of the Gulf of Carpentaria.

EXTENT: The length of the colony, from north to south, is about 1,300 miles, the breadth 800 miles, and the coast-line 2,550 miles. Its area is 668,000 square miles, or rather less than one-fourth of the continent.

Queensland is thus the largest of the three colonies of Eastern Australia, but both South Australia and Western Australia are one-third larger. And yet this vast and richly endowed colony, with an area 12 times that of England and Wales, has a population of only one-half that of Liverpool—the actual number, in 1891, being 393,718 or about one person to every 2 square miles of area. Included in this number are about 8,000 Chinese, 9,000 Polynesians, and about 1,000 belonging to other alien races. The aborigines are variously estimated at from 10,000 to 20,000, but they are probably much more numerous.

COASTS: The extensive seaboard of Queensland includes the whole of the eastern coast from Point Danger to Cape York—the northernmost point of the continent—and also the eastern and part of the southern shores of the Gulf of Carpentaria.

Along the eastern coast, at a distance of from 10 to 150 miles, the vast natural breakwater of the Great Barrier Reef makes sea-voyaging a pleasure for more than 1,200 miles. There are numerous openings in the reefs through which vessels may, in stormy weather, pass from the open ocean to the smooth water between the reefs and the coast, but there is only one really safe passage for ships. The long voyage, from Torres Strait as far south as Cape Capricorn, is entirely within the sheltered channel thus formed.

L

The coasts of Queensland are indented by a very large number of bays and gulfs, many of which form excellent harbours. Of these, the best known is **Moreton Bay**, which is about 40 miles long and 17 miles broad. Six navigable rivers—the **Brisbane**, **Logan**, &c.—enter this fine bay, in almost any part of which vessels may anchor safely, under shelter of the numerous shoals. **Port Curtis**, on the Pacific coast, and the recently discovered **Port Musgrave**—at the mouth of the Batavia and Ducie rivers, on the eastern side of the Gulf of Carpentaria—form splendid natural harbours. Other notable harbours are **Hervey Bay**, **Keppel Bay**, **Port Bowen**, **Port Denison**, **Rockingham Bay**, and **Port Albany** (near Cape York) on the eastern coast, **Thursday Island** harbour in Torres Strait, and **Investigator Roads** in the Gulf of Carpentaria.

Hundreds of islands stud the coastal waters of Queensland. Of these, the largest or most noteworthy are **Stradbroke**, **Moreton**, **Fraser** or **Great Sandy Island**, **Curtis**, and **Hinchinbrook Islands**, on the eastern coast; **Prince of Wales** and **Thursday Islands**, off the northern coast; and the **Wellesley Islands** in the Gulf of Carpentaria.

The principal headlands are **Point Danger**, **Sandy Cape**, **Cape Capricorn**, and **Cape Flattery** on the east, and **Cape York** on the north. Cape York, the extreme point of the long peninsula of Northern Queensland, is the northernmost point of the colony and also of the continent.

RELIEF : The main features in the relief of Queensland, as of New South Wales, are (1) the coastlands, (2) the tablelands and mountains, and (3) the vast interior plains.

Orographically, then, the general character of the surface of Queensland could be shown by three colours—one to mark out the comparatively narrow, low-lying lands between the sea and the mountains; another to show the long and irregular mountain ranges that rise from, or mark the seaward face of, the broad basaltic belt of tableland, which really extends from Cape York to the neighbourhood of Brisbane; and a third colour to distinguish the vast level plains that stretch away towards the western borders of the colony.

The **Coast Range** extends, under various names, from York Peninsula in the north to within a few miles of Brisbane. It runs nearly parallel with the coast, at an average distance of about 50 miles. The **Main Range**, which forms the northern portion of the long cordillera of Eastern Australia, runs inland of the Coast Range, and may be said to extend from the Macpherson Range, on the borders of New South Wales, to Cape York. The average elevation of these mountains is only about 2,000 feet; no summits in the Main Range exceed 5,000 feet. In the Coast Range the highest points are *Wooroonooran*, 5,400 feet, in the Bellenden Ker Range, and *Mount Dalrymple*, 4,200 feet, in the Mackay Range.

RIVERS : The Queensland rivers belong to four distinct systems : (1) those that flow eastward into the Pacific ; (2) those that form the headwaters of the Darling, and thus belong to the basin of the Murray ; (3) the streams that flow southwards, and many of those flowing westward from the Great Dividing Range, and are either lost in the sand or ultimately fall into the salt lakes of South Australia ; and (4) those that flow northward into the Gulf of Carpentaria.

Of the numerous rivers which flow eastward across the fertile coast plain into the Pacific, the principal, in order from south to north, are the Brisbane, the Mary, the Burnett, the Fitzroy, the Burdekin, the Herbert, the Johnstone, the Daintree, and the Endeavour.

The headwaters of the Darling River, which include the Warrego, the Condamine or Balonne, the Macintyre and its tributary the Dumaresq, drain a large portion of the tablelands in the southern part of the colony. Further west, the Victoria or Barcoo flows, under the name of Cooper's Creek, into Lake Eyre, while the Diamentina or Mueller River, though swollen during the rainy season by floods of many tributary "creeks," is lost in the stony desert which lies to the north of Lake Eyre.

To the tropical basin of the Gulf of Carpentaria belong the Gregory, the Flinders—one of the longest rivers in the colony—the Norman, the Gilbert, the Mitchell, and the Batavia. The Norman is navigable by sea-going vessels to Normanton. At the mouth of the Batavia is Port Musgrave, one of the finest natural harbours in Australia.

CLIMATE : Although the northern half of Queensland is within the Tropics, the heat is less oppressive than it is further south, while hot winds and sudden changes of temperature are unknown. "During a large part of the year, the weather is fine, the sky cloudless, and the air dry and exhilarating."

In such an extensive territory, stretching, as Queensland does, over 19° of latitude—from 10° to 29° S. lat.—there are, of course, great varieties of climate, but it may be said, generally, that the seaboard districts are hot and moist, while the interior plains are hot and dry, and that frosts and cold winds are known only on the elevated uplands in the south and west. In the summer, the heat is tropical, and along the coast it is felt the more on account of the heavy rains which then fall. For seven months in the year the heat is, however, tempered by the south-east sea breezes. The winter season is generally dry, and then it is most delightful. On the whole, although the Darling Downs in the south is the only district that can fairly be said to be within the Temperate Zone, the climate of Queensland, though hot, is healthy, and not unsuitable for Europeans. The colonists, both British and colonial born, pursue their ordinary avocations and enjoy their sports and pastimes all the year round as in England. On the cattle stations in the interior of the far north, the English immigrants are able to remain the whole day in the saddle, even in the height of summer. Boat races are rowed at Rockhampton even in the hottest weather ; and lawn tennis and cricket are carried on throughout the colony even in the summer time.[1] On the whole, it may be said that there is no country in the world at once so healthful and attractive, and certainly none where an Englishman can live an outdoor life in all seasons with such entire immunity from all physical ills.[2]

The rainfall[3] on the coastlands (and especially in the north) is, on the whole, very heavy, and in some parts excessive, frequently amounting in the south part of Cook district to over 160 inches in the year. But there is much variation even along the coast, the fall being moderate at Brisbane, scarcely 30 inches around the mouth of the Burdekin, and little more in the Port Curtis district. In the Darling Downs, the rainfall is generally sufficient for the farmers' needs, but in

1. Sir C. W. Dilke.
2. The *winter* months are May, June, and July—the *summer* months are December, January, and February. In North Queensland, as in other tropical countries, there are really only two seasons, the *wet* and the *dry*. The *average annual temperature* at Brisbane is 70° as against 51° in London.
3. As in other tropical and semi-tropical countries, the rain falls in summer ; in New South Wales and Victoria it falls chiefly in winter.

the western districts generally, the rains are scanty and irregular—amounting, in some parts of the *Mitchell* and *Gregory* districts, to as little as 10 inches a year, while severe droughts are not infrequent. But water may be found, even in the driest parts of the interior, by sinking wells, and irrigation on a large scale is practicable in many places. [1]

PRODUCTIONS : The chief commercial products of Queensland are **wool, gold, sugar, cattle, horses, sheep, tin, hides, and skins,** and the principal industries are connected with the production of these commodities.

The indigenous vegetable productions include, in addition to the eucalypti, acacias, cedars, and other characteristic Australian forms, some 500 species of plants identical with those of India and Malaya. The various *gum-trees* yield "hardwood," and many of the *acacias* grow to large trees and furnish excellent timber. The *red cedar*, the *Moreton Bay pine*, the *silky oak*, *beech*, and other trees all yield valuable timber. Queensland also possesses a few really good *indigenous fruits*, and no part of Australia is better supplied with *excellent grasses* and *nutritious fodder plants*, some of them, such as the "Mitchell Grass," capable of withstanding long droughts and of quick growth after slight rains.

The animals of Queensland are similar to those of the other colonies, and there are only one or two peculiarly tropical forms of mammals, but some of the birds on the northern coasts are allied to the 'birds-of-Paradise' of New Guinea.[2] *Snakes* are plentiful, and many of them poisonous. *Lizards* are numerous in all parts, and *alligators* are found in some of the northern rivers. *Fish*[3] swarm on the coast and in the rivers, and the largest *oyster* beds in Eastern Australia are in Moreton Bay and Sandy Strait. *Turtles* are caught in the Gulf of Carpentaria.

INDUSTRIES : The pastoral industry and mining are the chief pursuits in Queensland, and *wool* and *gold* are the staple products of the colony.

The broad plains of the interior afford the richest pasturage for sheep, which are fed almost exclusively on the native grasses, except on the Darling Downs and in some places near the coast, where portions of some farms are sown either with lucerne or with Californian prairie grass. In the coast districts generally, and especially in North Queensland, the rank natural vegetation makes excellent pasturage for cattle and horses, but is not suitable for sheep, which thrive best on the herbage in the drier climate of the interior. Over 18½ million sheep, 6½ million cattle, and over 429,000 horses are now depastured in the colony.[4]

AGRICULTURE : Agriculture is making steady progress, but only about a quarter of a million acres are as yet under cultivation. *Maize* and the *sugar-cane* are the chief objects of culture, but *wheat* and *rice, potatoes, bananas, pine-apples, oranges, grapes,* and other *fruits* are also grown.

1. The Government is vigorously promoting irrigation works in different parts of the colony, and mostly with great success. Artesian wells have been, and are being, sunk in numerous localities.
2. "Fossil remains, discovered in the Darling Downs, go to prove that in a distant age the Moa (*Dinornis*) existed in this part of the continent."
3. There are over 900 species of marine and freshwater fish, and over 300 of these are more or less edible.
4. "About one-half the area of the colony is natural forest, though little has been done hitherto to develop the forestry of the colony. A large proportion of the area is leased in squatting runs for pastoral purposes, amounting to 281,316,815 acres in 1893."—*The Statesman's Year Book*, 1895.

Maize, generally called "corn" in Queensland, is most successfully grown along the coast, and, on the northern seaboard, two crops can be grown every year. The sugar-cane is grown in the temperate and tropical valleys along the eastern coasts. The low-lying lands which border the lower courses of many of the eastern rivers are extraordinarily fertile, and the climate is most favourable for the cultivation of other tropical products, such as cotton, arrowroot, coffee, spices, rice, &c. The want of suitable labour and other causes have injured the sugar industry, but there were, in 1893, over 59,000 acres of land under sugar-cane, from which fully 76,000 tons of sugar were produced.

Sugar has been grown with great success along the whole coast-line of Queensland. In the south, the system adopted is cultivation on small holdings, the cane being crushed at mills owned by a company of from four to eight farmers, and very little coloured labour is here employed. In the middle districts, where the climate is still moderately cool, the same principle of farm cultivation is in practice, but the manufacture is undertaken by very large refineries. Further north, in the tropical regions, the cultivation and manufacture is carried on almost entirely by large capitalists.[1] Attached to each mill is an extensive area of cane, cultivated by labourers from the Pacific islands, brought into the colony by 'labour schooners' or *black-birders*, as they were called. In 1890, however, the importation of these islanders or *Kanakas* was prohibited—the traffic having become almost as bad as the slave-trade. Kanaka labourers have, however, been recently re-introduced under more stringent regulations.

Wheat, in favourable years, grows well, but it has not been largely cultivated except on the temperate Darling Downs, and even here, although the soil is suitable, the yield is uncertain owing to the prevalence of "rust" and occasional droughts. Rice is grown to some extent in the north, and there are large vineyards in different parts of the colony. Pine-apples, bananas, oranges, and other tropical fruits also come to perfection. Dairy-farming and market-gardening are carried on, chiefly in the vicinity of the towns.

MINING : Queensland is rich in minerals of all kinds, and mining, chiefly for gold, is the leading industry in many parts of the colony.

Gold has been found in nearly every part of the colony, from Warwick in the south to the Palmer River in the far north. Both the alluvial mines and the quartz reefs have proved as rich as those of New South Wales and Victoria, but, as in these colonies, the alluvial deposits are being worked out, while the reefs of auriferous quartz are practically inexhaustible. The Queensland gold-fields, which have produced over 32 million pounds' worth of gold, may be arranged in three groups :—

The Northern Goldfields include the rich and well developed field of *Charters Towers* —until recently the most productive in all Queensland—the *Etheridge, Ravenswood, Cloncurry, Croydon, Palmer River*, and other fields.

The Central Goldfields include the mines in the *Clermont* or Peak Downs, and in the *Gladstone* and *Rockhampton* districts. The Rockhampton fields include the famous *Mount Morgan Mine*, which is believed to be the richest deposit of gold in Australia.

The Southern Goldfields include the extensive and richly productive mines at *Gympie*, and less important fields at *Mount Perry* and *Warwick*. The Gympie fields are situated on the Upper Mary River, about 60 miles south of the port of Maryborough, with which they are connected by rail.[2]

Considerable quantities of copper, tin, and coal are also produced, while silver and lead, antimony, bismuth, manganese, iron, and quicksilver are found and worked to some extent.

1. *Queensland Handbook*, p. 8.
2. Most of the goldfields are now connected by rail with the nearest ports. See further, under railways. The student should trace the railways from the coast and note the position of the gold-fields.

The richest copper mines are at *Clermont*, *Mount Perry*, and *Cloncurry*—the lodes at Cloncurry being unusually rich and extensive. Wonderfully rich stream tin deposits are worked at *Herberton* in the north (about 60 miles south of Cairns), and at *Stanthorpe* in the south (about 180 miles south-west of Brisbane). Coal is extensively distributed throughout the colony, but the annual output is still under half-a-million tons. The principal coal mines now worked are those near *Ipswich* (24 miles from Brisbane), and those near *Howard* on the Burrum coalfield, 18 miles from the port of Maryborough. Marble and building stone are found near *Warwick*, and at *Gladstone* and *Rockhampton*.

The pearl fishery in Torres Strait (at Thursday Island, Prince of Wales Island, Somerset, &c.), and the bêche de mer fishery carried on at the Barrier Reef, Murray Island, &c., are both important industries, and a large number of " all sorts and conditions of men "—Aborigines, Malays, South Sea Islanders, Chinese, and even Japanese, Singhalese, Arabs, and Maories—are employed in diving for the lustrous pearl shells, or in collecting the *bêche de mer*, a kind of sea-slug, highly esteemed by Chinese gourmands. The oyster beds at Moreton Bay and Maryborough annually yield thousands of bags of the succulent bivalve.

COMMERCE : The commerce of Queensland is chiefly with the other *Australian Colonies*, and, next to them, with the *United Kingdom*. Annual value, 14 millions sterling — Imports, 4½ millions ; exports, 9½ millions.

The principal exports, the produce and manufacture of the colony, are *wool* —to the value of about 4 millions a year, *gold*—2 millions sterling a year, *raw* and *refined sugar*, *hides* and *skins*, *tin*, *silver-lead*, *copper*, *preserved meats*, *tallow*, *pearl shells*—from which mother-of-pearl is made—*bêche de mer* and *oysters*.

The imports include every description of *manufactured goods*, chiefly from the United Kingdom. *Agricultural implements*, *tools*, &c., are largely imported from the United States.

Five-sixths of the intercolonial trade is with New South Wales, and most of the rest with Victoria. There is a large overland trade with New South Wales in *live stock*.

The direct trade with the United Kingdom amounts to between 4 and 5 millions sterling a year. The exports to the United Kingdom amount to about 3¾ millions—seven-eighths of which consist of *wool*. The annual imports of British produce, chiefly *textile fabrics* and *iron*, wrought and unwrought, amount to over 1¼ million sterling.

PORTS : All the larger towns of Queensland are seaports, but almost all the foreign trade is centred in the four ports of Brisbane, Rockhampton, Townsville, and Maryborough.

RAILWAYS : From each of the more important ports a railway has been constructed, running, with one or two exceptions, almost due west into the interior, and will ultimately be extended to the western border of the colony, and most likely be united with each other by cross-lines along the coast and in the interior. Nearly 2,500 miles are now open for traffic, and about 500 miles are in course of construction. The principal lines are :—

The Southern and Western Railway, from BRISBANE, westwards for 483 miles *via* ROMA to CHARLEVILLE, and southwards for 232 miles *via* WARWICK to the frontier station of WALLANGARRA, where a junction is effected with the Northern Railway of New South Wales. There is thus through communication by rail between Brisbane and Sydney,

Melbourne and Adelaide.   Branch lines extend along the coast, north to GYMPIE, and south to COOLANGATTA near Point Danger.

The Maryborough Railway extends from the port of MARYBOROUGH to GYMPIE, a distance of 60 miles, and thence to BRISBANE.

The Bundaberg Railway, from BUNDABERG to MOUNT PERRY, and lines also run south to MARYBOROUGH, and north to GLADSTONE.

The Central Railway runs from the port of ROCKHAMPTON to LONGREACH, 55 miles west of BARCALDINE in the Mitchell District.   Branch lines connect CLERMONT on the north and SPRINGSURE on the south with the main line.

The Mackay Railway, from the port of MACKAY inland to MIRANI and ETON.

The Bowen Railway to WANGARATTA (48 miles).

The Northern Railway starts from the port of TOWNSVILLE, and is now open to HUGHENDEN (236 miles distant), with a branch to RAVENSWOOD.

Still further north are the Cairns to Herberton Railway ; the Cooktown Railway, from Cooktown towards the Palmer Goldfield ; and the Normanton to Croydon Railway, running from Normanton, at the head of navigation on the Norman River, to the Croydon Goldfield.

GOVERNMENT : The government of Queensland is vested in a Governor, aided by an Executive Council, and a Parliament of two Houses—the Legislative Council and the Legislative Assembly.

The Governor represents the Queen, and is appointed by the Home Government.   The 37 members of the Legislative Council are nominated by the Crown for life.   The Legislative Assembly now has 72 members, elected for 5 years.[1]

The annual Revenue, in 1893, was 3¼ millions sterling, and the Expenditure, 3½ millions sterling ; the Public Debt, incurred on account of immigration, railways, telegraphs, roads, bridges, and other remunerative public works, amounts to over 30 millions.

The defence of the colony is provided for, in addition to a strong militia and volunteer corps, by the Act of 1884, according to which every man between 18 and 60 is liable for military service in case of necessity.   There is also a small naval force, and a battery and torpedo defences protect the entrance to the Brisbane River.

Education is free, secular, and compulsory.   As in the other Australasian colonies, there is no State Church.

DIVISIONS and TOWNS : Queensland is divided into 12 large districts, some of which, and portions of others, are subdivided into counties.   All the larger towns are in the eastern or Pacific division of the colony ; there are few towns, and none of any considerable size, in the western districts.

These 12 districts may be arranged into two divisions—an Eastern or Pacific Division, and a Western Division.   Some of these districts are further subdivided, as Kennedy, in the Eastern Division, into *North Kennedy* and *South Kennedy*, and Gregory, in the Western Division, into *North Gregory* and *South Gregory*.

Only one town—BRISBANE—has above 50,000 inhabitants.   Seven other towns—Rockhampton, Maryborough, Gympie, Ipswich, Toowoomba, Charters Towers, and Townsville—have between 10,000 and 15,000 inhabitants.   Only a few of the western towns have a population of more than 1,000.

---

1. Members of the Queensland Parliament are paid £150 a year each, with travelling expenses.

The **Eastern or Pacific division** of Queensland includes the districts of *Moreton, Darling Downs, Burnett* and *Wide Bay*, in the south ; *Port Curtis, Leichhardt,* and *Kennedy*, in the centre ; and *Cook*, in the north.

MORETON, in the extreme south-east of the colony, is mainly an agricultural district, watered by the Brisbane and other rivers, and traversed by several railways, which radiate from the city of BRISBANE (52), the capital of the colony and the seat of Government, on the Brisbane River, about 25 miles above its outlet into Moreton Bay.   The river is navigable for vessels drawing over 21 feet, and the trade with England and the other Australasian colonies is very large.   IPSWICH (10), 24 miles west of Brisbane, is the centre of the productive coal mines in the basin of the Bremer River.

The DARLING DOWNS district, to the west of the Moreton district, contains the finest pastoral and some of the best agricultural land in the colony. Wheat and other cereals and all kinds of vegetables and fruits are largely grown, and good coal is found.   The chief towns are TOOWOOMBA, 100 miles west of Brisbane, on the summit of the Dividing Range ; WARWICK, 100 miles southwest of Brisbane, famous for its grapes; the "*tin-town*" of STANTHORPE, near the New South Wales border ; and GOONDIWINDI, also a border town, situated on the Macintyre River, about 200 miles south-west of Brisbane, and surrounded by productive orchards and fertile farm lands.

The BURNETT and WIDE BAY district, though principally occupied for pastoral purposes, has rich gold mines at GYMPIE, extensive coal deposits at BURRUM (Howard), and copper mines at MOUNT PERRY, all connected by rail with MARYBOROUGH, an important port on the Mary River, 25 miles above its mouth, and with BUNDABERG, a port for seagoing steamers, on the River Burnett, and the centre of an extensive sugar-growing district.

PORT CURTIS district is chiefly noted for its rich gold-mines, among them the famous Mount Morgan mine, which lies about 28 miles south-west of ROCKHAMPTON, the great gold and wool port on the Fitzroy River, 40 miles above its mouth.   GLADSTONE, founded in 1846, at the instance of Mr. W. E. Gladstone, has a fine harbour, and is connected by rail with Bundaberg.   This district is divided by a range of mountains from the

LEICHHARDT district on the west, a large tract of pastoral country, crossed by the *Central Railway*, from which branch lines run to the gold and copper mining centre of CLERMONT, to the north, and the pastoral centre of SPRINGSURE, to the south, of the main line.

The KENNEDY district extends from Cape Palmerston to Rockingham Bay, and thus occupies the middle portion of the Pacific coast of the colony.   It is well-watered by the Burdekin and its tributaries, and has splendid pastures for sheep and cattle, but is chiefly famous for its rich goldfields and productive sugar plantations.   CHARTERS TOWERS, RAVENSWOOD, and CAPE RIVER are the chief gold-mining centres, and MACKAY is the centre of one of the largest sugar-producing districts in the colony.   The goldfields and the pastoral country beyond them are connected by a railway with the port of TOWNSVILLE, on Cleveland Bay.   BOWEN, on Port Denison, and CARDWELL, on Rockingham Bay, are rising ports.

COOK district, in the extreme north of the colony, is like an immense wedge of land—as large as Ireland—between the Pacific Ocean and the Gulf of Carpentaria.   Sugar and rice are grown on the east coast, and large quantities of cedar are cut and exported, but immigrants have been attracted chiefly by the opening up of the Palmer River and other goldfields, while the tin mines are believed to be among the richest in the world. Silver and coal are also found. The only towns are COOKTOWN, a port on the coast, about 1,000 miles north-west of Brisbane and the outlet for the Palmer goldfield, towards which the railway is being extended ; and PORT DOUGLAS and CAIRNS, the outlets of the Hodgkinson and Etheridge goldfields and the Herberton and Tate tin mines.

The Western Division of Queensland includes the districts of *Maranoa* and *Warrego*, in the south, adjoining New South Wales ; *Gregory*, in the west, adjoining South Australia and the Northern Territory ; *Burke*, in the north, bordering on the southern shores of the Gulf of Carpentaria ; and *Mitchell*, almost in the centre of the colony.

MARANOA adjoins the Darling Downs on the east and New South Wales on the south. It is a pastoral district of tablelands and downs, though some land is under maize, oranges, vineyards, &c. Its chief town, ROMA, is a flourish-ing pastoral centre, on the *Western Railway*, which traverses this district as far as CHARLEVILLE, in the adjoining district of WARREGO, another pastoral district. but much drier than the districts to the east. The GREGORY district is also purely pastoral and not much known. South Gregory is traversed by the well-known Cooper's Creek, near which the heroic explorers—Burke and Wills — perished in 1861, and North Gregory, by the Diamentina and other rivers.

BURKE district adjoins the Northern Territory of South Australia on the west, and is bounded on the north by the Gulf of Carpentaria, into which the Norman, Flinders, Leichhardt, Albert, and other rivers flow. The vast plain which extends from the shore of the Gulf to the south is richly grassed, and, in the driest seasons, many of the rivers are running streams, while in many places the banks are so low and the country so level that large areas could be easily irrigated. The settled districts are as yet, however, mainly used for pastoral purposes, but there are two important mining centres—CLONCURRY, a rising town on the Cloncurry River, a tributary of the Flinders, in the centre of an extensive gold and copper-mining district in the south, and the new gold-mining field of CROYDON, in the north-east, which bids fair to be the most prosperous of the goldfields north of Charters Towers. Croydon is now con-nected by rail with NORMANTON, the chief port of the Gulf region, at the head of navigation on the Norman River.

The *Northern Railway*, from Townsville is now open as far as HUGHENDEN, in the south-east of this district, and will be extended to Croydon.

MITCHELL, the central district of the colony, is almost entirely pastoral. It is watered by the Thompson and the Barcoo rivers, and traversed by the *Central Railway* from Rockhampton to LONGREACH, near the Thompson River, 55 miles beyond BARCALDINE, a pastoral centre about 360 miles inland. The principal town is BLACKALL on the Barcoo River, the most central town in Queensland. It is some 60 miles south of the railway.

SOUTH AUSTRALIA.

SOUTH AUSTRALIA is, next to Western Australia, the largest colony on the continent, across which it extends from the shores of the Southern Ocean to the Gulf of Carpentaria and the Arafura Sea.

The name "South Australia" is somewhat misleading, as nearly the whole of Victoria is more to the south, while the Northern Territory extends almost as far north as Queensland.

BOUNDARIES : The colony is bounded on the *north* by the Gulf of Carpentaria and the Arafura Sea ; on the *south* by the Southern Ocean ; on the *east* by Victoria, New South Wales, and Queensland ; and on the *west* by Western Australia.

The colony thus includes the entire central division of Australia, and touches all the other colonies on the continent, having the three eastern colonies on the east and the immense western colony on the west. The northern boundary, which was originally the 26th parallel of latitude, was, by the annexation of the Northern Territory in 1863, shifted to the northern coast-line, so that the colony extends from the Southern to the Indian Ocean.

South Australia proper and the Northern Territory are divided from Western Australia by the 129th meridian East longitude. On the east, the boundary is marked by the 141st meridian, up to South latitude 26°, and thence north by the 138th meridian.

EXTENT : This vast territory measures over 1,800 miles from sea to sea, and has a total area of over 900,000 square miles, nearly one-third of the continent, and no less than 10 times the size of Great Britain, or 15 times the size of England and Wales.

The colony thus extends over 27° of latitude, and 12° of longitude in the south and 9° in the north. Its greatest length is 1,850 miles, its greatest breadth 650 miles, while the actual area is 903,425 square miles, or 578 million acres, of which only 10 million acres had been alienated up to March, 1890. The entire population of this vast territory, however, in 1891, only numbered 319,145, all of whom, with the exception of about 6,000 Aborigines, 4,000 Chinese, 10,000 Germans and 3,000 other foreigners, were British or of British origin.

South Australia proper, which extends from the Southern Ocean to the 26th parallel of South latitude, has an area of 379,825 square miles, and a population of 314,187, of whom about 6,000 are Aborigines.

The Northern Territory, which extends from the 26th parallel to the shores of the Indian Ocean, has an area of 523,600 square miles, but the population is only about 5,000, two-thirds of whom are Chinese.

COASTS : South Australia has a deeply indented coast-line of about 2,000 miles in length.

The principal *inlets* on the northern coast are Queen's Channel, Port Darwin, Van Diemen Gulf, and Arnhem Bay, opening into the Arafura Sea ; and Blue Mud Bay and Limmen Bight in the Gulf of Carpentaria.

The southern coast is indented by the much larger inlets of **Spencer Gulf** and **St. Vincent's Gulf**, in and on either side of which are numerous smaller openings. The **Coorong** is a long arm of the sea on the eastern side of **Encounter Bay**, from which it is divided by a narrow tongue of land.

The most important *straits* and *channels* are :—on the south, **Investigator Strait**, between Kangaroo Island and Yorke Peninsula, and **Backstairs Passage**, between the same island and the mainland; and, on the north, **Clarence Strait**, between Melville Island and the mainland, and **Dundas Strait**, between the same island and Coburg Peninsula.

The largest *islands* belonging to South Australia are **Kangaroo Island**, 85 miles long and about 30 miles broad, at the mouth of the Gulf of St. Vincent, on the south coast, and **Melville** and **Bathurst Islands**, off the northern coast, with **Groote Eylandt**, in the Gulf of Carpentaria.

The principal *headlands* are **Cape Northumberland, Cape Jervis, Cape Spencer**, and **Cape Catastrophe**, on the south coast ; and, in the Northern Territory, **Cape Van Diemen**, on Melville Island, **Cape Don**, on Coburg Peninsula, **Cape Wessel**, on the northernmost of the Wessel Islands, and **Cape Arnhem**, on the mainland in the north-east.

SURFACE : Though South Australia has no mountain ranges comparable to those of the eastern colonies, its surface is sufficiently diversified by **fertile plains**, several long **hill-ranges**, and **well-wooded valleys**, while the treeless, waterless, and arid districts often teem with mineral wealth.

The chief mountain ranges in South Australia Proper are the **Mount Lofty Range**, which runs north from Cape Jervis and forms the 'water-parting' between the Murray and the Gulf of St. Vincent, rising in *Mount Lofty*, 8 miles east of Adelaide, to 2,334 feet above the sea, and in *Mount Barker* to 2,331 feet ; the **Hummocks**, which run north from the head of St. Vincent's Gulf, nearly parallel with the eastern coast of Spencer Gulf ; the **Flinders Range**, which commences on the north-eastern side of Spencer Gulf and extends northward for some hundreds of miles to Lake Blanche, rising in *Mount Remarkable* and *Mount Brown* to upwards of 3,000 feet ; and the rugged **Gawler Range** in Eyre's Peninsula to the south of Lake Gairdner, extending from the head of Spencer Gulf towards Streaky Bay, with some peaks about 2,000 feet in height. In the south-eastern part of the colony are **Mount Gambier, Mount Schanck, Mount Terrible**, and other isolated peaks that were formerly volcanoes, their craters being occupied by beautiful little fresh-water lakes.

In the interior, the principal elevations are the **MacDonnell Ranges**, while in the coast region of the Northern Territory some of the mountains, which rise from "an exceedingly broken country, described as a mass of ranges" to the east of Blunder Bay, have an altitude of about 2,000 feet.

RIVERS : With the exception of the **Murray River** in the south-east, South Australia Proper has no large rivers, but the Northern Territory is (for Australia) exceptionally well-watered.

The lower course of the Murray is within South Australia. It is navigable from beyond Albury in New South Wales, and, during the season, an immense

amount of wool and other produce is brought down the river from the interior
of New South Wales and Victoria, After a course of 1,300 miles, the river enters
the shallow coastal Lake Alexandrina, and flows thence to the sea by a narrow
and scarcely navigable opening, called the Murray Mouth.

The **Gawler, Torrens,** and other streams which flow into St. Vincent's Gulf
are insignificant in length and volume, and are nearly or entirely dry for several
months every year.

A remarkable river in the interior is the **Finke,** a fine stream rising on the
southern slopes of the MacDonnell Ranges and flowing through fairly good
pasture land, but ending in sandy flats before reaching Lake Eyre, into which
the Barcoo or Cooper's Creek and the Warburton flow during the rainy season.

In the Northern Territory, several noble streams enable sea-going vessels
to penetrate into the interior for many miles. The **Roper,** which flows into
Limmen Bight in the Gulf of Carpentaria, is a fine deep river, navigable for
large ships and sea-going steamers for 100 miles inland. The Victoria is navigable for large vessels for 43 miles, and for smaller craft for 112 miles. Its
estuary is 20 miles wide, but the navigable channel is narrow. Large boats
have ascended the **Daly** River, which flows into Anson Bay, for 100 miles. The
Adelaide is a large river, and its lower course, like those of the Liverpool and
the South and East Alligator rivers, is navigable for sea-going vessels. There
are other large inland rivers, so that the facilities for water carriage are considerable.

LAKES : None of the other Australian colonies have so many or
such large lakes as South Australia, but, though some of the great
salt lakes are over a hundred miles in length, they are of no service
to the colony, as they are liable to be dried up, and are absolutely
unfitted for navigation.

One, and the largest, group of salt lakes occupies a vast area to the north of
Spencer Gulf. This group includes Lake Eyre, Lake Torrens, and Lake
Gairdner—each of them between 90 and 100 miles in length—with several
smaller basins, such as Lake Frome, Lake Blanche, &c. Considerably further
north, on the borders of Western Australia, are the still more extensive Lake
Amadeus and other salt lakes. All these lakes vary in outline and depth with
the seasons, and some of them are often mere swamps of saline mud.

Lake Alexandrina, at the mouth of the Murray, is fresh, but shallow and
difficult of navigation. The entrance from the sea, which forms the actual
mouth of the Murray, and is so called, is narrow and dangerous to pass through.

The craters of some of the extinct volcanoes in the south-eastern part of the
colony are occupied by beautiful fresh-water lakes. One of them, the Blue
Lake, on Mount Gambier, is 240 feet deep, and is surrounded by precipitous
walls, several hundred feet in height, and covered with luxuriant vegetation.

CLIMATE : The climate of South Australia is **hotter** and **drier**
than that of the other colonies, but for nine months in the year it
is agreeable. In the Northern Territory, the climate is **tropical,**
except on the tablelands, where the temperature is lower.

The climate generally resembles that of Southern Italy. There is no winter
in the English sense of the term, though slight frosts may be experienced on
the plains and ice seen on the hills in July and August, which are the coldest

months.   The three summer months—December, January, and February—are
very hot even in the south, and intensely so in the interior and far north, but
the dryness of the air renders even a temperature of over 100° F. bearable.   As
in Victoria and New South Wales, the heat is sometimes increased by hot
winds from the interior.   Fortunately, they are rare, occurring only in the summer, and then only for one, two, or at most three days, "lulling at night and
raging again in the forenoon."

The annual rainfall at Adelaide has averaged about 21 inches during the
past 60 years, but in 1839—the wettest year yet recorded—it rose to 31 inches.
In 1888, the whole of Australia suffered from one of the most disastrous droughts
ever experienced, and in the far north of South Australia the rainfall did not exceed an inch in the twelve months.   In the Mount Lofty and other hill districts
the fall is heavier, but on most of the sheep 'runs' in the interior the yearly
rainfall does not exceed 7 inches.   In the Northern Territory the rainfall is very
much heavier than in the interior and the south.   At Port Darwin, it has varied
from 45 inches in 1881 to 82 inches in 1885, and 61 inches in 1888.   The climate
of both divisions of the colony is, on the whole, **extremely healthy**, and the
death-rate is much lower than in England.[1]

PRODUCTIONS : Wheat and flour, wool, and copper, are the
staple products of South Australia, and both climate and soil are
extremely favourable to the extensive cultivation of the vine, the
olive, and the mulberry, and other plants that require dry heat in
order to come to perfection.   The soil and climate of the Northern
Territory are suitable for the cultivation of almost all tropical
plants.

The indigenous plants and native animals are similar to those of the adjoining
colonies.   Among the former, the most striking are gum trees, acacias, and
grass trees; while the latter include the ubiquitous kangaroo and other marsupials, some hundreds of species of birds, among which are the emu, the
laughing jackass, and other characteristic Australian birds, many species of
snakes, some of them venomous, and innumerable insects, among them the
troublesome mosquito.

The flora and fauna of the Northern Territory are the same as those of the
rest of the colony, but the plants include some tropical forms, while of the
crocodiles that infest the northern rivers, the largest are of the same species as
those of India.

The mineral productions include vast deposits of copper, iron, and silver-lead,
together with gold, cobalt, bismuth, marble, slate, granite, &c.

SOIL : In so vast a territory, with such a variety of soil, climate,
and rainfall, the character of the land varies exceedingly, ranging
from the extensive arable plains of the south and the rich alluvial
valleys of the north, to sterile expanses of sand, doomed to perpetual
aridity, in the interior.

In South Australia Proper, the "Adelaide Plains," to the east of the Gulf of
St. Vincent, are extremely fertile and well cultivated.   These plains are flanked
on the east by the Mount Lofty range of hills, the lower slopes of which are
dotted with gardens, vineyards, and villas, while portions of the range are

1. In 1893, the death-rate in South Australia was 13 44 per 1,000 of the population.

covered with dense forests. Beyond the mountains is a hilly tract of country, about 20 miles in width, beautiful in appearance, and including a large extent of the finest agricultural land. Thence to, and on either side of, the Murray, extend open grassy plains, surrounded by a vast waterless scrub. To the south-east of the Murray, the land is good, especially the rich agricultural and pastoral district of Mount Gambier. To the north, and towards and even in the interior, there is a large extent of good pastoral land, but the climate is too hot and dry for agriculture. The coast region of the Northern Territory is adapted to the cultivation of tropical products, while the less valuable country in the interior is suited for pastoral settlement.

AGRICULTURE : South Australia is pre-eminently the agricultural colony of Australia, and the **wheat** grown on the Adelaide Plains is the finest in the world.

In favourable years, the colony may well be called the "Granary of Australia," as sufficient wheat is grown on the extensive plains in the settled districts along the southern coast, and to the east of the gulfs, to supply the home demand and the wants of the neighbouring colonies, and also for export to England, where it is highly prized on account of its dryness and weight. The average yield per acre is certainly small compared to the heavy wheat crops of Manitoba and other countries, in fact, the yield is sometimes so small that without the employment of ingenious mechanical "aids," such as the "stripper"—an invaluable harvesting machine in all hot and dry climates, where the grain rapidly matures and must be quickly harvested—wheat-growing on the South Australian scale and system would be an impossibility.

Viticulture is also an important industry, and, as a wine-producing country, the colony already takes high rank, while **oranges** and other fruits are unsurpassed in size and flavour. The manufacture of preserved fruits and jams is rapidly becoming one of the staple industries of the colony, as is also the drying of raisins and currants.[1]

The **pastoral** industry, however, still yields the most valuable of all the staple products of the colony. The wool of South Australia is not, perhaps, so fine in quality as that of Victoria or of New South Wales, but the annual clip now amounts to 50 million lbs. Drought has destroyed a large number of sheep during the last few years, but there are over 7 million sheep, with about 320,000 cattle and 190,000 horses, in the colony. In the Northern Territory, cattle are much more numerous than sheep. Horses are bred there expressly for export to India, and large numbers are now also shipped to India from the southern ports.

Other stock are **camels**, used for transport in the interior, **Angora goats**, and ostriches. The largest ostrich farm is near Port Augusta.

MINERALS : The mineral resources include vast deposits of copper, iron, and silver-lead, and some gold, bismuth, and tin also exist. Of these, *copper* is by far the most important, and has been to South Australia what *gold* has been to Victoria, and *coal* to New South Wales.

1. The quantity of wine made in 1894 was 911,842 gallons.
The soil and climate of South Australia are particularly suitable for the production of all kinds of wine, and wine-making bids fair to become one of the main industries of the colony, and especially now that wheat-growing is becoming less and less profitable.

Copper was first discovered in one of the hills overlooking Adelaide. The first mine to be opened was at Kapunda, 50 miles north-east of Adelaide, in 1843, but this was eclipsed by the discovery, two years later, of the famous Burra Burra mine, one of the richest copper mines in the world. "The history of this mine, which lies about 90 miles north-east of Adelaide, is the history of the commercial progress of South Australia. Farms, land sales, immigration, rent, wages, have all rested on the yield of the Burra Burra." Even more extensive deposits of copper were discovered, in 1862, at Wallaroo and Moonta in Yorke Peninsula, and the mines there, although they have been worked for so many years, show no sign of exhaustion. The annual production of copper and copper ore now amounts to about £200,000. Some gold is also obtained from the *alluvial mines* at Teetulpa, 230 miles north-east of Adelaide, and other places, and from *quartz reefs* at Teetulpa, Echunga, Manna Hill, &c. Some of the Teetulpa reefs have proved to be rich, and all the north-eastern districts are being vigorously prospected. Silver mines are being worked at Glen Osmond near Adelaide, and near Beltana and Farina in the north.[1]

There are mines of bismuth near Mount Barker, and of cobalt near Blinman, about 350 miles north of Adelaide. Iron is known to exist in large quantities, and coal no doubt also exists.[2]

The numerous and important discoveries of gold, copper, tin, iron, lead, and other minerals, in the Northern Territory, indicate that it may have a great future as a mining country. Gold was first discovered there in 1869. Rich alluvial deposits and quartz mines are now worked. A gigantic copper lode has been found on the Daly River, and there are silver-lead mines on the Mary River about 30 miles from Pine Creek. Ores of iron are found everywhere, and tin occurs abundantly in the form of *reef tin*, and will eventually be, in all probability, the chief source of mineral wealth in the territory.

COMMERCE : Nearly the whole of the trade of South Australia, which is very large in proportion to the population, is carried on with the *United Kingdom, New South Wales,* and the other *Australasian Colonies.* Annual value, 16½ millions sterling—imports, 8 millions ; exports, 8½ millions.

The foreign trade of the colony consists principally in the export of its raw agricultural and mineral products ; the export of wine is, however, increasing. The imports are chiefly Manchester and Birmingham goods, farm implements, beer and spirits, &c.

Only about five per cent. of the trade of South Australia is done with foreign countries. "Of the remainder, on an average, about one-half of the imports are from the United Kingdom, and the other half from the other Australian colonies. Of the exports, about two-thirds go to the United Kingdom, and the bulk of the remainder to the Australian colonies."[3]

The trade with the United Kingdom amounts to about 4 millions sterling a year, and consists mainly in the export of *wool, wheat and flour,* and *copper ore,* and in the import of *iron and machinery, cotton and woollen goods, clothing, &c.*

The principal ports are PORT ADELAIDE, PORT AUGUSTA, and PORT PIRIE, in the south, and PORT DARWIN in the north.

1. The discovery of silver ore in the Barrier Ranges in New South Wales, on the eastern border of South Australia, has had an important influence on the commercial progress of the colony, as that district depends for its supplies from, and ships its products through, South Australia, and, besides, the argentiferous lodes may extend into the colony.
2. A bonus of £4,000 is offered by the Government of South Australia for the discovery of a payable coalfield.
3. *The Statesman's Year Book,* 1885. p. 280.

COMMUNICATIONS : There are nearly 5,000 miles of roads, nearly 2,000 miles of railways, and over 5,500 miles of telegraph lines.

The roads are, of course, all within the settled districts in the south—outside the settled area, and, in parts, within it, "most of the communication, even by mail coaches, is carried on along mere tracks over the sandy plains and through the bush, without the vestige of anything that can be called a road."

The **railways** radiate from Adelaide north, south, east, and west. The **Port Line** extends from ADELAIDE to PORT ADELAIDE and SEMAPHORE. The **North Line** runs from ADELAIDE to QUORN and PORT AUGUSTA, connecting at Quorn with the **Great Northern Line**, open for traffic from PORT AUGUSTA to OODNADATTA or ANGLE POLE, to the north-west of Lake Eyre and 688 miles from Adelaide. A branch from ROSEWORTHY, on the North Line, runs to MORGAN, an important river port on the Murray. The **Southern Line** extends from ADELAIDE to PORT VICTOR, at the head of Encounter Bay. At Mount Barker Junction a line branches off, connecting at *Serviceton* with the Great Western Railway of Victoria, while another line runs south from the adjoining station at WOLSELEY, and curves to the coast at KINGSTON, and connects further south with the **Mount Gambier and Rivoli Bay Railway.** A branch starting from HAMLEY BRIDGE on the North Line gives through railway communication between ADELAIDE and PORT WAKEFIELD, KADINA, and WALLAROO, whence there is a tramway to MOONTA. The railway from PORT PIRIE to *Broken Hill* and *Silverton*, in New South Wales, crosses the North Line at PETERSBURGH. In the Northern Territory, a railway 150 miles in length connects PALMERSTON with PINE CREEK, at one time an important gold-mining centre.

The **Overland Telegraph Line** to *Port Darwin* was commenced in August, 1870, and completed in August, 1872, at a cost of nearly half-a-million sterling, or twice as much as the original estimate. This stupendous work crosses the continent almost along the route followed by McDouall Stuart in 1860. The distance from Adelaide to Port Darwin is 1,973 miles, and, although the difficulties and hardships encountered in making the line were very formidable, only 7 men died, out of some hundreds employed, which proves that the climate, though hot, must be extremely healthy. South Australia has also constructed a line through the even more inhospitable and desert region along the southern coast to *Eucla* at the head of the Great Australian Bight, where it connects with the overland line from Albany and Perth.

GOVERNMENT : A **Governor**, appointed by the Home Government, is at the head of the executive. The legislative power is exercised by a Parliament, which consists of a **Legislative Council** and a **House of Assembly.**

The **Legislative Council** and the **House of Assembly** are both elected, the former by voters holding property qualifications, the latter by manhood suffrage.

The Northern Territory is governed by a **Resident**, appointed by the authorities at Adelaide, and assisted by a small staff.

The **Revenue** and the **Expenditure** for 1894 each amounted to about 2¼ millions sterling. The **Public Debt** amounts to rather more than 21½ millions, one-half of which was spent on railways and tramways, and the rest on water-works and water conservation, harbour improvements, telegraphs, roads, school buildings, and other reproductive public works.

For the defence of the colony there is an efficient volunteer and militia force, and a permanent force of artillery, numbering in all about 2,200 men. A powerful gunboat—the *Protector*—is stationed off Port Adelaide, which is also defended by two strong forts.

Education is compulsory up to a certain standard, and free to all who are unable to pay the school fees. There is a training college for teachers, and a university at Adelaide authorised to grant degrees in arts, law, music, medicine, and science.

DIVISIONS : The settled portion of the colony is divided into Counties, Hundreds, and District Councils. There are also three Pastoral Districts—the Western, Northern, and North-Eastern.

The Counties, 44 in number, serve chiefly for electoral purposes. The division into District Councils is more important, as it confers the powers of a municipality, the ratepayers having the power of levying rates, granting licenses, &c., and of applying the funds so raised to making and repairing the roads, &c. The Hundreds are blocks of country thrown open for agricultural settlement.

The Pastoral Districts cover immense areas, the *Western* being 152,623 square miles, the *Northern* 121,650 square miles, and the *North-Eastern* 28,139 square miles.

In the Northern Territory, four counties—*Palmerston*, *Malmesbury*, *Disraeli*, and *Rosebery*—have been formed in the north-west, and one—*Gladstone*—in the basin of the Roper River on the Gulf coast.

TOWNS : Being principally a pastoral and agricultural country, South Australia contains very few towns of any considerable size. Adelaide, the capital, with a population, including its suburbs, of 120,000, is the only large town, and Port Adelaide is the only other town with a population of over 5,000.

ADELAIDE, the capital of South Australia and the seat of government, stands in a plain on the small River Torrens, about midway between the shores of the Gulf of St. Vincent and the Mount Lofty range of mountains. The city is surrounded by "Park Lands," a strip of which also extends between the residential part of the city—North Adelaide—and the business portion—South Adelaide. The streets in both divisions are built at right angles, and many of them are planted with trees. South Adelaide is built nearly in the form of a square, and is bounded by four terraces which face the park lands on the north, south, east, and west. The Torrens, an insignificant stream, has been, by means of a dam, converted into a fine expanse of water about 2 miles in length. Adelaide is now in direct railway communication with Melbourne, the journey occupying about 18½ hours. Most of the trade of the city, however, passes through Port Adelaide, which is situated on a fine natural harbour, formed by an inlet of the Gulf, and now deepened so as to admit all but the largest steamers. Port Adelaide is separated by a sandy tongue of land, about 2 miles in width, from the shore of the Gulf on which is situated the outport of Semaphore. Between the Semaphore pier and the mouth of Port Adelaide River is Larg's Bay, where a pier has been built to form an outer harbour, accessible to vessels of the largest tonnage in all weathers. Mail steamers also call at Glenelg, a beautiful watering-place, 6½ miles south-west of Adelaide, with which it is connected by two railways, trains running each way about half-hourly.

M

Other important ports are **Port Victor**, at the head of Encounter Bay, connected by rail with Adelaide and also with the Murray river port of **Goolwa**, on the western side of Lake Alexandrina ; **Beachport**, on Rivoli Bay, connected by rail with **Mount Gambier** (3), and the chief outlet of the rich agricultural district in the extreme south-east of the colony ; **Port Wakefield**, near the head of the Gulf of St. Vincent, about 60 miles north of Adelaide, with which it communicates by rail and by sea ; **Wallaroo** (2), the seaport of the famous Wallaroo copper mines, near which is the small town of **Kadina** (2).   A tramway, 12 miles in length, runs from Wallaroo Bay to **Moonta** (2), another great copper-mining centre and port, on the shores of Spencer Gulf.   About 100 miles further north, on the same side of the Gulf, is **Port Pirie** (2), the principal wheat port of the colony, and now busily engaged in forwarding supplies to, and shipping produce from, the silver-mining district of the Broken Hill and Barrier Ranges. Still further north, and also on the eastern side of Spencer Gulf, at the head of navigation, is **Port Augusta** (1), which has a fine natural harbour, and which, as the starting point of the *Great Northern Railway* (now open to Oodnadatta, and is being continued further north, and may, in time, span the continent) and as the outlet of the vast pastoral and large agricultural districts of the north, must become a still more important centre of trade in *wheat* and *wool*, and, perhaps, in mining products.

Of the inland towns, the most noteworthy are **Gawler** (3), 25 miles north-east of Adelaide, in a large wheat-growing district ; **Kapunda** (2), a copper-mining centre, about 50 miles north-east of Adelaide, on the *Kapunda and North-West Bend Railway*, which branches off from the main *North Line* at Roseworthy, where the Government has established an Agricultural College, and terminates at **Morgan**, an important river-port on the north-west bend of the Murray, and the chief place for the shipment of goods to, and the discharge of produce (principally wool) from, the interior ; and **Kooringa** or Burra (3), the town of the famous Burra Burra[1] copper mine, which, since 1844, has yielded over 4 million pounds' worth of copper ore.

In the *Northern Territory* the principal place is **Palmerston**, the capital and chief port, on Port Darwin.

PALMERSTON,[2] the metropolis of the Northern Territory, is situated on the eastern side of a splendid natural harbour, the well-known *Port Darwin*. A magnificent jetty of wood, faced with copper, has been built ; it runs out to a depth of 38 feet, so that vessels of the largest tonnage can lie alongside and load or discharge at all states of the tide.   For many years a good bush road, with substantial bridges over dangerous creeks, has connected Palmerston with the goldfields, which extend to **Pine Creek**, 150 miles to the south, and a railway is now being made to the same place and is practically completed.   The principal stations on the line between Palmerston and Pine Creek are **South-port Road Station, Adelaide River, Port Darwin Camp**, and **Burrundie**, the latter being the Government and the Police headquarters at the Reefs.   Southport, about 24 miles south (by water) of Palmerston, on a river debouching into one of the arms of Port Darwin, was formerly the principal port and depôt for the gold mines, but it is now deserted.

---

1. The Burra Burra mine was actively worked until a few years ago, when, owing to the low price of copper, operations had to be suspended.   The diminution in the yield of copper from the South Australian mines is, in fact, due to the fluctuations in price and not to any failure of the supply, which is practically inexhaustible.   The Burra Burra and other mines are still being worked on a small scale, but the mines at Wallaroo, Matooroo, and Moonta are the only important mines now working.

2. "Palmerston has many advantages as the site of a large town.   It is accessible to ocean-going vessels of the largest draught ; a natural site exists for a dry dock, and the projections of the coast are admirably suited for the erection of lighthouses and forts."—*Australian Handbook.*

PROGRESS: "South Australia can now point to the result of half-a-century of colonial enterprise and labour—to the energy, patience, and sagacity that, out of a wilderness occupied by a few wandering savages who did not cultivate a rod of ground, have built cities and towns, established harbours, constructed two thousand miles of railway, and thousands of miles of macadamized roads; spanned the continent with electric wire, raised corn in abundance for a considerable population, and shipped a large surplus to distant lands; planted orchards and vineyards; worked valuable mines that are known throughout the world; stocked the country with millions of sheep; built up a trade that, in proportion to the population, is hardly equalled by that of any other people; founded a commonwealth with the institutions of a free and Christian people, rejoicing in their privileges; and, notwithstanding the defects and inequalities belonging to every human society, possessing the comforts, luxuries, and refinements of older and larger communities."

WESTERN AUSTRALIA.

WESTERN AUSTRALIA includes the whole of the continent to the west of the meridian of 129° E., and is the largest, but the least populous, of all the Australian colonies.

Though founded in 1829, Western Australia[1] has, until lately, made but slow progress, yet the material elements of prosperity are by no means wanting. "She has been the Cinderella of the Australian family, and while her more fortunate sisters have got on in the world, have been gay and prosperous, and have received much company in the shape of immigrants, she has led a solitary and unnoticed existence." Now, however, the tide has turned—the long neglected resources of this vast colony are being actively turned to good account, and there is every reason to believe that the whole of the habitable region, which may be said to include most of the land within about 200 miles of the seaboard, will be settled by prosperous agriculturists, dairy farmers, and fruit growers, while the large areas of land admirably adapted for pastoral purposes will be fully utilized, and although mining and prospecting are yet in their infancy, it is now certain that belts of rich mineral country extend from one end of the colony to the other, and that, in the near future, Western Australia will prove to be one of the richest provinces of the continent.

BOUNDARIES: Western Australia is bounded on the *north* and *west* by the **Indian Ocean**, on the *east* by **South Australia** and its Northern Territory, and on the *south* by the **Southern Ocean**.

The colony is thus surrounded by the sea on all sides except the east, where it adjoins South Australia and the Northern Territory, the boundary from sea to sea being formed by the meridian of 129° E.

The *extreme points* of the mainland are **Cape Londonderry**, on the north; **Steep Point**, on the west; and **Peak Head**, on the south.

---

1. Western Australia was originally known as the "Swan River Settlement"—that settlement, however, was confined to the south-western corner of the present colony.

EXTENT : This vast colony has an area of over **one million square miles,** fully one-third of the continent, and no less than 18 times the size of England and Wales.

The greatest length, from Cape Londonderry on the north to Peak Head on the south, is 1,490 miles ; and the greatest breadth, from Steep Point on the west to the 129th meridian on the east, is 850 miles. The total area is estimated at **975,920** square miles, or, inclusive of the contiguous islands, over one million square miles. The population, in 1895, amounted to 90,000.

The occupied portion of the colony is confined to a belt of land between ALBANY, on King George's Sound, on the south, and WYNDHAM, on Cambridge Gulf in East Kimberley, on the north, a distance of about 1,400 miles from north to south, with an average breadth of from 100 to 200 miles.

COASTS : The coasts of Western Australia are indented by many inlets and estuaries, and are fringed by numerous islands.

The coast-line measures about 3,000 miles in length, an average of about 1 mile of coast to every 300 square miles of area. A line of coral reefs extends along a considerable portion of the coast, and protects it from the full force of the Indian Ocean. Between these reefs and the shore there is, in many places, safe anchorage.

Although some portions of the coast are deeply indented, Western Australia is rather deficient in good harbours, the only inlets deserving mention are King **George Sound** on the south coast, **Sharks Bay** on the west coast, **King Sound** on the north-west coast, and **Cambridge Gulf** on the northern coast. The other inlets are of little value, owing to the direction of the prevalent winds, the currents, their shallowness, or to bars at their entrance.

Of the numerous islands along the coast, only two are as yet of any importance. These are **Dirk Hartog's Island** off Shark's Bay, and **Rottnest Island** off Fremantle, both on the west coast.

The principal headlands are **Cape Londonderry,** the most northerly point of the colony ; **Cape Lévêque,** at the entrance to King Sound ; **North-West Cape,** at the entrance to Exmouth Gulf ; **Steep Point,** at the False Entrance from Sharks Bay ; **Cape Naturaliste,** on the western side of Géographe Bay ; **Cape Leeuwin,** the extreme south-western point of the colony ; **Cape Howe,** on the western side of Tor Bay ; **Peak Head,** the southernmost point ; **Bald Head,** at the South Channel entrance to King George Sound ; and **Cape Arid,** on the western side of the Great Australian Bight.

RELIEF : The surface of Western Australia is less diversified than that of any of the other Australian colonies, and the whole country is virtually a **vast plain,** often undulating, but generally flat, and broken only by the hill ranges, which stretch along the seaboard, and by isolated elevations and depressions in the interior.

The whole of the settled district,[2] nearly the size of France, is usually level, occasionally undulating, but never mountainous. The western seaboard, generally, is comparatively flat and of a sandy character, being composed chiefly of the detritus of old coral reefs, which has been again deposited by the

---

1. Recent discoveries of rich goldfields have attracted a large influx of miners, &c., and the population now amounts to about 90,000.
2. The account here given of the natural features of the colony is largely based on papers and speeches by two ex-governors—Sir F. A. Weld and Sir F. N. Broome—but principally upon the reports of Mr. H. P. Woodward, the Government Geologist of Western Australia, as given in the *Australian Handbook* (1895), p. 377-378.

action of water. The whole country from north to south, except the spots cleared for cultivation, may be described as one vast forest ; sometimes, but comparatively seldom, the traveller comes upon an open sand plain, covered with shrubs and flowering plants in infinite variety and exquisite beauty, and often, especially in the northern and eastern districts, low scrubby trees and bushes fill the place of timber ; but, taking the word 'forest' in its widest sense, as wild, woody, and bushy country, the colony, for long distances inland, is covered with one vast forest, stretching away into regions yet unexplored. A large portion of the western seaboard is very heavily timbered, and immense tracts of land in the south-west are covered by great forests of the extraordinarily durable jarrah, the karri and tuart (both eucalypti of enormous size), sandal-wood, and other valuable trees.

MOUNTAINS : The principal ranges are the **Stirling Range** in the south, the **Darling Range** in the west, and the **King Leopold Range** in the north.

The mountains of Western Australia are not remarkable for their height, none of them exceeding 4,000 feet, and few of them attaining an elevation of more than 2,000 feet. But many of them present a striking appearance, inasmuch as they rise abruptly from level plains. The **Stirling Range**, near Albany on the south coast, is the loftiest range in the settled district, and, "being perfectly isolated, and rising from a dead level plain, it is visible for an immense distance." In this range *Ellen's Peak* rises to a height of 3,420 feet, and *Mount Kyenerup* to 3,500 feet. The much longer and more important **Darling Range** extends for about 300 miles along the western coast almost due north and south, at a distance of 18 to 20 miles from the sea, "towards which it presents a steep face, and although it has no peaks over 1,500 feet in height, yet it has a more imposing appearance than the Roe Range, which runs parallel to it, but further east, and of which the highest peak, *Mount William*, reaches 3,000 feet above sea-level." In the partially explored districts in the north and north-west are numerous hill ranges and many isolated hills, one of which, *Mount Bruce*, rises near the sources of the Fortescue River in the north-west, to a height of 3,800 feet. The highest peaks in the **King Leopold Range** and other ranges in the north may prove to be still loftier.

RIVERS : The rivers of Western Australia are, with few exceptions, simply storm water channels, which carry off immense floods in the rainy season, but are dry, or consist only of occasional pools of water, during the rest of the year.

The **Swan** is the only river in the settled districts in the south-west which is capable of being navigated to any extent. Daily river steamers run from FREE-MANTLE, at its mouth, to PERTH, the capital of the colony, 12 miles up the river, and a steamer conveys goods and produce to and from Guildford, 9 miles above Perth. The Blackwood, in the south-west, falls into Flinders Bay, a short distance east of Cape Leeuwin. Many other small streams fall into the sea between Cape Leeuwin and Cape Arid, but not a single river or creek breaks the desolate shore-line of the Great Australian Bight.

The northern and north-western rivers have much longer courses, but few of them run throughout the year, and navigation is generally limited to the estuaries of the larger streams. The Murchison has a channel of some hundreds of miles in length, but, though in times of flood an immense amount of water rushes through it to the sea, it is generally dry, or consists of a mere chain of pools, at other seasons.   The **Gascoyne**, also a large river when in flood, flows into Sharks Bay through an arid and waterless sandstone plain, which extends along the coast from the Murchison River to the Ashburton.   With the valley of the Ashburton, the pastoral district of the North-West Division commences, and extends for 300 miles to the De Grey River.   The principal rivers of this division – the **Ashburton** and the **Fortescue**—are each about 200 miles long and 100 yards wide at their mouths, and flow through good alluvial land and some well-grassed plains.   There is also some good land on the **De Grey** River and its tributary, the **Oakover**.   The Kimberley Division, in the extreme north of the colony, is watered by numerous running streams, their banks generally covered with acacias, palms, small bamboos, and other tropical vegetation. The longest is the **Fitzroy**, which rises in the interior beyond the King Leopold Range and falls into King Sound.   Its chief tributary, the **Margaret River**, and the **Ord River**, which runs north into Cambridge Gulf, flow through the well-known *Kimberley goldfields*.   These rivers are generally running streams, and usually contain, in the driest seasons, some water in rock-holes and pools, then a welcome supply indeed to the hardy miners, who, in spite of the hot climate and hard fare, and no little danger from hostile natives, have made their way thither from Derby on the west or from Wyndham on the north.   Several of these northern rivers flow through alluvial plains of great fertility, which it is hoped may be utilized for the cultivation of tropical products.[1]

LAKES :   The so-called lakes are really immense salt pans or marshes, perfectly dry "except after heavy rains, when they may be covered with a few inches of water."

Lake Amadeus, on the eastern border, is mainly within South Australia.   A great number of "lakes"—Lakes **Moore**, **Austin**, **Barlee**, &c.—are marked on the map in what is called the " Lake District," between the Darling Range and the Great Desert, but they are little more than shallow expanses of salt or saline water, and in the hot season become wholly dried up.

CLIMATE :   The climate of Western Australia is one of the most healthful and enjoyable in the world.   In the north, there is a true tropical climate ; in the central portions of the coast region, the climate is like that of **Southern Italy** ; while in the south-west it is like that of the **South of England**, but the summer is much hotter, and the winter brighter and not nearly so cold.

The south-western division is also the most salubrious; it is very seldom too hot, and very rarely too cold.   Snow is never seen, and ice only early in the morning in the depth of winter.   At Perth, the temperature[2] may occasionally reach 100° F., but it never falls below 35°.   In this division, the year consists of two seasons, the wet and the dry—the former lasting from April to October, the latter from November to March.

Further north, about the Gascoyne and Murchison rivers, heavy rains fall in summer—the rest of the year is dry and healthy.

The entire region north of the Ashburton is within the Tropics ; here the summer heat is very great, especially in the Kimberley District, but it is dry and

---

1. The West Australian Government gives a bonus of 500 acres to anyone who produces, in the Kimberley Division, a certain amount of cotton, tobacco, sugar, tea, &c.

2. In 1889, the highest temperature at Perth was 107° F. in the shade in January, and the lowest 36° in May.   The mean temperature in the same year was, at Albany 58°, Bunbury 59°, York and Perth 61°, Freemantle 64°, Geraldton 66°, Carnarvon 72°, Onslow 75°, Derby 85°, and Wyndham 87°.

tempered by cool sea breezes, and the air is free from the moistness and malaria which characterize tropical climates generally.

The rainfall[1] varies from about 40 inches on the coast, from Albany to Fremantle, to less than 20 inches in the hills 50 miles inland, while in the interior there are only occasional showers during thunderstorms.   The severe droughts and heavy floods prevalent in Eastern Australia are almost unknown in the south-western districts, though common in the northern and central parts of the colony.

The north-western districts are sometimes swept by violent hurricanes,[2] which often cause great loss of life and property.

The healthfulness of the climate is proverbial—epidemic diseases are almost unknown, and it is particularly favourable to consumptive people.   Exposure to all weathers scarcely ever produces any ill effects,[3] and the mortality of the whole colony is said to have averaged only one per cent. since its formation, that of Great Britain being two and a half per cent.

PRODUCTIONS : Wool is the staple product, and the annual clip exceeds in value the pearls and pearl-shells, the timber and sandalwood, the gold, and other exportable produce of the colony.

The indigenous plants are generally similar to those of the other colonies, and comprise immense forests of valuable timber trees, such as the indestructible *jarrah*, the *tuart* or white gum, the *karri*, the *red* and *blue gum*, and other eucalypti, together with *shea oaks* and *wattles*—the bark of the latter being almost as valuable as oak-bark for tanning—while in the comparatively bare and sterile districts are found the most varied and beautiful plants.   "Some of the sandy plains, too poor to support a forest growth, are yet covered with *shrubs* and *flowering plants* in infinite variety and of exquisite beauty, among them being many of the choicest adornments of English greenhouses."

The native animals are also similar to those of the rest of the continent.   Among them are the *kangaroo, wombat, bandicoot*, and other marsupials, with the *dingo* or native dog—the terror of sheep farmers—and the *native cat*, an untameable carnivorous marsupial.   *Parrots* and *cockatoos* are numerous in the forests, and *emus, brush-turkeys, eagle-hawks*, and other large birds are met with in the interior.   From the red-beaked *black swan*, found on the Swan River, both the river and the settlement were named.   The West Australian waters swarm with *fish, alligators* abound in the northern rivers, there are several kinds of *snakes*, many of them poisonous, and *insect life* is prolific everywhere.

INDUSTRIES : Notwithstanding recent discoveries of large and productive *goldfields*, the large increase of land under *cultivation*, the valuable *pearl-fisheries*, and the large and increasing *timber trade*, the pastoral industry is by far the most important occupation, and the chief source of wealth to the colony.

Over 100 million acres of land are held for pastoral purposes, and, since the opening up of the pasture-lands of the north-west and north for sheep and cattle-farming, the colony has made great progress as a producer of wool and live

---

1. The average rainfall at Perth, from 1876 to 1889, was 33 inches. In 1890, the rainfall at Fremantle and Perth was 46 inches; at Derby, 31; at Geraldton, 29; at York, 23; at Carnarvon, 10; at Ashburton, 11; and at Cossack, 6.

2. The hurricane of April, 1887, almost annihi-

lated the pearling fleet, and caused great loss of life on the north-west coast.

3. "Throughout Western Australia, the traveller camps out at night, generally without any covering but a rug, and never seems to receive any injurious effects."

stock, and now possesses some 2½ million sheep, 180,000 cattle, and about 45,000 horses. Nearly a third of the sheep and cattle are in the Kimberley district in the north.

Agriculture, though now extending, has been retarded by the want of labour and the difficulty of transporting produce to market. The making of new roads and railways has, naturally been followed by a large increase in the area under cultivation, which even yet, however, only amounts to 180,000 acres—an insignificant acreage indeed in a country of over 1,000,000 square miles in extent, even granting that vast areas are sterile and uninviting and doomed to perpetual aridity and barrenness. Although the agricultural land available for settlement is limited, compared to the vast area of the colony, there is a large extent of good, and in many places excellent, soil suitable for European grains, fruits, and vegetables, most of which can be cultivated and brought to a high state of perfection. But *wheat, barley, oats,* and *potatoes,* though grown to some extent, are largely imported; and, although many districts are well adapted for dairy-farming, large quantities of *butter* and *cheese* and other farm products have to be imported from the other colonies. In the alluvial river plains of the north, both soil and climate are suited for the cultivation of sugar, cotton, and other tropical products. In the south, strawberries, apples, pears, &c., grow well, while grapes, oranges, apricots, figs, and bananas come to perfection in the warmer districts. The number of vineyards and olive gardens is increasing, and *wine* and *oil* of good quality are produced.

The pearl-fishery is an important industry along the north-west coast, especially in *Sharks Bay* and round *Cossack.*

Timber-cutting is a great source of wealth in the south-western districts, and a number of short railways and tramways have been laid down from the forests[1] to the shipping ports on the coast. The famous *jarrah*—a species of eucalyptus, but better known as the West Australian mahogany—is in demand all over the world for all purposes requiring durability and imperviousness to the white ant and the *teredo navalis.* The *karri,* which abounds along the south coast to the west of Albany, is even stronger and larger, growing sometimes to a height of 300 feet. The *sandalwood* tree abounds principally in the York district.

MINERALS: The mineral resources of Western Australia are as yet imperfectly known, but when the deposits of gold, copper, lead, tin, and coal, now worked or known to exist, are fully developed, the colony may prove to be as richly endowed in this respect as any of the other colonies.

Gold is said to have been discovered on the north-western coast, in 1688, by the bold buccaneer, Dampier, and in some old Dutch charts this part of the coast is marked "Provincia Aurifera." Exactly two centuries later, in 1888, rich alluvial deposits were found in the same district, and much gold has been already obtained from the Pilbarra field, on the Yule River, 80 miles east of Roebourne. The new goldfields opened in 1890, on the Ashburton River, are most promising, but more gold has been taken from the field at Nullyagine, on the De Grey River, than from any in the colony. In 1887, rich gold-bearing quartz reefs were found in the Yilgarn Hills, about 200 miles east of Perth, and the precious

---

1. The forests of Western Australia cover an area of 37,000 square miles, and include *white gum,* 10,000 square miles; *jarrah,* 14,300 square miles; | *karri,* 2,300 square miles; *red gum,* 8,000 square miles; and *York gum,* 2,400 square miles.

metal has been found to the north of Lake Austin, and at other places, but the greatest rush has been to Southern Cross and Coolgardie, the chief centres of the Yilgarn Goldfields. The **Kimberley goldfield**, in the north of the colony, is traversed by the headwaters of the Ord and the Margaret rivers, and lies about 350 miles east of Derby, on King Sound, and 300 miles south of Wyndham, on Cambridge Gulf, the two ports of the district. Nearly 60,000 ozs. of gold, valued at £226,000, were exported from the four fields of Kimberley, Murchison, Pilbarra, and Yilgarn in 1892, and about 100,000 ozs., value £420,000, in 1893.

Fine lodes of **lead** and **copper** have been worked near NORTHAMPTON and at other places, but, owing to unremunerative prices, the mines are now closed. Good coal has been discovered at WYNDHAM on Cambridge Gulf in the north, and coal seams also crop out on the upper branches of the Irwin River, to the north-east of Geraldton. Rich deposits of **stream tin** are now worked on the Blackwood River near Bridgetown, in the extreme south-west, and this valuable mineral has also been found on the goldfields at Roebourne on the north-west coast.

COMMERCE : About half the trade of Western Australia is carried on with the *United Kingdom*, and the rest mainly with the other *Australian Colonies*. Value, in 1894-5, 3¼ millions sterling ; imports, 2 millions ; exports, 1¼ millions.

The trade of the colony consists in the export of raw produce, such as **wool** and **gold**—which amount in value to four-fifths of the total exports—**timber** and **sandalwood**, **pearl-shells** and **pearls**, **hides and skins**, **guano**, **bêche de mer**, and **horses**, and in the import of **manufactured goods**, principally from the mother country. The exports to Great Britain consist almost entirely of gold, wool, pearl-shells, and timber.

The chief ports are FREMANTLE and ALBANY in the south ; and DERBY and WYNDHAM in the north.

COMMUNICATIONS: The P. & O., the Orient, and the Messageries Maritimes steamers call at Albany, whence the mails are sent to Perth by rail. Coasting steamers run regularly from port to port, and a line of steamers runs direct to London, *via* Singapore. There are about **500** miles of *railway* open for traffic, and several hundred miles are under construction. The principal line, 354 miles in length, runs from Fremantle to Perth, and thence by Guildford, York, and Beverley to Albany on King George's Sound. A branch line is now open from Spencer's Brook to Southern Cross. Other lines open or in course of construction are (1) from Geraldton northwards to Northampton and southwards to Guildford on the *Eastern Railway*, and eastwards to Millewa, the present terminus for the Murchison goldfields ; (2) Bunbury north to Perth, and south to the timber district ; (3) a tramway from Cossack to Roebourne ; and (4) some 50 miles of railways and tramways constructed by the Timber Companies, for bringing timber from the forest ranges to the sea. It is proposed to construct a railway under a Land Grant from York right across the Lake District to the port of Eucla, at the head of the Great Australian Bight, and other lines are projected in the northern half of the colony. All the chief centres are connected by telegraph, and the Western Australian system is connected with that of South Australia at Eucla, and has also been extended northward from Perth to Roebourne and Derby, and thence to the Kimberley goldfields and Wyndham.

GOVERNMENT : Western Australia now possesses full responsible government.

Western Australia has been the last of the Australian colonies to obtain the privileges of complete self-government.   As in the other colonies, the **Governor** represents the Queen and holds the executive power, while the legislative authority is vested in a nominated **Legislative Council** and an elected **Legislative Assembly** representing the 33 constituencies into which the colony has been divided.

The **Revenue**, in 1894-5, amounted to £1,000,000, and the **Expenditure** to £930,000, while the **Public Debt** was 4 millions sterling.

For the **defence** of the colony there is a small force of artillery and rifle volunteers, but there are no regular forces or military works, with the exception of two small forts at the entrance to King George's Sound.

Education is compulsory, but not free.   Both the Government and the Assisted Schools are under inspection.   There is a High School at Perth and a Grammar School at Fremantle.

DIVISIONS and TOWNS : There are six land divisions, namely, the **South-West Division,** the greater part of which has been sub-divided into *counties*, and the **Gascoyne, North-West, Kimberley, Eucla,** and **Eastern Divisions.**

The **SOUTH-WEST DIVISION** includes the settled districts, and is by far the best part of the colony. It has three well marked natural divisions, (1) the *coastlands*, which are admirably adapted for agriculture and pasture, (2) the *great forest belt*, extending from the ranges to within from 10 to 15 miles of the sea, with inexhaustible supplies of the finest timber, and (3) the *uplands*, with a mean elevation of about 1,000 feet above sea-level, between the forests and the unoccupied interior, and along which the railway from Albany to Beverley (and Perth) runs.

The chief towns in this division are **PERTH** (9), the capital of the colony, prettily situated on a fine lake-like reach of the Swan River, about 12 miles above its port, **Fremantle** (5), at the mouth of the river ; **Guildford,** also on the Swan, 9 miles above Perth, a charming little town surrounded by fields and vineyards ; **York** (3), 80 miles east of Perth, on the *Eastern Railway,* which connects at **Beverley** with the *Great Southern Railway,* which runs thence to the principal port of the colony, **Albany** (2), an important port of call on King George's Sound.   From the port of **Bunbury,** on the western coast, much timber, sandalwood, tin (from the Greenbushes Tin Fields on the Blackwood River), and many horses are exported.

The **Victoria District,** in the northern part of this division, contains good pasture lands, and there are large wheat farms between the Irwin River and **Geraldton** (1), the capital and chief port of the district.   From Geraldton a short railway runs north to the lead and copper mining town of **Northampton,** and south to **Walkaway,** whence a line 300 miles in length runs to Guildford on the Eastern Railway.   Another line runs eastwards to **Millewa.**

The **GASCOYNE DIVISION** extends from the Murchison River to the Ashburton, and takes its name from the Gascoyne River, which flows through it into Sharks Bay.   There is much good pastoral land about **Carnarvon,** the principal port at the mouth of the Gascoyne.   The basin of the Murchison River is rich in gold.   There is an important pearl fishery in Sharks Bay.

The **NORTH-WEST DIVISION,** another pastoral and mining district, extends along the coast for over 400 miles.   It includes the basins of the Lower

Ashburton, the Fortescue, the De Grey, and other rivers, and, though the rainfall is uncertain, large areas are well grassed and provided with water for stock. Much gold has been obtained from the Pilbarra and the Nullyagine goldfields. Roebourne, the chief town in the district, is connected by a tramway, 8 miles in length, with the port of Cossack, a great centre of the pearl and pearl-shell fisheries.

KIMBERLEY DIVISION, in the extreme north, is already very largely occupied by sheep and cattle farmers, and the *goldfields* in East Kimberley may ultimately prove very productive. The fields are about 350 miles east of Derby, the capital and chief port of the division, picturesquely situated on King Sound, and some 200 miles south of Wyndham, a rising port on Cambridge Gulf.

The vast EASTERN DIVISION is as yet imperfectly known, but it appears to be, for the most part, a waterless desert, covered here and there with spinifex or with mulga scrub. In the lake district, on the western side of this division, are numerous shallow depressions, which are filled with water after the rains, but are generally perfectly dry. The *Yilgarn goldfields*, in the south-west, are about 200 miles east of Perth. The chief gold-mining centres in this district are Southern Cross and Coolgardie.

The EUCLA DIVISION includes the south-eastern part of the colony, and, though the Hampton Plains are now being utilised for pastoral purposes, most of the division is sterile and uninhabited. At Eucla, a station on the border of South Australia at the head of the Great Australian Bight, the telegraph line of Western Australia connects with the South Australian system.

TASMANIA.

TASMANIA is an island, nearly as large as Ireland, situated to the south-east of Australia, from which it is separated by Bass Strait, a broad channel of from 80 to 150 miles in width. It is by far the smallest of the Australian colonies, but it is, in many respects, the most interesting and certainly the most beautiful of the "New Englands" under the Southern Cross.

The "Garden of the South," as Tasmania is justly called, is a "beautiful and well-watered island, rich in harbours and inlets, traversed by high mountain chains, full of crags, glens, and ravines of commanding appearance. Everywhere on the coast there are good anchorages and many excellent harbours. Altogether, the coast offers the most charming scenery, being for the most part bold and rocky. The interior, especially, is delightful, and here are united, so to speak, the climate of Italy, the beauty of the Apennines, and the fertility of England. Mountain and valley, hill and dale, crowned with high forests and rich pasture grounds in the plains, afford the most pleasing variety." And yet, with an area of more than half that of England, it has a population of less than 150,000, or one-tenth that of Wales.

Tasmania takes its name from the Dutch navigator, Tasman, who discovered it in 1642, and named it Van Diemen's Land, in honour of the Governor of the Dutch East Indies, under whose orders he had sailed to explore the "Great South Land." A penal settlement was established on the present site of Hobart in 1803. Free settlers followed, and, in 1825, the island was formed into an independent colony. Transportation was abolished in 1853, and, in 1856, a Constitutional Government was granted, and the name was then changed from Van Diemen's Land to Tasmania.

BOUNDARIES: Tasmania is bounded by **Bass Strait** on the *north*, by the **Tasman Sea** on the *east*, and by the **Southern Ocean** on the *south* and *west*.

Bass Strait separates Tasmania from the coast of *Victoria*, a line drawn from the southernmost point of that colony, Wilson Promontory, would pass through the middle of the island. Tasman Sea is the name given to that part of the *South Pacific Ocean* which lies between Australia and Tasmania on the west, and New Zealand on the east. The Southern Ocean, to the west of Tasmania, is, strictly speaking, a part of the *Indian Ocean*.

EXTENT: This heart-shaped island is about 200 miles in length from north to south, and a little less from east to west, while the total area, including the lakes and islands, is over 26,000 square miles, or more than half that of England.

The main island has an area of about 24,330 square miles; the smaller islands, most of which lie off the northern and south-eastern coasts, cover in all about 1,800 square miles.

Tasmania is roughly triangular in shape, the base being to the north and the apex to the south. The three **extreme points** are *Cape Grim* on the north-west, *Cape Portland* on the north-east, and *South Cape* on the south.

COASTS: "The comparatively smooth **north coast** is broken by the long estuary of the Tamar; the **west coast** is a line of cliffs with one great inlet, Macquarie Harbour, about the middle; but the southern outcurve and the **east coast** are split into a labyrinth of long inlets, irregular peninsulas, and rocky islands like Western Scotland."

Although the Tasmanian coast is, on the whole, bold and rocky, many of the numerous estuaries and bays form excellent harbours. Even on the inhospitable *west coast* there are at least three accessible ports—**Port Davey**, formerly much frequented by whaling vessels, **Macquarie Harbour**, a large inlet running inland for about 25 miles, and the estuary of the **Pieman River**. On the *north coast*, besides **Port Dalrymple** at the mouth of the Tamar, there are several harbours such as those of **Port Sorell**, **Devonport** at the mouth of the Mersey, **Emu Bay**, and **Stanley** or **Circular Head**. On the *east coast*, the chief openings are **Oyster Bay** and **George's Bay**. The *south* and *south-east coasts* are studded with safe bays and harbours, the principal being **Port Arthur**, in Tasman Peninsula, **Storm Bay**, leading into the estuary of the Derwent (on which stands Hobart, the capital of the colony), **Frederick Henry Bay**, and **Norfolk Bay**; with **Récherche Bay**, **Southport**, and **Port Esperance**, on the western side of D'Entrecasteaux Channel.

Of the 55 *islands* which belong to Tasmania, the largest are **Flinders Island** and **Cape Barren Island**, in the **Furneaux Group**, at the east end of Bass Strait, and **King's Island** at the western entrance; with **Hunter's Island** off the north-west coast, **Schouten and Maria Islands** on the east coast, and the double **Bruni Island** on the south. "The 100-fathom line round Australia includes all these islands, and Tasmania itself indicates the former union of the two countries."

The three chief *peninsulas* are **Freycinet Peninsula** on the east coast, and the double **Tasman and Forestier Peninsula**, with **Ralph Bay Peninsula**, on the south-east coast.

NATURAL FEATURES : Tasmania is a mountainous country, and high ranges of hills and isolated peaks, rocky precipices and tortuous ravines, mountain lakes, rushing streams, and picturesque waterfalls, alternate with beautiful valleys, fertile plains, and grassy uplands.

Tasmania has been called the "Switzerland of the South," and is perhaps the most thoroughly mountainous island in the world. There are scarcely any continuous mountain ranges, the entire surface of the island being a most irregular and picturesque succession of mountains and valleys, peaks and glens, and presenting every variety of the most pleasing scenery, which, in many places, is thoroughly English and reminds one of the finest parts of Kent and Surrey. "It is England all over. Everywhere you descry lovely country houses, surrounded by fine gardens, extensive shrubberies, verdant parks and lawns, fields in pasture or under the plough, and dense woods sloping down from the hills. There are lanes here than which there is nothing more thoroughly English even in the beautiful county of Kent itself, but, amid all this English outlook, the new comer is reminded that he is not at home by the appearance of gaudily-coloured parrots and other birds unknown in the mother country."

MOUNTAINS : On either side of the deep valley or glen which runs right across the island, from the estuary of the Tamar on the north to that of the Derwent on the south, are several irregular mountain ranges and extensive tracts of high tableland, which culminate in **Cradle Mountain**, 5,069 feet, on the west, and **Ben Lomond**, 5,010 feet, on the east.

The **Eastern Range** winds in the form of an irregular Z, at an average distance of 40 miles from the coast. It has an average height of about 3,000 feet, and several peaks attain an elevation of about 4,000 feet, but only its highest point, **Ben Lomond**, rises above 5,000 feet.

The **Western Highlands** are rather loftier and much more extensive than the Eastern Range. The central tableland is traversed and edged by several ranges, of which the **Great Western Range** rises in Table Mountain to 3,600 feet, and in Miller's Bluff to nearly 4,000 feet. Further north, several peaks rise over 4,000 feet, and further west, Cradle Mountain, the loftiest summit in the colony, rises to a height of over 5,000 feet above the level of the sea.

RIVERS : Tasmania is well-watered by numerous rivers, some of them of considerable size, the largest being the **Derwent** in the south and the **Tamar** in the north.

The **Derwent**, which issues from Lake St. Clair, and receives the overflow of the three other alpine lakes of the central tableland, has a course of 130 miles, and is the longest river in the colony. HOBART, the capital, stands on the western side of its estuary, which forms one of the finest harbours in the Southern Hemisphere.

The **Tamar**, the chief river of the north, is a tidal river 45 miles in length, formed by the confluence of the North and South Esk at LAUNCESTON, to which it is navigable for vessels drawing 16 feet of water. Its mouth is known as Port Dalrymple.

The Davey and the Huon rivers in the south are also navigable streams, and nearly all the rivers which discharge into Bass Strait on the north are navigable at their mouth for medium-sized craft.   There are several river harbours for small vessels on the east coast.   On the west, the chief rivers are the Gordon, which flows into Macquarie Harbour, and the Pieman River, which is fed by numerous streams from the Du Cane Range and the Surrey Hills.

LAKES : "Tasmania shows itself to be a truly alpine region by the possession of numerous mountain lakes near the sources of its rivers."

The largest are the Great Lake, 12 miles in length and covering an area of 44 square miles, Lake St. Clair, Lake Echo, and Lake Sorell, all drained into the Derwent.   The much smaller Woods Lake and Arthur Lake are drained by the Lake River into the South Esk and the Tamar.   Most of these beautiful lakes are very deep, and are analogous in formation and scenery to the charming mountain lakes of Switzerland and Scotland. [1]

CLIMATE : The climate of Tasmania is admirable.   It is hardly ever hot or unpleasantly cold, and the weather is less variable than it is in England. [2]

Tasmania is undoubtedly one of the healthiest countries in the world, and the cause assigned for most of the deaths is always "old age."   The hot winds of Australia rarely reach Tasmania, and, when they do, they are never of long duration.   It is naturally cooler in summer than any of the adjoining colonies, and the winters are as mild as those of the south of France. [3] Snow rarely falls at Hobart, but Mount Wellington, which overlooks the town, is sometimes covered with it even in the summer months.   The rainfall varies greatly, not only in different parts of the island, but also at the same place.

The western half of the island is much wetter than the eastern portion.   The western and south-western counties are always very wet, Mount Bischoff having a fall of over 80 inches, and Corinna, on the Pieman River, over 70 inches, while at Macquarie Harbour it often amounts to over 100 inches a year.   Hobart and the east coast have a range of from 14 to 40 inches—the average being about 24 inches.

The winds are often violent, but thunderstorms are rare.   "The atmosphere is rich in ozone, and epidemic diseases are almost unknown.   The climate is highly favourable to infant life, and especially restorative to constitutions enfeebled in warmer countries."

". "Tasmania has for some years become the summer resort of large numbers of visitors, who come from the hotter climates of Australia to enjoy the comparatively cool and health-restoring breezes.   They, for the most part, flock to Hobart, where, during the months of January and February,[4] every hotel and lodging-house is crowded.   For many years the Australian Squadron has also spent some weeks in the harbour at Hobart during this season.   The visitors find abundant occupation in excursions on the river, in driving along the slopes of Mount Wellington to the Huon River, through forests and romantic scenery ; in ascending Mount Wellington, and enjoying a walk in one of its fern valleys, by a rippling stream, under the shade of fern trees, sassafras, and eucalypti, and in collecting flowers and berries of every hue.   Dances, pic-nics, and other entertainments are of daily occurrence." [5]

<hr/>

1. All these lakes lie at an altitude of between 3,000 and 4,600 feet above the level of the sea.
2. Sir Edward Braddon.
3. The average summer temperature is about 62°, and that of winter 45°—the mean annual temperature being 55° F.

4. The seasons are, of course, nearly the opposite to what they are in England ; summer commences in December, so that a Tasmanian Christmas is a very different thing from its prototype in the old country.
5. Sir W. L. Dobson.

PRODUCTIONS : With some remarkable exceptions, the *indigenous plants* and *native animals* of Tasmania are similar to those of Australia. Both climate and soil are extremely favourable to the cultivation of English cereals and fruits, and sheep, cattle, and horses thrive on the luxuriant pastures, while the rich mines of tin, gold, and coal are a great source of wealth to this prosperous colony.

The flora of Tasmania is essentially Australian, and closely resembles that of the uplands of Victoria. The forests abound with valuable timber trees—the blue gum, which, in some of the southern counties, attains a height of 350 feet and a girth of 100 feet and rivals the giant eucalypti of Victoria, while the celebrated Huon pine supplies the finest timber for shipbuilding. Beautiful flowers carpet the lovely glens and grassy uplands, and large tree-ferns are plentiful in the mountain gullies and deep ravines.

The kangaroo, wallaby, wombat, and opossum of Tasmania are the same as those of the mainland, and that ornithological curiosity, the duckbilled platypus and its relative, the *echidna setosa* or porcupine ant-eater, are common throughout the island. But the most interesting, as well as the most formidable, of Tasmanian animals—the beautiful tiger-wolf and the fierce little Tasmanian devil—are both quite unknown on the mainland, although their bones have been found in a fossil state in the caves of New South Wales. These savage animals often cause great loss to the sheep-farmers in the outlying districts, but they are becoming scarce, as also is another carnivorous marsupial, known as the native cat, which is said to have a most decided *penchant* for chickens.

Owing to the salubrity and comparative coldness of the climate, together with an abundant supply of food and water all the year round, the domestic animals of Tasmania are, on the whole, superior to those of the mainland, and the stud sheep, cattle, and horses exported to Victoria and New South Wales often command fabulous prices. The wool is also highly esteemed. [1]

AGRICULTURE : Most of the European grains, fruits, and vegetables can be cultivated and brought to perfection in this colony, and some tropical plants also thrive in certain localities. [2]

The soil and climate are all that could be desired for the cultivation of all the English cereals, fruits, trees, and plants, and, "on account of the mildness of the winters and the greater amount of sunshine, their growth is more rapid, and the production of fruit especially is much more certain and abundant than in England." But out of a total acreage of 17 millions, only about half a million acres are as yet under cultivation. Wheat, oats, and barley are largely grown, and *gardens* and *orchards* cover no less than 10,000 acres. About half a million bushels of the finest apples and 30,000 bushels of pears are annually produced, besides large quantities of grapes, figs, raisins, currants, strawberries, and other fruits, all of which come to perfection. *Fruit-growing* and *jam-making* are already important industries, and green fruit and jam are largely exported. On Maria Island, on the east coast, grapes are grown for *wine-making*. Hops are also largely grown, and much beer is brewed for export to the adjoining colonies.

---

1. There were, in 1883, over 1¾ million sheep, and 174,000 cattle, and 30,000 horses in the colony. | 2. T. C. Just (*Handbook of Tasmania*).

MINERALS: Mining, principally for tin and **gold**, is the most important industry in the colony, but many rich mines of coal and **silver-lead** are also worked, and excellent **slate** and **stone** are quarried. *Iron ore* exists in abundance, and *copper, zinc, bismuth, antimony, asbestos,* and *precious stones* are also found.

Tin : The extensive and extremely rich deposits of tin ore at *Mount Bischoff*,[1] in the north-west of the island, were discovered in **1872**, and this famous mine, together with the productive workings round *Ringarooma* and *Portland,* in the extreme north-eastern corner, have yielded in all about 6 million pounds' worth of tin—the annual output of ore now amounts to about a quarter of a million sterling.

Gold occurs throughout the northern and western districts, both in alluvium and in quartz veins. The principal gold mines are at *Beaconsfield*, on the western side of the Tamar, which produce twice as much gold as the north-eastern and the west coast mines. The west coast diggings are at *Mount Lyell* and along the *King River* and around CORINNA on the *Pieman River.* Important discoveries of silver and silver lead were made in 1889, at *Zeehan* and *Dundas,* on the west coast, just south of the Pieman River. Rich deposits of both copper and silver have also been found at Mount Lyell in the same district.

Coal is widely distributed, and mines of excellent coal are worked at *Fingal* and *Mount Nicholas* (or MILLBROOK), in the north-east ; at the *River Mersey,* on the north ; at *New Town,* near Hobart ; and elsewhere. The total output, however, does not exceed 50,000 tons a year. Fine **slate** is quarried at *Selby,* in the north-east, and there are large **stone** quarries round Hobart and Launceston.

COMMERCE: The commerce of Tasmania is carried on almost entirely with the adjoining colonies and the mother country. Value in 1893, 2½ millions sterling, the exports being slightly larger than the imports.

More than half the trade of the colony is carried on with Victoria, and the rest principally with the United Kingdom and New South Wales, and to a much less extent with New Zealand and South Australia. Wool and minerals, principally tin and gold, comprise more than half the *exports*—the rest include **green fruit** and **jam**, **potatoes**, **timber** and **bark**, **hops**, **hides** and **skins**, **sheep** and **horses**, &c. The *imports* are chiefly textile fabrics and other manufactured goods, and articles of food and drink.

The direct trade with foreign countries is practically *nil*, and it is estimated that fully one-half of the intercolonial trade of the island is really with England, goods from, and produce for, the United Kingdom, being received or sent through Melbourne and Sydney, to which the large steamers of the *Tasmanian Steam Navigation Company* run regularly from HOBART and LAUNCESTON, the two chief ports of the colony.

There are about 6,000 miles of good roads and over 400 miles of railways. Telegraph lines traverse all the settled districts, and a submarine cable connects the island with Melbourne.

GOVERNMENT: The **Parliament of Tasmania** consists of a Legislative Council and a House of Assembly. The **Governor** is aided by an **Executive Council**.

---

1. The Mount Bischoff ore is extremely rich, and | is sent by rail to Emu Bay for shipment.
y elds from 70 to 80 per cent. of pure tin. The ore |

Both the **Legislative Council** and the **House of Assembly** are elected by duly qualified voters. The executive power is vested in the **Governor**, aided by a **Cabinet** of four responsible ministers—the Chief Secretary, the Treasurer, the Attorney-General, and the Minister of Lands and Works—who, with other Ministers of the Crown, form the **Executive Council**.

The **Revenue**, in 1893, amounted to a little under three-quarters of a million ; the **Expenditure** to a little over three-quarters of a million. The **Public Debt**, in 1893, was about 7½ millions sterling, the whole raised for public works.

For the **defence** of the colony there is a small volunteer force. The Derwent and the Tamar are defended by strong batteries. Hobart is the summer station of the Australian squadron.

Education is compulsory and unsectarian. The elementary schools are under Government control. There are many grammar schools and private colleges, and an Act of 1889 authorised the establishment of a university at Hobart.

DIVISIONS : Tasmania is divided into **18 counties**, and these are again subdivided into **parishes**.

Only 4 of the counties are entirely inland. Of the rest, 3 are on the north coast, 2 on the east, 4 on the south, and 5 on the west.

The **Northern Counties** are *Wellington, Devon,* and *Dorset.*

The **Eastern Counties** are *Cornwall* and *Glamorgan.*

The **Southern Counties** are *Pembroke, Monmouth, Buckingham* (the metropolitan county), *Kent,* and *Arthur.*

The **Western Counties** are *Montgomery, Franklin, Montagu,* and *Russell.*

The **Inland Counties** are *Lincoln, Westmoreland, Somerset,* and *Cumberland.*

TOWNS : The largest towns are **Hobart**, on the Derwent, in the south, and **Launceston**, on the Tamar, in the north.

HOBART (36), the capital and seat of government, is picturesquely situated at the foot of Mount Wellington, on the River Derwent, about 12 miles from its mouth. Mount Wellington is often snow-capped even in the middle of summer, and the Derwent, which forms one of the finest natural harbours in the Southern Hemisphere, is here 2 miles wide and has sufficient depth and capacity for almost any number of vessels of the largest tonnage. The country around Hobart is delightful, and the city is a favourite place of residence and summer resort of wealthy Australians. Mr. Anthony Trollope thus speaks of this attractive city :—" It is beautifully situated, just at the point where the river becomes sea, and is surrounded by hills and mountains from which views can be had which would make the fortune of any district in Europe. And the air of Hobart is perfect air. The summer weather is delicious. All fruits which are not tropical grow to perfection, and grapes ripen in the open air. . . . . So much in regard to the gifts bestowed by nature on the Tasmanian capital. Art has made it a pretty, clean, well constructed town, with good streets and handsome buildings."

Launceston (22), the only other large town, is a fine city on the Tamar, about 40 miles from its mouth (Port Dalrymple), and at the confluence of the North and South Esk rivers. Though not so populous as Hobart, its trade is as important and even greater in value than that of the capital, its nearness to Australia giving it a great advantage as an outlet for the mineral and other products of the colony.

N

The most important of the smaller towns are **Beaconsfield**, a gold-mining centre on the west bank of the River Tamar, in the county of Devon ; **Waratah**, the township at the foot of the famous Mount Bischoff, the tin from which is conveyed by rail to Emu Bay (**Burnie**) and shipped thence to Launceston for smelting ; **Ringarooma**, the shipping port for the tin mines in the north-east ; **Georgetown**, a watering-place at the mouth of the Tamar ; **Devonport**, which includes **Formby**, with **Torquay** (Devonport East) at the mouth of the Mersey, and **Latrobe**, at the head of the estuary ; the agricultural centres of **Deloraine** and **Westbury**, both on the railway which leaves the Hobart and Launceston Main Line at Evansdale Junction and is now open to Ulverstone, on the northern coast, and is to be extended to Emu Bay ; **Stanley** or Circular Head, the chief port in the north-west ; **Corinna**, the centre of the Pieman River gold-fields ; **Zeehan** and **Dundas**, the chief centres of the north-west silver fields—Zeehan is the terminus of a railway to **Strahan**, a rising port on the northern shores of Macquarie Harbour ; **Franklin**,[1] on the Huon River, famous for its apples, pears, and jam fruits ; **New Norfolk**, in the hop-growing district north of Hobart ; **Longford**, on the Norfolk Plains, the "Garden of Tasmania" ; and **Fingal**, a coal-mining town on the South Esk River, 120 miles north-east of Hobart, and 70 miles south-east of Launceston. There are a large number of other delightful little towns and pretty villages in this prosperous and pre-eminently British colony.

NEW ZEALAND.

THE COLONY OF NEW ZEALAND consists of two large islands known as the **North Island** and the **South Island**, together with a much smaller island called **Stewart Island**, to the south of South Island, and a number of outlying islands collectively known as the **Off Islands**—the whole group being situated in the South Pacific Ocean, about 1,000 miles to the south-east of Australia.[2]

New Zealand was discovered in December, 1640, by the famous Dutch navigator, Tasman, who gave it the name, first of all, of *Staaten* or *Staatenland*, in honour of the States-General or Parliament of Holland, afterwards altering it to *Nova Zeelanda*, after his native province of Zeeland in Holland. Tasman did not land on any part of the islands, and no European is known to have visited the islands until 1769, when the celebrated Captain Cook landed at Poverty Bay, on the east coast of North Island. The account which the natives themselves gave of their impressions of Cook's arrival, is recorded by Mr. Polack, who had it from the mouths of their children in 1836. "They took the ship at first for a gigantic bird, and were struck with the beauty and size of its wings, as they supposed the sails to be. But on seeing a smaller bird, unfledged, descending into the water, and a number of parti-coloured beings, apparently in human shape, the bird was regarded as a houseful of divinities. Nothing could exceed their astonishment. The sudden death of

1. Named after the famous Sir John Franklin, who was Governor of Tasmania from 1837 to 1843. Under Sir John, assisted by his noble wife, the colony made great progress.

2. The three islands of New Zealand were named *New Ulster*, *New Munster*, and *New Leinster* respectively, by the first Governor of the colony, an Irishman, "because New Zealand, like Ireland, had no roads." Subsequently they were distinguished as *North Island*, *Middle Island*, and *South Island* respectively, but the two main islands are now known as the *North Island* and the *South Island*—the third island being always called *Stewart Island*, after the settler who first discovered that it was a separate island, and not, as had been supposed, a part of South Island.

"The Maori name of North Island was *Te Ika a Maui*, 'the fish of Maui,' a native hero or deity. South Island was called *Te Wahi Pounamu*, 'the land of the Greenstone.' Greenstone, or jade, was the substance from which their weapons of offence, as well as their symbols of authority, were made, and supplied the natives with 'stone implements,' and being formerly to them as valuable as metal is to ourselves."—*Bowden's Manual of New Zealand Geography*, p. 3. (London : George Philip & Son).

their chief (it proved to be their great fighting general) was regarded as a thunderbolt of these new gods. To revenge themselves was the dearest wish of the tribe, but how to accomplish it with divinities who could kill them at a distance was difficult to determine." Cook took formal possession of the islands and spent altogether 327 days in surveying the coasts, &c., quitting it for the last time in 1777. Soon after, the islands became a favourite resort of British, French, and American whalers, whose stations were scattered along the southern coasts and on both sides of Cook Strait. Australian traders then began to visit the country, and, in 1814, the first missionary station was established, but the first actual settlement was not made until 1839, and in the following year New Zealand became a British Colony, Captain Hobson having concluded the *Treaty of Waitangi*, by which the native chiefs ceded the sovereignty of the island to Great Britain. The progress of the colony was greatly checked at various times by wars with the natives, whose power was not finally broken until 1881, and, though some disturbances have since occurred, no further trouble is probable or even possible.

BOUNDARIES: The **South Pacific Ocean** is the boundary of New Zealand on all sides. That part of it which lies between New Zealand and Australia is now distinguished as the **Tasman Sea,** in honour of the first discoverer of New Zealand and Tasmania.

The position of New Zealand in the South Pacific Ocean is almost at the very antipodes of the British Isles in the North Atlantic. A line drawn from Greenwich through the centre of the globe and continued to the surface on the opposite side, would reappear near *Antipodes Islets*, which are only a few hundred miles to the south of South Island. But, although the geographical position of the "Britain of the South" is almost the counterpart of that of the Britain of the North, its environment is entirely different—the British Islands being situated in the middle of the greatest extent of land on the globe, and New Zealand almost in the midst of the greatest extent of water; further, while Great Britain is, as it were, moored alongside the largest of the land masses, New Zealand is divided by over 1,000 miles of sea from the nearest, and that the smallest, of the continents. But, "looking at the extent, climate, fertility, abundant coast-line and harbours, adaptation for trade, and the bright future that awaits it in connection with the development of Australasia," New Zealand justly merits the title of the "Britain of the South."

In shape, New Zealand, as seen on the map, resembles a top-boot, turned upside down, broken in two just above the instep, and having the toe pointing towards Australia—the *North Island* representing the foot, the *South Island* the top or leg, and *Stewart Island* "the torn loop." New Zealand thus resembles Italy in shape, as it also does in size, climate, and natural conditions generally, and, "if Italy were insular and surrounded by vast tracts of water, the resemblance would be complete."

EXTENT : With the exception of the northern portion of North Island, which bends towards the north-west, the islands extend in a south-west to north-east direction for nearly 1,200 miles, but a straight line from the *North Cape*, in North Island, to the *South Cape*, in Stewart Island, does not exceed 900 miles in length. The **breadth** varies from a few miles, as at Auckland, to 250 miles, the average being about 120 miles. The **total area** of the colony is over 100,000 square miles,[1] or considerably more than that of Great Britain.

North Island is 550 miles long, and has an area of about 44,500 square miles, or one-tenth less than that of England.   Its breadth varies from a few miles, as at Auckland, to 250 miles between Cape Egmont and East Cape.

South Island is also about 550 miles in length.   Its breadth varies from 150 to 200 miles, and it has an area of 58,500 square miles, so that it is almost equal in extent to England and Wales together.

Stewart Island is much smaller, being only 30 miles long, 25 miles broad, and with an area of not more than 668 square miles.

The Off Islands include the Chatham Islands, 500 square miles, the Auckland Islands, 300 square miles, and other smaller groups and islets.

COASTS : The coasts of New Zealand nearly equal in extent the coasts of Great Britain, but though they are, in parts, **deeply indented** by numerous inlets, they are not so rich in **harbours and navigable estuaries** as the British coasts.

No part of New Zealand is more than 75 miles from the sea, an important fact in connection with the development of the country, and which would be still more so but that the harbours are very unequally distributed, and really safe and commodious harbours are not numerous; while long stretches of coast, especially on the western side of South Island, are destitute of a single natural harbour.   North Island is much more irregular in shape and more deeply indented than South Island, the coasts of which, except in the north and south-west, are remarkably bold and unbroken.   Almost the whole of the western coast of this island is open and exposed—Westport, Greymouth, and Hokitika being the only available harbours.   The "Sounds" on the south-western coast are long, narrow, fiord-like inlets, hemmed in by lofty cliffs, and afford some of the grandest and most picturesque coast scenery in the world.   Other parts of the coast of South Island and portions of that of North Island are very beautiful. The symmetrical cone of Mount Egmont, on the west coast of North Island, is a striking feature, and, viewed from a little distance from the shore, it appears to rise from the sea.

INLETS : The chief inlets in the *North Island* are the **Bay of Islands, Hauraki Gulf,** and the **Bay of Plenty,** on the north-east ; **Poverty Bay** and **Hawke Bay,** on the east ; **Palliser Bay** and **Port Nicholson,** on the south ; and the **North** and **South Taranaki Bights,** with **Kawhia, Manukau, Kaipara,** and **Hokianga Harbours,** on the west coast.

The principal openings in the *South Island* are **Golden Bay** and **Tasman Bay,** on the north ; **Cloudy Bay,** on the north-east ; **Pegasus Bay,** with **Port Lyttelton,** and **Akaroa Harbour,** on the east ; **Otago Harbour** and **Molyneux Bay,** on the south-east ; **Bluff Harbour, New River Harbour,** and **Tewaewae Bay,** on the south ; and **Chalky Inlet, Dusky Bay,** and **Milford Sound,** on the south-west ; together with **Canterbury Bight** on the east coast, and **Westland** and **Karamea Bights** on the west coast.

In *Stewart Island* the only large inlets are **Port Pegasus** on the south, and **Paterson Inlet** on the east.

---

1. The official estimate of the area is 105,340 square miles.

STRAITS : The principal straits are **Cook Strait**, a navigable channel, from 15 to 80 miles in width, between North and South Island ; **Foveaux Strait**, 15 miles in width, between South Island and Stewart Island ; **Coromandel Channel**, between Great Barrier Island and the Coromandel Peninsula, on the eastern side of Hauraki Gulf ; and **French Pass**, between D'Urville Island and the north coast of South Island, on the eastern side of Tasman Bay.

CAPES : The principal headlands in the *North Island* are **Cape Maria Van Diemen**, the most westerly point ; **North Cape**, the most northerly ; **East Cape**, the most easterly ; and **Cape Palliser**, the most southerly point of the island. **Cape Egmont** is the extreme point of the great outcurve on the west coast.

In the *South Island* the chief capes are **Cape Farewell**, the most northerly point ; **Cape Jackson** and **Cape Campbell**, on the north-east ; **East Head** and **Cape Saunders**, on the east ; **The Bluff** and **Windsor Point**, on the south ; with **West Cape**, **Cascade Point**, and **Cape Foulwind**, on the west.

At the south of *Stewart Island* is **South Cape**. South-West Cape is the extreme point of an adjacent islet, and is the southernmost point of New Zealand.

ISLANDS : There is a considerable number of islands and islets on the coasts of the main islands, such as the **Three Kings** off the extreme northern coast, the **Great Barrier** and other islands on the north-east coast of North Island, **D'Urville** and **Arapawa Islands** on the north-east coast of South Island, and **Resolution** and other islands on the south-west coast. **Kapiti Island** is in Cook Strait, and **Ruapuke Island** in Foveaux Strait.

The *Off Islands* of New Zealand include several island-groups and islets situated some hundreds of miles to the north, east, and south of the main islands. They include the **Chatham Islands**, about 536 miles to the east of Lyttelton ; the **Auckland Islands**, 180 miles, and **Campbell Island**, about 320 miles to the south of South Island ; the **Bounty Islands** and the **Antipodes Islets**, about 470 miles east of Stewart Island ; and the **Kermadec Islands**, a group 600 miles north-east of Auckland. All these and a few other small islands belong to New Zealand, but none of them, with the exception of the Chatham group, have a permanent population. They are occasionally visited by whalers, and on many of them supplies of food, &c., are stored, in case of vessels being wrecked on them.

RELIEF : The surface of New Zealand is agreeably diversified by lofty mountains, wooded hills, well-grassed plains, fertile valleys, beautiful lakes, and swiftly-flowing rivers. In the South Island, the snow-covered "cloud-piercing" Southern Alps, with their huge glaciers and alpine lakes, rival those of Switzerland, while the lofty volcanoes and the wonderful lakes and hot springs of the North Island are among the most marvellous physical phenomena on the globe.

The **scenery** in North Island has all the grace and charms of Southern Italy, with volcanoes that surpass Vesuvius and rival Etna in altitude, with a brilliant sky and a marvellously luxuriant vegetation, while the Hot Lake district, in the centre of the island, prior to the terrible eruption of Mount Tarawera in 1886, and the destruction of the famous Lake Rotomahana, with the boiling springs and the marvellous Pink and White Terraces, was, and to some extent still is, a veritable wonderland.[1]	In South Island, the massive lofty mountains, extensive snowfields, huge glaciers, snow-capped peaks, foaming torrents, mist-

---

1. For a description of *the Wonderland that remains*, see Dr. Moore's *New Zealand*, p. 132 et seq.

crowned waterfalls, placid lakes embosomed in deep mountain-valleys, recall and rival the grandly-picturesque scenery of Switzerland. Compared with Australia, nothing can be more complete than the contrast between that vast country and New Zealand. "Marcus Clarke has told us that weird melancholy is the dominant note of Australian scenery, which is true enough, for the Australian landscape is as lonely, as melancholy, and as solemn as the Roman Campagna, with the added weirdness of strange bark-shedding trees, and of uncouth birds and beasts. New Zealand is wholly different—severe and frowning in the south, open and alluring in the north, with a bright Polynesian loveliness. Australia is, in summer, a land of dry rivers, brown grass, yellow, lurid glare, and brassy sun ; and, in the greater part of winter, a land of blue sky and soft, smoky haze. New Zealand, in summer, may resemble parts of Australia in winter, but she has a real winter in the South Island and a wet winter in her extreme north. The west of the Middle or South Island, whence come the New Zealand coal and gold, is a country of constant rain, of glaciers, and of tree-ferns and chattering paroquets, inexpressibly distinct from the dried-up Australian goldfields of Bendigo. South Central Australia has the climate of Greece, while New Zealand, owing to its enormous length from north to south, has, like Japan, and for the same reason, all the climates of the world, except the dry and intense brilliancy of Australia or of Greece. New Zealand, which is all but tropical in the Bay of Islands, is Scotch at Invercargill."[1] But, apart from its alpine, volcanic, and tropical features, the general character of the New Zealand scenery is not very different from that of the British Isles, and Mr. Trollope thinks that "in New Zealand everything is English, and that the scenery, the colour, and general appearance of the waters, and the shape of the hills, are altogether un-Australian, and very like that with which we are familiar in the west of Ireland and the Highlands of Scotland. The mountains are brown, and sharp, and serrated, the rivers are bright and rapid, and the lakes are deep and blue, and bosomed among the mountains. If a long-sleeping Briton could be set down among the Otago Hills, and, on awaking, be told that he was travelling in Galway, or the west of Scotland, he might be easily deceived, though he knew these countries well ; but he would feel at once that he was being hoaxed, if he were told in any part of Australia that he was travelling among Irish or British scenery."

MOUNTAINS : With the exception of a few **lofty volcanic peaks,** the mountains of *North Island* are of **moderate elevation,** and do not vie in grandeur or magnitude with the **great ranges** which traverse the *South Island,* and rise, in the massive **Southern Alps,** far above the snow-line.

The most striking feature in the relief of New Zealand is the long mountain range, which runs through both islands, in the direction of south-west to north-east, from Windsor Point in the south-west of South Island to East Cape in the north-east of North Island. This range, consisting of "up-heaved zones of stratified and massive rocks of different ages, constitutes **the powerful backbone of the country,**" and is broken only by the Strait and by occasional passes.

In the North Island, the **main range** extends, under various names, from the north-eastern outcurve to the shores of Cook Strait—the two principal sections being the **Ruahine Range** and the **Tararua Mountains.** The former has an average height of about 4,000 feet, and is cleft by a gorge through which the Manawatu River flows. The latter extends to the south of this river and is prolonged southwards, ultimately terminating in the bold headlands which rise on either side of Port Nicholson.

A series of minor ranges extends between the main range and the south-eastern coast from the shores of Hawke Bay to Cape Palliser, while to the west and north of the Ruahine Range rise the **Kaimanawa** and other ranges, beyond which lies the *Wonderland of New Zealand*. To the south of Lake Taupo rise the lofty cone of **Tongariro**, an active volcano, 6,500 feet in height, and the still loftier and more massive **Ruapehu**, an extinct volcano, 9,195 feet above the sea. Between Lake Taupo and the Bay of Plenty is the remarkable *Hot Lake District*, famous for its hot lakes, mud volcanoes, boiling springs, and the exquisite terraces—which, alas, were destroyed in 1886, when Mount Tarawera, till then believed to be a wholly extinct volcano, broke out with terrific violence, and devastated what had been, and to some extent still is, one of the most wonderful regions in the world.

There are several other ranges, such as the **Coromandel Range**, &c., in the northern half of North Island, but they seldom exceed 1,500 feet in height, with the exception of a few of the numerous isolated volcanic peaks, none of which, however, approach in altitude or grandeur the symmetrical snow-capped cone of **Mount Egmont**, an extinct volcano, standing in solitary state in the centre of a rounded promontory on the south-west coast of the island. This remarkable mountain is almost a perfect cone, and its sides, which are clothed with magnificent forests, "curve off so gently and gracefully into the general slope of the country, that, viewed from a little distance from the shore, it appears to rise from the sea."

Three distinct lines of craters occur in North Island, one at the *Bay of Islands*, in the extreme north ; one at *Auckland*, near which over 30 craters may be counted ; and one from *Mount Egmont* to *White Island*, an active volcano in the Bay of Plenty, about 30 miles from the shore. Numerous other islands on the coast are of volcanic origin, and indicate a submarine continuation of the volcanic areas. Strangely enough, no volcanoes are known to exist in the South Island, although slight earthquakes occur in both islands.

In the South Island, the **main range** is, as in the North Island, known by different names in different parts, but, unlike those of the north, the southern ranges run nearer the western than the eastern coast. The central and loftiest portion of the chain is known as the **Southern Alps**, which rise far above the limit of perpetual snow. The culminating point, *Mount Cook*, rises to a height of 12,350 feet, and is thus no mean rival to Mont Blanc, the monarch of the European Alps. This massive mountain, with *Mount Hochstetter*, 11,200 feet, and many other heights, are covered with perpetual snow, while the higher valleys are filled with immense glaciers, which feed the alpine lakes, the basins of which were formed by ancient glaciers of still greater extent.

"Towards the north of the island, the principal ranges diverge like the legs of a compass, forming an eastern branch, which terminates at the promontory of Cape Jackson, and a western branch called the **Tasman Mountains**, terminating at Cape Farewell. Above the fork, whence these two principal chains diverge, rises the lofty peak of *Mount Franklin*, 10,000 feet in height, clothed in dazzling snow, and surmounting, like a watch-tower, all the northern region of the island. In the western branch are the conspicuous peaks of *Mount Arthur*,

Mount Snowdon, and *Mount Peel,* the latter being 6,000 feet in height ; and in the eastern branch are *Ben Nevis, Mount Rintoul, and Mount Richmond."*

Between Mount Franklin and the west coast are the **Paparoa Mountains,** a detached range between the Buller and the Grey Rivers ; between the east coast are the **Kaikoura Mountains,** which rise in *Mount Odin* to a height of 9,700 feet, and the **Lookers On Mountains,** so named by Captain Cook from a number of natives whom he observed watching the ships.

To the south of the Southern Alps, the country is very mountainous, and some of the mountains attain a considerable height, the loftiest being *Mount Earnslaw,* 9,165 feet, at the head of Lake Wakatipu.

Stewart Island is also mountainous, but its highest peak, *Mount Anglem,* is only 3,200 feet.

∴ The most extensive plains in New Zealand are the **Canterbury Plains,** which extend from Banks' Peninsula to the Southern Alps, a distance of about 100 miles. Narrower plains also extend along the western coast of South Island, and there are several level or slightly undulating tracts of considerable extent in the North Island.

RIVERS : New Zealand abounds in rivers, and running streams are numerous everywhere. Though some of the rivers are of considerable length, none are navigable for more than a short portion of their course. The longest rivers are the **Waikato,** in North Island, and the **Clutha,** in South Island.

The mountains of New Zealand altogether surpass those of Australia, and rival those of almost any mountain region on the globe, but its rivers, owing to the direction of the mountain ranges and the narrowness of the country, are comparatively small and generally so rapid in their course as to be useless for navigation. The Waikato, the chief river in the North Island, is 170 miles in length ; it rises on the northern slopes of Mount Ruapehu and flows through Lake Taupo, and thence north through a remarkable series of hot springs, which extend along its banks for more than a mile. "The river here plunges through a deep valley, and its floods, whirling and foaming around rocky islets, dash with a loud uproar through the defile. Along its banks white clouds of steam ascend from hot cascades falling into the river, and from basins full of boiling water shut in by white masses of stone. Steaming fountains rise at short intervals, sometimes two or more playing simultaneously, and producing endless changes, as though experiments were being made with a grand system of waterworks. Dr. Hochstetter counted seventy-six separate clouds of steam visible from a single station, and among them were numerous intermittent geyser-like fountains, with periodical water eruptions." The most graphic description, however, conveys but a faint idea of the peculiar grandeur and beauty of these natural wonders. About 100 miles from Lake Taupo the river bends sharply west and enters the North Taranaki Bight by an estuary, accessible to large vessels, while small steamers can ascend the river itself for 80 miles.

Other considerable rivers in the North Island are the **Thames** or Waiho, and the **Piako,** both flowing north into the Firth of Thames—as the southern extension of the Hauraki Gulf is called—and both navigable for small vessels ; the **Wairoa,** in the north, which enters Kaipara Harbour ; the **Wanganui,** 120 miles in length, which rises on the slopes of Mount Tongariro and falls into South Taranaki Bight ; and the **Hutt,** which flows into Port Nicholson.

In South Island, the largest river is the **Clutha** or Molyneux, which receives the overflow from Lakes Hawea, Wanaka, and Wakatipu, and has a southerly course of 170 miles into Molyneux Bay. Several rivers rise among the glaciers of the Southern Alps, and flow eastwards into Pegasus Bay and Canterbury Bight, and westwards into Westland Bight. The eastern rivers of South Island are much longer than those which drain the short western slope, but all alike are unnavigable and subject to heavy floods. The **Waitaki**, which drains three of the alpine lakes and flows into the sea at the southern end of Canterbury Bight, is a considerable stream ; the two largest of the western rivers, the **Buller** and the **Grey**, drain the country to the west of Mount Franklin and are both navigable to some extent. A number of other rivers are of some local import- ance, their estuaries in many cases forming small but useful harbours, while the streams, when unnavigable even for boats or canoes, supply abundant water-power for working the gold mines and for other purposes. Altogether, New Zealand is favoured far above Australia in the abundance of perennial rivers and running streams, and in an entire immunity from the droughts, which prove so disastrous in many parts of Australia.

LAKES : The lakes of New Zealand are doubly interesting— those of North Island being of *volcanic origin*, while the alpine lakes of South Island have been formed by *glacial action.*

The largest lake in North Island is **Lake Taupo**, a beautiful expanse nearly in the centre of the island, with a diameter of some 20 miles and an area of about 200 square miles, and of unknown depth. It is surrounded by a tableland from which rise numerous volcanic cones, while in the vicinity are a number of hot springs, many of them of an extremely high temperature, and some actually boiling. To the south of Taupo, and midway between its shores and the active volcano of Tongariro, is the small lake of **Roto-aira**, at an elevation of over 1,700 feet. But *the* lake district—a region which forms one of the wonders of the world—lies to the northward of Taupo, about midway between that lake and the shores of the Bay of Plenty. Here a belt of country, some 30 miles in width, is occupied by a succession of hot lakes, mud volcanoes, solfataras, fuma- roles, and hot springs throwing up jets of boiling water and comparable only to the famous geysers of Iceland and the Yellowstone. Of the sixteen lakes in the district, the largest is the picturesque **Lake Tarawera**, on the eastern side of which rises the well-known Mount Tarawera. Next in size is **Lake Rotorua**, fringed by numerous hot springs ; but the most wonderful of all, previous to the terrible eruption of 1886, was the **Rotomahana**[1] or Hot Lake, which was then, however, considerably altered, while the famous Pink and White Terraces, the boiling springs, and the resplendent basins of warm or hot water, which formed a series of exquisite natural baths, were destroyed.

On the slopes of the Southern Alps, in South Island, are a number of true alpine lakes, the largest of which, both in size and depth, are found towards the south. **Te Anau** (132 square miles), **Manipori**, **Wakatipu** (114 square miles), **Wanaka**, and **Hawea** are the largest ; further north are **Pukaki**, **Tekapo**, and **Oahau**, fed by the Mount Cook glaciers and drained by the Waitaki River. "The bottoms of many of these lakes are far below the sea level, though they must have been filling up for ages by the sediment carried into them, a proof that either the entire land was once much higher, or that the lakes have been

1. *Roto*, in the Maori language, means "lake."

ground out by glaciers." The lakes to the north of Mount Cook are smaller in size, but there is an extensive sheet of fresh water—Lake Ellesmere—on the east coast, close to Banks' Peninsula. Many of the inlets along the coast are so nearly enclosed by land as to form salt-water lakes or lochs.

CLIMATE : The climate is temperate and healthy, and differs only from that of Great Britain in being **warmer and more equable,** while the air is drier and more elastic.[1]   **High winds and gales are** frequent, and **rain falls all the year round.**

In the extreme north of North Island, the climate is sub-tropical and decidedly warm in summer, and over the whole of North Island frost and snow are unknown except on the uplands and mountains ; while, in the South Island, the frost is occasionally severe at sea-level on the east coast, and the heat in summer is very great.   In the extreme south, severe frosts and deep snow on the uplands are common in winter.   Stewart Island is subject to violent winds and frequent fogs.

The climate as a whole is agreeable, having neither extreme heat nor intense cold, but, of course, it varies considerably in different parts of a colony that extends over a thousand miles from north to south.   "The changes of weather and temperature are very sudden, the transition from heat to cold, from sunshine to rain, from calms to gales, being so frequent and marked as to defy calculation, and to prevent its being said that there is any uniformly wet or dry season in the year."

Dr. Moore, in his interesting work on New Zealand,[2] says that the bright and bracing climate of New Zealand stimulates the faculties of the whole nature of the immigrant, so that he who has been slow, unintelligent, and depressed in England, becomes quick, lively, hopeful, and energetic in the "Britain of the South."[3]   The Briton, he further remarks, who travels for health, and looks round the world for a region where pure and bracing or mild air, interesting natives, beautiful scenery, good water for drinking, mineral baths, his accustomed food, convenient excursions, and pleasant society among people of his own language, will find all these, together with an unsurpassed climate, in New Zealand ; and at Auckland, the Bay of Islands, Napier, Rotorua, or Nelson, he will discover a new world of calm delight in a balmy yet invigorating atmosphere.   And it is really astonishing how cheerfully one, fresh from the gloom of wet days in London, Liverpool, Manchester, and Glasgow, or other huge smoky cities, can wait indoors in New Zealand for the cessation of a rainstorm. There is almost always a glimpse of blue sky somewhere—and what a blue ! Read Froude's eloquent description in "Oceana." There is an exhilaration in the air, only temporarily veiled by the transient dull weather.

The prevailing winds are from the north-west, and the rainfall is much heavier on the western than on the eastern coasts—the amount falling at Taranaki, on the west coast of North Island, being more than double the fall at Napier, on the opposite side of the island ; while in the South Island nearly five times as much rain falls on the west coast as on the east.   The snow-line is at a height of about 7,500 feet, and thus the summit of Ruapehu, the highest mountain in

---

1. "The mean annual temperature of the North Island is 57° F., and of the South Island 52°, that of London and New York being 51°.   There is about 20° difference between the warmest and the coldest months.

2. *New Zealand for the Emigrant, Invalid and Tourist* (London: Sampson Low).

3. The average death-rate in New Zealand is lower than that of any other of the Australasian colonies, and considerably lower than that of Great Britain.   The excess of births over deaths, which in England is about 57 per cent., is in New Zealand over 200 per cent.

the North Island, and the higher portions of the great ranges in the South Island, are covered with perpetual snow.[1]

PLANTS and ANIMALS : With one or two doubtful excep-tions, there are scarcely any truly indigenous animals, but the native plants are wonderfully peculiar, and most of them are found nowhere else.

There are about 120 indigenous forest trees, all of them evergreen, and many of them yielding valuable woods. The well-known kauri pine, occasionally exceeding 15 feet in diameter and 150 feet in height, furnishes not only excellent timber, but also the valuable gum or resin known as *kauri gum*. Other species of native pines produce good lumber for all ordinary purposes, while the wood of the native beech is specially adapted for shipbuilding. Ferns grow in great variety and in endless profusion almost everywhere, and the extensive plains and hill sides are clothed with native grasses, which support millions of sheep. Another vegetable product of considerable importance is the New Zealand flax, which is now exported to the value of half a million a year, and is largely used in rope-making, as it is very tenacious and durable.[2] None of the characteristic eucalypti and acacias of Australia are found in New Zealand, neither are there any of the marsupials or snakes so common in that continent. When the islands were explored by Captain Cook, the only land mammals were a native *dog* and a native *rat*—then used as food by the Maoris, but now extinct—and two species of the *bat*. Birds are comparatively numerous, and about 70 species are not found in any other country. There are handsome parrots and pigeons, and some good singing birds, but the most remarkable of all is the kiwi or apteryx, a wingless and tailless bird, which is now rapidly disappear-ing and will doubtless soon become extinct. The moa, a gigantic representa-tive of the same class, which must have been common at one time, has long since disappeared.[3] Fresh-water fish are not numerous, with the exception of *eels*, which are both large and abundant, but the neighbouring seas teem with many kinds of edible fish, and excellent oysters abound on the coasts. There are no noxious reptiles. Snakes are unknown, but there are a few varieties of lizards, and frogs are occasionally seen.[4]

INHABITANTS : With the exception of about 42,000 Maoris, 4,500 Chinese, and about 15,000 Germans, Danes, Scandinavians, French, Americans, &c., the people of New Zealand are British or of British origin—more than one-half of them born in the colony, the rest being settlers from Great Britain and Ireland.

The Maoris (*i.e.*, the aborigines or natives), as the natives of New Zealand call themselves, are the finest in physique and the highest in intelligence of all the Polynesian peoples. They are in every respect far superior to the natives of Australia, and the best type of Maori is not much inferior to the average Euro-pean. Although not a pure Polynesian race, there being evidently an admixture of Papuan and Melanesian elements, their language is a Polynesian dialect, differ-ing but slightly from the Hawaiian and other similar languages, and easily understood by almost all the Polynesian islanders. They are supposed to have

---

1. "New Zealand never suffers from drought. Its natural formation and its situation effectually prevent that. In fact, in seasons when the drought is most cruel in Australia, New Zealand is at its prime; the unbroken weather of a dry summer bringing to perfection the crops which have been well nourished by the unfailing rains of winter and spring."

2. The prepared article is known under the name of *phormium tenax*.

3. Remains of many species of this bird have been found. Some of them must have attained to the extraordinary height of 12 feet, and they seem to have formed the staple food of the natives, until they were exterminated.

4. For the domestic animals and cultivated plants *see* p. 177.

colonized the country about 500 years ago, coming, according to their own traditions, from " Hawaiki," which may denote Hawaii, the principal island in the Sandwich group, or Savaii, in the Navigator Islands.[1]   All the original settlements of the Maoris were probably made on the shores of the North Island[2]—the South Island, together with Stewart Island and the islets, being peopled by migrations from thence, and by subsequent conquest.   When the islands were first settled by Europeans, the Maoris were much more numerous than at present, and their numbers are still diminishing, and there is but little doubt that the beginning of the end has come for the Maori race.[3]   In character, the Maoris are "warlike, courageous, quick at learning, good at imitation, fond of oratory, and susceptible of strong religious feelings ; but they are vain and proud, revengeful and jealous, though not devoid of good qualities, and are now on good terms with the colonists."

The British colonization of the islands may be said to date from 1814, when the first missionary settlement was established at the Bay of Islands.   Traders from New South Wales then came and established agencies, and Sydney merchants formed numerous whaling and lumbering stations on the coasts of both islands.   But nothing of much importance was done until 1839, when Colonel Wakefield selected a site for a settlement on Port Nicholson (now Wellington) under the auspices of the New Zealand Company, and, in the following year, numerous vessels arrived with hundreds of immigrants from Great Britain and a few from Australia.   The total population in 1893 (exclusive of the Maoris) amounted to 672,000, an average of 6 per square mile.

The order in which the various settlements were formed is thus given in the *Australian Handbook*:—(1) **Wellington**, as already stated, founded by the New Zealand Company in 1840 ; (2) **Auckland**, established by the first Governor, Captain Hobson, in the same year, who also made the first treaty with the natives—the Treaty of Waitangi—by which the sovereignty of the island was transferred to Great Britain ; (3) **New Plymouth**, also founded by the New Zealand Company, in September, 1841, after a preliminary expedition the year before ; (4) **Nelson**, founded by the Company in October, 1841 ; (5) **Otago**, founded in March, 1848, by a Scotch Company working in connection with the New Zealand Company, and under the auspices of the Free Church of Scotland ; (6) **Canterbury**, similarly founded in December, 1850, in connection with the Church of England ; (7) **Hawke's Bay**, originally part of Wellington Province, was formed into a separate province in 1858 ; and (8) **Marlborough**, originally a part of Nelson, was separated in the same manner in 1860.

INDUSTRIES : New Zealand is, first and foremost, a **sheep-farming** and therefore a *wool* producing country, also exporting large quantities of *frozen mutton;* secondly, a **cattle-rearing** country, exporting *hides* and some *frozen beef;* thirdly, an **agricultural** and **fruit-growing** country, exporting *grain* and *farm products* largely ; and, lastly, a mining country, producing *gold, coal, silver,* and other minerals.[4]

1. ' Both of these names would be pronounced by a Maori as *Hawaiki*, but some think that the word has only a mythical signification and reference, being used much as " home " is by ourselves.'
2. About 40,000 of the Maories dwell in the North Island. Less than 2,000 are found in the South Island, about 1,000 on Stewart Island, and about the same number in the Chatham Islands.
3. The natives are said to be conscious of their approaching fate, a fate in which not only the people themselves, but also the native fauna and flora, seem involved. Hence the Maoris rightly say:—' As the white man's rat has extirpated our rat, so the European fly is driving out our fly. The

foreign clover is killing our ferns, and so the Maori himself will disappear before the white man.' Dr. Moore, however, thinks that the wonderful vigour and tenacity of life of the Maori race is such that, in time, they will blend into the mixed nation called "young New Zealand," just as the Celts have blended into our Anglo-Saxon-Danish-Norman nation, and that the ranks of this young nation will furnish to the world orators, politicians, poets, merchants, and warriors equal in bravery, ability, and energy to any of those born of a purely white race.
4. See further the summary of agricultural and pastoral information in the *New Zealand Handbook*, issued by the Emigrants' Information Office.

The pastoral industry is by far the most important, and the available land is mainly used for rearing sheep and cattle. But although the sheep and cattle "runs" cannot be compared in size to those in Australia, they are, owing to the fact that much larger areas are covered with *sown grasses*, more productive and capable of supporting a comparatively much larger number of stock. There are now in the colony about 18½ million sheep, 830,000 cattle, and 211,000 horses.

Wool is the staple product, and about 110 million lbs. are now annually exported, nearly the whole of it going to London, "whence about two-fifths of it is re-exported to France, Belgium, and America. Much of the best wool used in the carpet factories abroad comes from New Zealand." The trade in frozen meat only dates from 1882, but the exports have risen from 15,000 cwts. in that year to 903,000 cwts. in 1893. The export of hides, skins, and tallow is also large, but comparatively few horses or cattle are sent out of the colony.

But although the colony is mainly pastoral, it is also very largely agricultural, and produces more wheat, oats, and barley than any other Australasian colony; the yield per acre of all the grain crops, including maize, is also larger. The annual production of cereals is now nearly 20 million bushels, and their cultivation is rapidly advancing throughout the colony, which bids fair to be, in the near future, the great agricultural centre in the Southern seas. "Not only grain and bread-stuffs, but potatoes, hay, chaff, roots, vegetables, butter, cheese, bacon, hams, even meat and fish, preserved or frozen, are sent over in immense quantities, and are eagerly bought at Sydney, Brisbane, and Melbourne, at highly remunerative prices. This trade is steadily growing, and there is no doubt that in time a great part of Australia will be entirely supplied with food staples from New Zealand." All kinds of English fruits and vegetables thrive in almost all parts of the colony, and grapes and oranges come to perfection in the warmer parts of North Island.

MINING : The mineral resources of New Zealand are almost as rich as those of any other Australasian colony. They include rich deposits of gold, extensive coalfields, almost every variety of iron ore, as well as immense quantities of iron sand, which abounds on the sea coast, some silver, tin, copper, and other useful metals and minerals.

Gold—about 50 million pounds' worth of which has been produced in the colony—was first discovered in 1842, but it was not practically worked until 1852, at Coromandel, on the northern coast of the North Island. But little of the precious metal, however, was obtained until, in 1860, rich deposits were discovered in Otago, and subsequently in Westland, Nelson, and Marlborough, in the South Island, and in the Thames Valley in the North Island. The distribution of gold is one of the most puzzling phenomena of nature, and nowhere is it more remarkable than in New Zealand. There "it is found in all sorts of places and conditions; pure, in lumps, loose among the gravel; in scales or particles in the sand of rivers; in nuggets or rough pieces in holes among stones or huge boulders; mixed pellmell with the spoil of rivers backed up by the sea; in fine dust, mixed with black steel sand, thrown up on the beach from the bottom of the sea during storms; in veins, or specks, or needles, in the very substance of quartz rocks, often invisible to the naked eye when most plentiful; in ragged patches in rotten stone crumbling to the touch;

combined with silver and all sorts of other minerals ; lying on the surface of the ground as if somebody had just spilt it there out of his pocket ; fixed in the rifts of rocky gorges of thundering torrents ; cropping out in rough ridges of quartz reefs on the tops of hills ; hidden half a mile in the bowels of a mountain, and 500 feet below the surface of the earth."

Extensive coalfields exist in New Zealand, and coal-mining is rapidly becoming an important industry in both the North and South Island. The annual output is now considerably over half a million tons. Iron ore is abundant, but the workings are limited to the rich iron sand which occurs plentifully along the coast, and which is found to yield a metal equal to the best Staffordshire iron. Copper exists at D'Urville's Island and elsewhere, and silver has been found in various localities. Large quantities of tin ore have been recently discovered in the Remarkable Mountains in Stewart Island, and are now being worked. There are productive springs of petroleum or rock oil at Taranaki, on the west coast of North Island, and at Gisborne, on Poverty Bay, on the east coast. Building stone and marble are regularly quarried, and graphite, antimony, and manganese are also found and worked. The kauri gum industry yields an important article of export, the annual value being about half a million. The gum is found, chiefly in the Auckland district, by digging on the sites of old kauri pine forests.

COMMERCE : The commerce of New Zealand is mainly carried on with the United Kingdom and the other Australasian colonies— the trade with foreign countries, chiefly the *United States* and *China*, is very small. As in the other Australasian colonies, the volume of trade in proportion to the population is very great, but the exports are largely in excess of the imports in New Zealand. Total value in 1893—exports, 9 millions sterling ; imports, 7 millions.

The annual turnover, therefore, amounts to no less than 16 millions sterling, an amount which, in proportion to the population—672,000—is very large, being over £23 per head.

The chief exports (excluding specie) in 1893, were, in order of value, *wool, frozen meat, gold ; grain, pulse*, and *flour ; kauri gum ; hides, skins*, and *leather ; butter* and *cheese, phormium* or New Zealand hemp, *tallow, timber, grass seed, preserved meats, live stock*, and *bacon* and *hams*.

The chief imports (excluding specie) in 1893, were also, in order of value, *clothing* and *clothing materials ; iron* and *steel goods, machinery, &c. ; sugar ; paper, printed books*, and *stationery ; spirits, wine*, and *beer ; tea, fruit, bags* and *sacks, tobacco* and *cigars, coal, oils*, and *fancy goods*.

Over one-half of the imports are from, and the bulk of the exports go to, Great Britain, the annual value of the trade between the colony and the mother country being now considerably over 11 millions sterling. The principal exports to the United Kingdom, in 1893, were *wool*, £4,478,000 ; *frozen meat*, £1,757,000 ; *hemp*, £33,800 ; and *kauri gum*, £510,000. The chief imports from the United Kingdom, in the same year, were *iron* (wrought and unwrought), £392,000 ; cottons, £434,000 ; apparel and haberdashery, £420,000 ; and woollens, £257,000.

PORTS : The principal ports for vessels entering and clearing for the United Kingdom are Auckland and Wellington in the north

Island, and **Lyttelton** for *Christchurch*, and **Port Chalmers** for *Dunedin*, in the South Island.

These ports carry on an extensive trade with the Australian colonies, as well as with the mother country, as also do the ports of Napier, in the North Island, **Bluff Harbour** (the port for *Invercargill*), **Oamaru**, **Timaru**, and **Nelson**, in the South Island.

COMMUNICATIONS: In addition to about **2,000** miles of railways, there is an extensive **coaching system** between the railway termini and other important centres, and constant communication by **steamers** between all the principal ports in the colony, and also with the Australian colonies, England, and America.

With the exception of 164 miles of private lines, the **railways** of New Zealand belong to and are worked by the Government, which, up to the present, has spent in their construction over 16 millions sterling. In 1894, the Government lines open for traffic were 1,948 miles in length—744 miles in the North, and 1,204 miles in the South, Island. The railway system of the colony is as yet incomplete, but the main lines are gradually being extended, and it will be possible, before long, to travel by rail from Auckland in the north to Invercargill in the south—the only break being at Cook Strait.[1]

The principal lines in the North Island are (1) from AUCKLAND north to some distance beyond HELENSVILLE and south to MOKAU, and also to MORRINSVILLE, TE AROHA, and LICHFIELD; (2) from TAURANGA, on an inlet of the Bay of Plenty, to the *Hot Lake District*; (3) from NAPIER by WOODVILLE and EKETAHUNA to WELLINGTON; (4) from WELLINGTON to NEW PLYMOUTH; and (5) from WELLINGTON to MANAWATU (private line).

In the South Island, there is a continuous line from CULVERDEN, about 60 miles north of CHRISTCHURCH, along the eastern and south-eastern coast to and beyond Invercargill, passing through TIMARU, OAMARU, PORT CHALMERS, DUNEDIN, and other coast towns and ports, with branches from CHRISTCHURCH to LYTTELTON (its port), and from INVERCARGILL to BLUFF HARBOUR and to KINGSTON, on Lake Wakatipu. An important line also runs from NELSON, on the northern coast, to GREYMOUTH and HOKITIKA, on the west coast, and there is also a short line along the coast from WESTPORT to the coal mines north of that town. On the north-east coast, a short line runs from PICTON through BLENHEIM to AWATERE, and will be extended southward to connect with the main line of South Island. Several branches strike inland from the main line, most of them in the Otago and the Canterbury districts. The Midland Railway, which will connect Greymouth and Christchurch, is being built.

The **coaching system** is very extensive and complete, and passengers and cargo are conveyed from port to port by the vessels of the **Union Steamship Company** and other lines, and to and from the minor ports by small coasting steamers. The **New Zealand Shipping Company's** steamers sail every fourth Thursday from LONDON, calling at *Teneriffe*, *Cape Town*, and *Hobart*, with cargo and passengers for all New Zealand ports. The steamers leave New

1. In 1893-4, the number of passengers carried [5,192,245] and the quantity of goods carried [including allowance for season-ticket holders] was [amounted to 2,216,740 tons].

Zealand every fourth Thursday for LONDON, *via Rio de Janeiro* and *Teneriffe*. The Shaw, Savill, and Albion Company's mail steamers also run once a month from LONDON, calling at *Teneriffe, Cape Town*, and *Hobart*, to New Zealand, returning *via Rio* and *Teneriffe*. The steamers of the Union Steamship Company and the Oceanic Steamship Company also maintain a monthly mail service between AUCKLAND and SAN FRANCISCO, *via Honolulu*. Although WELLINGTON is 16,000 miles from LONDON, the passage between the two ports has been frequently made in less than 40 days.

A submarine cable connects New Zealand with Sydney, and telegraphic lines unite all the chief centres of population in the colony, while the telephone is in general use in the larger towns.

GOVERNMENT : The general government consists of a **Governor**, appointed by the Crown, a **Ministry** who form the executive, and a **Parliament** of two Chambers.

The New Zealand Parliament consists of a **Legislative Council** of 46 nominated members (two of whom are Maoris) ; and a **House of Representatives** of 74 members (4 of whom are Maoris), chosen by duly qualified electors.[1]

The **Revenue**, for 1893-4, amounted to 4½ millions sterling, and the Expenditure to nearly the same amount, while the **Public Debt** amounted to nearly 39 millions sterling, or over £57 per head of the population.

For the defence of the colony there is a *volunteer* force of 8,000 men, and a small artillery force. The approaches to the principal forts are defended by *heavy batteries*, supplemented by *torpedo boats* and *submarine mines*.

Elementary education is free, secular, and compulsory. Higher education is provided for in a large number of endowed colleges and grammar schools. The **University of New Zealand** has power to confer degrees, but, like the London University, it is solely an examining body. To it are affiliated the *Otago University* at Dunedin, the *Canterbury College* at Christchurch, and the *University College* at Auckland.[2]

There is no State Church, and no State aid is given to any form of religion.

DIVISIONS : Up to the year 1876, New Zealand was divided into **nine provinces**—four in the North Island, and five in the South Island—but in that year the provincial governments were abolished, and the colony was then divided into **Counties**. The nine provinces are now termed **Provincial Districts**.

The Four Provincial Districts in the North Island are *Auckland*, in the north ; *Taranaki*, in the west ; *Hawke's Bay*, in the east ; and *Wellington*, in the south.

The provincial district of Auckland occupies nearly the whole of the northern half of the North Island, and is characterized by its warm and pleasant climate, its valuable kauri forests,[3] and the absence of open pastoral country. It is,

---

1. Until 1875, each of the provinces into which the colony was, until then, divided, had its own separate Government, which consisted of a *Provincial Council*, presided over by a *Superintendent*.

2. There are 70 village schools and 4 boarding schools for Maori children, about 2,500 of whom are under instruction. There is also a Native College (Te Aute, Hawke's Bay) for the higher education of Maori youths.

3. The kauri pine, the most famous of New Zealand trees, is entirely confined to Auckland, and almost wholly to its northern extremity. The valuable kauri gum is found on the site of ancient kauri forests, and lies, in many cases, at a considerable depth, and is indeed largely mingled with the strata of tertiary coal, which abounds in the province.— *The Australian Handbook*.

however, "richly watered, and presents clusters of fertile valleys running inland from the numerous ports, estuaries, and river harbours, which are scattered along the extensive coast-line." The hot lakes and springs, the active and extinct volcanoes and volcanic islands of Auckland, are among the most remarkable in the world.

Taranaki, formerly called New Plymouth, is a small district on the western side of the island, and its soil and climate are so fine that it is called the "Garden of New Zealand," while the magnificent volcanic cone of Mount Egmont forms, perhaps, the most striking feature in the scenery of the whole island. The district of Hawke's Bay, on the eastern side of the island, consists of rich alluvial plains and undulating hills, rising gradually from the sea coast to the Ruahine Mountains.

The district of Wellington, which includes the southernmost part of the North Island, contains some of the finest open country and undulating forest-land in the colony, but chiefly owes its importance to the geographical and political position of the city of Wellington, with its magnificent harbour—Port Nicholson —on the great waterway of Cook Strait.

The Five Provincial Districts in the South Island are *Nelson* and *Marlborough*, in the north; *Westland*, in the west; *Canterbury*, in the east; and *Otago*, in the south.

Nelson and Marlborough occupy the northern end of the South Island. Both districts are rugged and mountainous, but there is much cultivable and grazing land in the valleys, and their coasts are deeply indented—many of the inlets being very beautiful, and some of them, such as Pelorus Sound, almost unequalled for variety and romantic grandeur, while, in the interior, the prevailing scenery is bold and grand—lofty mountains alternating with rich and fertile valleys, dense forests, and beautiful lakes. The climate is delightful, and agriculture is successfully carried on in the valleys, but the great wealth of these districts is in their minerals, chiefly coal and gold.

The provincial district of Westland includes a long narrow strip of country on the western side of the South Island, between the Southern Alps and the sea. The climate is moist, and much of the land is covered with dense forests, while the rivers abound in fish; but the district is chiefly famous for its rich goldfields, even the sands on the seashore being impregnated with the precious metal. The much more extensive district of Canterbury includes the open country which slopes down from the Southern Alps to the eastern coast. The valleys, even in the hilly Banks' Peninsula, are in a high state of cultivation, while sheep and cattle rearing and wheat-growing are extensively carried on on the celebrated Canterbury Plains.

Otago occupies the southern part of South Island, and, though the youngest, it is now one of the most important provinces in the colony, and has the largest population. Lofty mountains, with huge glaciers and alpine lakes, and an abundance of running streams, vast forests, lofty downs suitable for sheep and cattle grazing, and fertile lowlands well adapted for agriculture, distinguish this district, but Otago is chiefly remarkable for its rich goldfields, which have yielded nearly 20 million pounds' worth of the precious metal.

Stewart Island also has much mineral wealth. The black iron sand on its shores is equal to that of Taranaki, numerous quartz reefs have been found at Pegasus Bay and Port William, while extensive deposits of tin have been recently discovered.

COUNTIES: These Provincial Districts are, for purposes of local government, divided into **Counties**, which, in 1876, were 63 in number—32 in the North Island, 30 in the South Island, and 1 in Stewart Island. The number of Counties has since been increased to 79 by the subdivision of some of the larger counties.

The Counties in the North Island, 46 in number, are: Mongonui, Whangaroa, Bay of Islands, Hokianga, Whangarei, Hobson, Otamatea, Rodney, Waitemata, Eden, Manukau, Coromandel, Thames, Ohinimuri, Waikato, Raglan, Waipa, Piako, Kawhia, West Taupo, East Taupo, Rotorua, Tauranga, Whakatane, Cook, and Waiapu, *in the Provincial District of Auckland;* Wairoa, Hawke's Bay, Kaikora North, Waipawa, and Patangata, *in the Provincial District of Hawke's Bay;* Clifton, Stratford, Taranaki, Hawera, and Patea, *in the Provincial District of Taranaki;* and Waitotara, Wanganui, Rangitikei, Oroua, Manawatu, Horowhenua, Pahiatua, Wairarapa North, Wairarapa South, and Hutt, *in the Provincial District of Wellington.*

The Counties in the South Island, 33 in number, are:—Sounds, Marlborough, and Kaikoura, *in the Provincial District of Marlborough;* Collingwood, Waimea, Buller, Inangahua, Grey, Amuri, and Cheviot, *in the Provincial District of Nelson;* Westland, which includes the whole *Provincial District of Westland;* Ashley, Selwyn, Akaroa, Ashburton, Mackenzie, Geraldine, and Waimate, *in the Provincial District of Canterbury;* and Waitaki, Maniototo, Waihemo, Waikouaiti, Taieri, Peninsula, Vincent, Lake, Fiord, Wallace, Southland, Tuapeka, Bruce, Clutha, and Stewart (Stewart Island), *in the Provincial District of Otago.*

TOWNS: All the chief towns of New Zealand are on or near the coast, and although two-fifths of the people live in towns, there is no such concentration in one large town as in Victoria, where one-half of the people live in Melbourne, or as in New South Wales, where two-fifths of the population reside in Sydney. Recent Returns show a large increase in the town population, but even yet there are only four towns with over 10,000 inhabitants, namely, Auckland and **Wellington**, in the North Island, and **Dunedin** and **Christchurch**, in the South Island.

WELLINGTON (30), the capital of the colony and the seat of Government, stands on the shores of a splendid natural harbour—Port Nicholson—on the northern side of Cook Strait. It lies about 1,200 miles south-east of Sydney and 1,400 miles east of Melbourne, and with both these ports, as well as with London (via Rio de Janeiro and Teneriffe on the homeward route, and via Cape Town and Hobart on the outward route), and with San Francisco via Honolulu, there is regular steam communication. The extension of the direct steam service and the completion of the railway system of the colony, will hasten the progress of the city, and cause it to "steadily grow into her true position as the commercial and maritime as well as the political and geographical capital of the colony."

Wellington is now connected by rail with **Foxton** (2), a rising port on the Manawatu, a tidal river navigable for small vessels for 50 miles inland, **Wanganui** (5), the centre and outlet of a rich agricultural and pastoral district, and **New Plymouth** (3), the chief town in the provincial district of Taranaki.

on the west coast, near Mount Egmont. Another line of railway runs from Wellington through Masterton (3), the largest inland town in the North Island, by Eketahuna and Woodville to Napier (9) on Hawke Bay, the chief port on the east coast of the island, and the sole port for a large grazing and timber district. Gisborne (3) is another rising port, on the shores of Poverty Bay— memorable as the scene of Cook's first landing in New Zealand.

AUCKLAND (50), the largest city in the North Island, and formerly the capital, is still the leading seaport. The "Corinth of the South Pacific," as Auckland is called, is picturesquely situated on the eastern side of a narrow isthmus, about 6 miles in width, and its harbour, which opens out into the beautiful Gulf of Hauraki, has sufficient depth of water for the largest ocean steamers.[1] Auckland is connected by rail with Onehunga (3) on Manukau Harbour, an outlet for much timber and agricultural produce, and with Helensville on the Kaipara River, and other places to the north. At Onehunga there are ironworks for smelting the iron sand which abounds on the coast. To the south, the main line has been extended up the Waikato Valley and southwards to the Taraka Plain, which lies about midway between Lake Taupo and the west coast, and the main trunk line, which will ultimately connect Auckland and Wellington, is now being constructed. A line of railway also runs from Tauranga (2), a seaport on an inlet of the Bay of Plenty—the only safe port for shelter for large vessels between Auckland and Wellington—to the heart of the famous Hot Lake District.

In the South Island, by far the largest towns are Dunedin and Christchurch. DUNEDIN (48), the "Edinburgh of New Zealand," was founded by members of the Free Church of Scotland in 1848, but did not make any great progress until the discovery of the rich goldfields in the Otago district attracted thousands of diggers, and now the city is perhaps the most important commercial centre in the colony. This "remarkably handsome town," as Trollope calls it, stands on the shores of a fine bay, about 9 miles above its outport—Port Chalmers—at the entrance to the same inlet (Otago Harbour). It is well built, and, like its prototype in the Northern Hemisphere, is environed by the most diversified and romantic scenery. CHRISTCHURCH (40), the chief city of the provincial district of Canterbury, is "eminently English in its appearance, architecture, and surroundings." It stands on the banks of the Avon, about 6 miles from the port of Lyttelton (4), with which it is connected by a railway tunnelled through the hills. This "City of the Plains" is the centre of the largest agricultural and pastoral district in the colony, and its port, Lyttelton, has a very large shipping trade.

Christchurch and Dunedin are on the main line of the South Island Railway, which has been extended on the north to Culverden, 70 miles north of Christchurch, and on the south to Invercargill, about 150 miles south-west of Dunedin. Invercargill is the chief town in South Otago, and does a large trade in timber, frozen meat, wool, grain, &c. It is situated on an estuary, 17 miles north of its outport, Campbelltown, on Bluff Harbour, which opens into Foveaux Strait, and with which it is connected by rail. Queenstown, on the eastern shores of Lake Wakatipu, is a favourite tourist resort, and is reached by steamer from Kingston, the present terminus of the railway from Invercargill and from Dunedin, at the south end of the lake.

1. It has been proposed to connect the port by a ship canal with the harbour of Manukau on the | western side of the isthmus, which is about 100 miles nearer to Sydney than Auckland Harbour.

Oamaru (5) and Timaru (4) are two important ports on the east coast, between Dunedin and Christchurch. At both places, artificial harbours, rendered necessary by the large export of agricultural produce, especially cereals, have been constructed.

Blenheim (3), the chief town of the district of Marlborough, in the northeast of South Island, is connected by rail with Picton, the chief port of the district, at the head of Queen Charlotte Sound, the favourite rendezvous of Captain Cook, and about 50 miles distant from Wellington, on the opposite side of Cook Strait. Nelson (11) is prettily situated on the south-eastern shores of Tasman Bay, and is the terminus of the railway which runs to Greymouth (4), the " Newcastle of New Zealand," and Hokitika (3), the capital of Westland. Hokitika owes its rise to the discovery of productive goldfields in the vicinity in 1865. Its harbour, though greatly improved, is not so good as that of Greymouth, or that of Westport (3), another coal port and gold-mining centre, about 40 miles to the north of Greymouth. Westport is by far the best port on the west coast, and extensive harbour works are in progress. The breakwaters and training walls at Greymouth are also being completed, so that the port will be available at all times.

THE OFF ISLANDS.

The Off Islands of New Zealand include the Auckland Islands, Campbell Island and Macquarie Island to the south, the Antipodes Islets and Bounty Islets to the south-east, the Chatham Islands to the east, and the Kermadec Islands to the north, of the main islands.

The Auckland Islands are a group of one large and several small islands, about 200 miles south of New Zealand. The largest of them—Auckland Island—is about 30 miles in length and 15 miles in width. Along its western coast, the cliffs rise perpendicularly from the water to a height of several hundred feet, but its eastern shores are deeply indented, several of the inlets forming extensive and well-sheltered harbours. The islands are mountainous throughout, but the soil is rich and the vegetation luxuriant ; the climate, though mild and healthy, is extremely wet and stormy, and, since the whale-fishing settlement, established in 1850 by the Messrs. Enderby (to whom the British Government had granted the islands in recognition of their efforts to develop the whale fishery in the South Polar Seas), was abandoned, the group has remained unpeopled, except when they have formed the temporary refuge of shipwrecked mariners. Many disastrous wrecks have taken place upon their coasts.[1]

Campbell Island lies about 145 miles south-east of the Auckland Islands, and over 300 miles south of South Island. It is about 36 miles in circumference, and has some good harbours, but is usually uninhabited.

Macquarie Island is as far again from New Zealand as the Auckland Islands. It is about 20 miles long, and is covered with vegetation, but is only visited by a few vessels during the seal-fishing season.

The Antipodes Islets, about 500 miles south-east of New Zealand, are remarkable only as being the land nearest to the antipodes of London.

The Bounty Islets are an uninhabitable group of rocky islets, about 180 miles north of the Antipodes group.

---

1. For a detailed account of these and the | *Zealand Geography* [London: George Philip & other Off Islands, see Bowden's *Manual of New* | Son.

The **Chatham Islands** are situated about 536 miles to the east of Lyttelton. The largest of the three islands—Chatham Island—is 38 miles long and 25 miles broad. The first discoverer of these islands—Captain Broughton (1791) —found an indigenous race, called Morioris, who, however, were nearly exterminated by the Maori settlers from New Zealand. The present population is small, only about 500, but "excessively mixed, and is said to include Morioris, Maoris, Kanakas, Negroes, Chinese, Spaniards, Portuguese, Danes, Germans, English, Irish, Scotch, and Welsh!" They support themselves by stock-rearing and seal-fishing, and by supplying whaling ships and other vessels, which often call at the ports of WAITANGI or WHANGAROA, for provisions, &c. The climate is mild, and the soil generally fertile.

The **Kermadec Islands**, a group of islands some 600 miles to the north-east of Auckland, were formally annexed in 1887. Raoul or Sunday Island, the largest of the group, is a rugged and wooded island, about 12 miles in circumference.

NORFOLK ISLAND—LORD HOWE ISLAND.

Between the Kermadec group and Australia are two other islands —**Norfolk Island** and **Lord Howe Island**—both of which belong to the colony of New South Wales.

Norfolk Island is situated about 1,100 miles north-east of Sydney, and nearly midway between New Zealand and New Caledonia. This beautiful island was discovered by Captain Cook in 1774, and was subsequently used as a penal settlement. The convicts, who had brought the island into a high state of cultivation, were removed in 1855, and the British Government handed it over to the Pitcairn Islanders—the descendants of the mutineers of the *Bounty*— some of whom, however, returned to Pitcairn Island. The rest have now increased to about 500. They support themselves by the cultivation of the fertile soil and by the whale fishery.

Lord Howe Island is a mountainous and well-wooded island, about 5½ miles in length, nearly midway between Norfolk Island and the Australian coast, and about 830 miles north-east of Sydney. The few inhabitants are connected with the whale fishery, and whaling ships call here for supplies.

PITCAIRN ISLAND.

Between the easternmost group of the Polynesian archipelagoes and the South American coast, a distance of 4,000 miles, there are only a few solitary islets, one of which—**Pitcairn Island**—possesses special interest, as the refuge of the mutineers of the *Bounty*.

Pitcairn Island is an isolated mountainous island, about 2 miles long and three-quarters of a mile wide, with a fine climate, a fertile volcanic soil, covered with palms and fruit trees. It was in 1790 that the mutineers of the *Bounty* settled here. Their descendants were removed first to Tahiti and then back again to their island, only to be again removed, in 1856, to Norfolk

Island, but two years later many of them returned to their island home. They have since increased to about 200, and there is probably no healthier, happier, or more contented and comfortable a community in the world than the isolated islanders of Pitcairn.

BRITISH NEW GUINEA.

BRITISH NEW GUINEA, which includes the south-eastern part of the island, is a Crown Colony.   It has an area of 90,000 square miles, and a population of perhaps half a million.

"The territory was first taken over, in 1884, as a Protectorate, and was then placed under the management of a Special Commissioner ; but, on the colonies of Queensland, New South Wales, and Victoria undertaking to guarantee £15,000 a year for the cost of administration, which was further secured by the British New Guinea (Queensland) Act of 1887, the Queen's sovereignty was formally proclaimed in 1888, and the territory was constituted as a colony under the name of British New Guinea."

The colony is governed by an Administrator, aided by a nominated Legislative Council.   Each of the four divisions of the territory is in charge of a Magistrate.

The seat of government and the chief trading centre is **PORT MORESBY**, a small settlement conveniently situated on the shores of a land-locked harbour on the eastern side of the Gulf of Papua.   The coast region around Port Moresby is said to be beautifully diversified by forest, hill, and valley, and considerable parts of it are industriously cultivated by the natives, whose little groups of pile-raised huts are surrounded with thriving plantations of yams, sweet potatoes, fruit trees, sago palms, and sugar-cane.   There are, undoubtedly, large areas well suited to the production of an infinite variety of tropical products, but the hot, moist climate and malarious exhalations from the flooded forest-growths in the rainy season, are often fatal to Europeans, and must always have a deterrent effect on immigration and settlement.   Then there is the native question, more formidable here than in almost any other British possession.   In some parts of the colony, the natives are reported to be singularly amiable, honest, and pleasant to deal with ; in other places, they have been found savage and treacherous, while not a few tribes are fierce cannibals and inveterate head-hunters.   But another bar to settlement of European cultivators is the fact that—as Sir William Macgregor, the Administrator, says—the natives are agriculturists, each man planting for himself and his family, and, as a rule, they derive the means of existence from the soil, supplemented irregularly on the coast by a few fish, and inland by an occasional wallaby or pig.   They differ altogether from the aborigines of Australia.   The Papuan is an agriculturist, and is therefore a settler, with clear ideas as to proprietary rights in the soil he requires for his support ; the Australian aboriginal is nomadic, with vague hunting rights over ill-defined, great areas.   The Papuans are present in great numbers in large and permanent towns ; the Australian is a wanderer, and possesses no hereditary hearth or foundation, and their numbers are very small.   The missionaries have made very considerable progress, but the Papuan possesses no religious enthusiasm, and cannot in this respect be compared to the Polynesian ; but missionary influence diminishes cases of murder and massacre, and otherwise does good.

The trade is very small and appears to be decreasing, although valuable timber abounds and the coco and sago palms are plentiful. Traces of gold have been found on the mainland, but the only payable deposits discovered were at Sudest and St. Aignan, two islands of the Louisiade Archipelago, off the south-eastern peninsula, and even these seem now to be almost, if not entirely, exhausted. The principal exports are *gold, bêche de mer, copra, bird-skins, pearl-shells, rattans,* &c., but the annual value is under £20,000. Commercially, the prospects of British New Guinea are not very bright, and, apart from the discovery of gold in payable quantities, there is nothing that would make it a valuable commercial possession, nor will it ever be the home of white men, such as Australia has become.

∴ Both the **D'Entrecasteaux Islands** and the **Louisiade Archipelago** form part of the colony of British New Guinea. The exceedingly fertile Murray Island, at the eastern entrance to Torres Strait, about 100 miles north-east of Cape York, is now annexed to the colony of Queensland.

THE FIJI ISLANDS.

The charming archipelago of Fiji, or more properly *Viti*, embraces in all 255 islands and islets, scattered over an ocean area of 300 miles from west to east, and 200 miles from north to south, between the parallels of 15° and 22° S. latitude, and about 1,250 miles north of Auckland, 1,860 miles north-east of Sydney, and nearly 5,000 miles south-west of San Francisco. Two of the islands are of considerable size—the largest, **Viti Levu** (4,112 square miles), being considerably larger than Cyprus, while the second largest, **Vanua Levu** (2,432 square miles), is about three times the size of Mauritius. The other islands range from an area of 217 square miles to mere rocks. The total area of the inhabited islands is **7,740 square miles**, or slightly larger than that of Wales, and the population, including the 2,300 inhabitants of **Rotumah**, a small island annexed to the colony in 1880, amounts to about **123,000**, of whom 105,000 are native Fijians, 9,000 Indian and 4,500 Polynesian immigrants, 2,500 **Europeans**, and over 1,100 half-castes.

The Fijian Islands are the most easterly of the Melanesian Islands, and perhaps also the most diversified and attractive of all these tropical archipelagoes.[1] All the islands are of volcanic origin, and are therefore mountainous, some peaks in Vanua Levu rising to about 5,000 feet above the sea level, the smaller as well as the larger islands are abundantly watered by numerous rivers—through almost every valley flows a running stream, from which an ample supply of water for irrigation and other purposes can be obtained all the

---

1. The scenery in all parts of the archipelago is beautiful, and in some places even grand. " In passing over the mountains of Vanua Levu," says Mr. Horne, the Government botanist of Mauritius, " many magnificent views present themselves. Here forest and woodland, with valley opening into valley in oft-repeated succession ; there on one side the open grass country ; on the other, the blue sea studded with islands—with spots and lines of white foam where the sea is breaking on the reefs ; all these, seen from a considerable elevation, combine in forming a panorama of which words can convey a very faint idea."

year round.   Many of the rivers in the two larger islands are even navigable for canoes and good-sized boats.   The longest river is the **Rewa Rewa**, in Viti Levu ; it is navigable for 50 miles from its mouth, and its volume, owing to the copiousness of the rains—the mean annual fall is over 100 inches, and even that is only one-half the amount registered at some places situated in the path of the monsoon—is such that it deserves its other name of Via Levu, or Great Stream.

Every island in the group is almost encircled by a barrier reef, which forms an admirable breakwater, and, once through the opening, vessels ride at anchor in perfect safety.

The **climate**, though hot,[1] is remarkably healthy, and the larger and some of the smaller islands afford a delightful tropical residence.   Hundreds of English people have lived continuously in Fiji for years in the enjoyment of excellent health.

The soil is everywhere fertile, and there is hardly any land that is not capable of being profitably cultivated.   Many tropical products, such as *bananas, maize, cotton, sugar, tea, coffee, yams, pineapples*, &c., are already grown to some extent ; the *coco-nut palm plantations* afford a certain income, while the *numerous forests* contain a great number of valuable timber trees.   And now that the Fijians, who are by far the most intelligent of all the Melanesian peoples, have, mainly as the result of the labours of the Wesleyan missionaries, become civilized and Christianised[2]—it is said that there are no more devoted Christians in all Polynesia than the erewhile treacherous and savage cannibals of Fiji—the wonderfully rich resources of these ' gardens of the tropics ' are ripe for that development which British energy and capital alone can effect.   The Fijian himself is by no means an idler, but prefers to work on his own land, and thus the European planter is forced to depend upon imported labour.   Anyone, however, who is tempted by the genial climate and productive soil of Fiji to make that distant colony his home, will, with prudence and energy, succeed, and will find the calling of a tropical horticulturist or agriculturist one of peculiar interest, and one presenting a field of profit and research, without the ordinary monotony associated with farming in the extra-tropical portions of the empire.[3]

The **geographical position** of these islands, relative to the Australasian colonies and the Pacific Coast of America, their multifarious resources, and fine harbours, so easily rendered impregnable, combine to make Fiji one of the most valuable and important of the smaller British possessions.

The group was discovered by the famous Dutch navigator, Tasman, in 1643 —a year after his discovery of Tasmania and New Zealand.   Early in the present century, the island became the resort of South Sea traders, but, until the advent of the missionaries, the Fijians were savage cannibals, and horrible

---

1. The absolute minimum temperature at Suva, in 1889, was 63° F., and the absolute maximum was 91° F.; the mean minimum at the same place and in the same year, was 69° F., and the mean maximum 81° F.

2. Out of a population of something like 121,000, more than 15,000 are regular attendants at Wesleyan churches, and the remaining 23,000 are not

heathens, but, for the most part, members of other Christian Churches.   The people of Fiji are now a Christian people.   The education of the native Fijians is almost entirely conducted by the Wesleyan Mission, in whose schools 40,000 children are taught.

3. *The Handbook of Fiji*, by the Hon. J. E. Mason (Cowes).

human sacrifices attended most of their religious and other ceremonies. Old and sick people were buried alive, and " when a chief died, a whole hetacomb of wives and slaves had to be buried alive with him. When a chief's house was built, the hole for each post must have a slave to hold it up and be buried with it. When a great war canoe was to be launched or to be brought home, it must be dragged to or from the water over living human beings tied between two plantain stems to serve as rollers." The first Wesleyan missionaries arrived in 1835, and their success in civilizing such a people has been most remarkable. In 1858, the native king, Cakobau (pronounced Thakombau), offered the sovereignty of the islands to Great Britain, but it was declined. Owing to the desire of the European settlers, who were becoming more numerous, for a settled government, a constitution was framed, which provided for a partly European and partly native government under Cakobau, but this lasted for two years only, and in March, 1874, the sovereignty was again offered to, and again declined by, the British Government. In October of the same year, however, by an **Act of Cession,** the islands were ceded to Britain, and all lands in Fiji passed to the Crown. Soon after, a charter was issued, constituting the group into a separate colony.

The Fiji Islands form a **British Crown Colony** under a Governor, appointed by the Home Government, assisted by an Executive and a Legislative Council.

The members of both Councils are either official or nominated by the Crown. The natives are governed by their own chiefs, under the Governor's supervision. The two larger islands and Rotumah are each under a European Commissioner. The other islands are grouped in **14** provinces, each of which is placed under a superior native chief. Justice is administered by 10 European and 30 native magistrates. The Revenue is a little over, and the **Expenditure** under £80,000, and there is a small Public Debt. The **imports** amount to about £270,000, and the **exports** to over £350,000. The direct trade with Great Britain is small— most of the imports from, and exports to, Great Britain passing through Australian and New Zealand ports. There is regular steam communication with Sydney twice a month, with Auckland once a month, and with Melbourne every five weeks. There is a subsidised inter-island steamer trading regularly in the group. Seven-eighths of the vessels trading with the colony are British.

Fiji possesses but two towns, namely, **Suva,** the present capital, on the island of Viti Levu, and **Levuka,** the former capital, on the island of Ovalau.

SUVA, selected in 1880 as the capital, is situated on the south coast of Viti Levu, the largest and most populous island in the colony, and has a good harbour, and a white population of about 700. Levuka, the former capital, is on the coast of Ovalau, a small island off the eastern coast of Viti Levu, and perhaps the most centrally situated of the whole group. The harbour, formed by a barrier reef about a mile from the shore, though smaller than that of Suva, is excellent, and vessels can pass in and out with any wind.

∵ The little island of **ROTUMAH,** which lies about 400 miles N.N.W. of Levuka, and about five times that distance from Sydney, was annexed to the colony of Fiji in 1880. It is a mountainous island of purely volcanic formation, inhabited by about 2,300 copper-coloured natives, a peaceable, law-abiding, sober, and reliable race. The British Commissioner appointed by the Fiji Government is, at the same time, arbitrator, adviser, and conciliator, and even medical adviser, to these interesting people.

∴ The Governor of Fiji also acts in the capacity of Her Majesty's **High Commissioner for the Western Pacific**, and, as such, his jurisdiction extends over all islands in the Western Pacific, not being within the limits of the colonies of Fiji, Queensland, or New South Wales, and not being within the jurisdiction of any civilized power.

THE SOLOMON ISLANDS.

About 500 miles to the eastward of New Guinea is a large group of islands known as the Solomon Islands. They extend for 700 miles in a north-west and south-east direction, and consist of a chain of islands—the four northern islands, Bougainville, Choiseul, Ysabel, and Malayta, being separated by channels varying in width from 15 to 50 miles from the southern chain, which includes three large islands—**New Georgia, Guadalcanar, and San Christoval**—and several smaller islands and islets. The group has a total area of perhaps 15,000 square miles, or considerably more than twice as large as Wales, while the savage Melanesian inhabitants may number 150,000.

All the islands are **mountainous** and generally **volcanic**, and most of them are girdled by coral reefs. There is an active volcano on Guadalcanar and several quiescent and extinct volcanoes, and earthquakes are not infrequent.

These islands, says Mr. Woodward, an accomplished naturalist who visited them in 1886, 1887, and 1888, are for the most part clothed from coast to summit with the densest tropical forest, in which the immense *ficus* trees, of several species, are often conspicuous objects. In the neighbourhood of native villages the beach is fringed with **cocoa-nut palms**. The natives of the coast villages are constantly at war with the villages of the interior, while even on the same island a walk perhaps of five miles along the sea coast would bring one not only to a hostile village, but to a tribe speaking a distinct dialect from that of the village started from. Like all savages, they are suspicious of strangers, and treacherous when they see their opportunity, and a long list of murders and massacres darkens the history of the intercourse of the white man with the islanders, from their first discovery to the present time, the fault being sometimes with the native and sometimes with the white man.[1]

These islands were first discovered by a Spanish navigator, Mendaña, in 1568. He gave them the name of the Islands of Solomon, in order that his countrymen, supposing them to be the islands whence King Solomon got his gold, might be induced to colonise them. The group remained unvisited until about the end of the last century, and, since then, they have been occasionally visited by whalers, and in recent years by ships engaged in the South Sea Island labour-trade, recruiting natives to work on the plantations in Fiji and Queensland. During the last 20 years, traders from Sydney have visited, and in a

1. The above description is based on that given in Mr. Woodward's interesting work—"A Naturalist among the Head Hunters," an account of his three visits to the Solomon Islands, in the years 1886, 1887, and 1888 (London: George Philip & Son).

few instances temporarily settled in, the group, while it is now annually visited by the missionaries of the Melanesian Mission.

By an arrangement between Great Britain and Germany, the group was divided, the northern islands remaining within the German sphere of influence, and the southern islands within the British sphere.[1]

In 1886, Germany definitely annexed Bougainville, Choiseul, and Ysabel, the three largest islands in the group, and attached them to her New Guinea Protectorate. In 1893, the rest of the group was formally declared a British Protectorate.

THE SANTA CRUZ ISLANDS AND THE NEW HEBRIDES.

THE SANTA CRUZ ISLANDS and the NEW HEBRIDES form a long chain of islands between the **Solomon Islands and New Caledonia.** The Santa Cruz Islands, though independent, are within the *British sphere of influence*, while the New Hebrides are under the *joint protection* of France and England.

These islands form three distinct groups—(1) the Santa Cruz Islands, (2) the Northern New Hebrides, and (3) the Southern New Hebrides. These groups are divided from each other by about 60 miles of sea, and the Santa Cruz group is about as far from the southernmost of the Solomon Islands as the Southern New Hebrides are from New Caledonia.

The **Santa Cruz Islands** are small, but fertile, and bear a population of some 5,000. *Vanikoro*, one of the group, is famous as the scene of the shipwreck of the French navigator, La Perouse.

The **Northern New Hebrides** include a few large islands—*Espiritu Santo*, the largest, is 75 miles long by 40 wide—and many smaller islands and islets. All are mountainous, and culminate in the volcanic peak of *Lopevi*, which rises to a height of 5,000 feet. Espiritu Santo contains beautiful mountains, charming rivers, and extensive woods. **Banks' Islands** form the northern part of the group.

The **Southern New Hebrides** consist of five islands, the most northerly of which, **Erromango**, is memorable as the place where the courageous and devoted missionary, John Williams, was killed by the savages. In the southernmost island of the group—**Aneiteum**—Scotch missionaries have long laboured, with the result that the people are all Christians, and have become civilized and docile, although they were formerly as savage and degraded as most of the people of **Tanna**, a large island between Aneiteum and Erromango, still are. The two other islands—**Aniwa** and **Fotuna**—are to the east of Tanna. Both are small.

The New Hebrides and the larger islands of the Santa Cruz group are of volcanic origin, and some of the volcanoes are still active, and there are boiling springs on one of the Banks' Islands. The appearance of almost all the islands

---

1. In April, 1886, diplomatic arrangements were entered into between the British and German Governments, which provided that, for political purposes, the Western Pacific should be considered as embracing that part of the ocean lying between the 8th parallel North and the 30th parallel South, and between 165° West longitude and 130° East. A conventional line of demarcation between the British and German spheres of influence within the area thus defined, was then laid down. This line, starting from the north-east coast of New Guinea, at a point near Mitre Rock, on the 8th parallel South, runs due east to the intersection of 8° S. with 154° East. Then it bears north-east to a point south of Bougainville Island, at the intersection of 7° 15′ S. with 155° 25′ E. From there it is carried south-east to the south of Choiseul and Ysabel islands, and from the intersection of 8° 50′ S. with 159° 50′ E. it holds due north-east to the south-east corner of the Marshall Islands, where it turns directly north, running to 15° N.

is very beautiful—steep hills, covered with fine forest trees, rising abruptly from
the sea, while groves of cocoa-nut palms surround the villages.   None of the
Pacific islands " have suffered more than the New Hebrides by the reckless
action of the 'black-birding' schooners seeking labourers for the Queensland
plantations, and carrying away the natives by force."   In some years, 6,000
islanders left their home, and not more than one-half or two-thirds ever returned.
No wonder that the population of all the islands has decreased, but, although
the labour trade is now conducted under more stringent regulations, the epidemic diseases introduced by Europeans, and even the adoption of more
civilised habits, are everywhere diminishing the population.   " There must
surely be something wrong in the method of civilisation, which, throughout all
Polynesia, has this one invariable and disastrous effect."

THE GILBERT ISLANDS.

The **Gilbert Islands,** an equatorial group of 16 little atolls, annexed by Great Britain in 1892, lie to the north-east of the Solomon
Islands.   The aggregate area is about 170 square miles, but the
population amounts to nearly 40,000, an average of about 230 per
square mile.

This group, which is cut by the Equator and the 175th meridian E., is one
of the most remarkable of all the Pacific archipelagoes.   The islands are so
small—the largest covering only 15 square miles--and the hard coral rock is
covered with so little soil, that nothing can be grown beyond a little taro,
while the cocoa-nut is almost the only spontaneous plant product, and yet
these barren atolls are more densely peopled than the most fertile islands in
all Oceania.   The smallest atolls have a population of 1,500 to 2,000 ; one of
the largest islands, **Taputeouea,** with an area of only 10 square miles, has no
less than 7,500 inhabitants, an average of 750 per square mile !   The climate
is admirable, and the people, who are industrious fishermen and skilful canoe
builders, are strong and healthy, and were formerly much sought after by the
labour contractors from Queensland and Fiji.

THE ELLICE ISLANDS.

The **Ellice Islands,** which lie mid-way between Fiji and the
Gilbert Archipelago, are within the British sphere of influence.

The eight islands of this group are, like those of the more northerly groups,
purely coralline, and are, in fact, called *Lagoon Islands*.   They have an area
of about 170 square miles, and a population of only 2,500, a small number
compared with the swarming population of the adjoining Gilbert Islands.

THE PHŒNIX, UNION, AND MANIHIKI ISLANDS.

The **Phœnix, Union** or Tokelau, and **Manihiki Islands,** together
with several isolated islets—**Malden, Starbuck, Christmas, Fanning,** &c.—all belong to Great Britain, with the exception of two
islands in the Phœnix group and two of the Union Islands, which
are occupied by the Americans.

The **Phœnix** Islands are all within about 5° of the Equator ; the widely-
scattered **Union** and **Manihiki** Islands lie further south, between the Ellice Islands
and the Marquesas.   All of them are low coral islands, very seldom visited.   Some

of them contain deposits of guano, which have been exploited by the English or American traders, and one of them—a small atoll called *Penrhyn Island*—is " interesting as being the extreme eastern outlier of the Melanesian race. The inhabitants are tall, of a dark brown colour, have wavy hair sometimes frizzled into mops, and prominent noses and brows. They are described as being excessively noisy and quarrelsome. They fish for food or dive for pearl-shell all day, come home by sunset, eat, and begin to talk. They soon quarrel; the women join; they wrangle and storm; the children even join in; and this keeps on all night. It all ends in nothing, they never fight, but bluster, and shout, and scream, night after night."

The natives of **Manihiki** or **Humphrey Island**, which gives its name to the group, are Christians, and can read and write English. Further to the south-west are the **Suwarrow Islands**, now also British.

Two other isolated islands to the north of the Equator—**Christmas Island** and **Fanning Island**—also belong to Great Britain.

THE COOK ARCHIPELAGO.

The **Cook Archipelago**, so named in honour of its discoverer, the famous Captain Cook, consists of a group of nine islands, situated about 700 miles south-east of Samoa, in Southern Polynesia.

These islands, also called the Hervey Islands, are either volcanic or coralline, and are all encircled by dangerous coral reefs. Their total area is only about 300 square miles, and the population is not more than 8,000. The largest island is the well-known **Rarotonga**, the scene of the missionary labours of John Williams, the apostle of the Pacific, and one of the pioneers of missionary enterprise in the South Sea Islands. The islanders, formerly fierce cannibals, are now in an advanced state of civilization, but they are rapidly diminishing in numbers. Their petition for annexation to Great Britain was unheeded until 1888, when a British Protectorate was proclaimed.

APPENDIX A.

THE INTERNATIONAL STRUGGLE FOR THE NIGER.

By Sir George Taubman-Goldie, K.C.M.G.

When the Lower Niger was discovered in 1829, it was hoped that Central Africa would be at once opened to trade. Yet the efforts of the first thirty or forty years, in spite of the generous support of Parliament, were absolutely fruitless. Of the repeated Government attempts to open up the Niger, those of 1832 and 1841 are best known, from their deplorable waste of money and life. At last, about a quarter of a century ago, the British Government wisely resolved to withdraw their resident Consul, to grant no more subsidies, and to abandon the task as hopeless. The field was thenceforth clear to independent adventure. Several petty firms gradually established permanent trading stations among a few tribes in the Niger regions at their own cost and risk. By continual subsidies to the chiefs of these tribes, the traders obtained tolerable security from plunder in their factories, and the local natives gradually learnt the advantages of commerce and peace; but this security was purely local, and constantly disturbed by the attacks of neighbouring tribes; while the trading steamers were fired upon, and sometimes seized and plundered, on their transit up and down river. In 1876, one of these steamers had forty-three holes knocked in her hull with round shot.

The British Government occasionally despatched gunboats to chastise piratical tribes; but as gunboats, owing to their deep draught, could only ascend the river during three months of the year, and as they never left the main river, nor took part in inland expeditions, nor (for fear of the climate) proceeded far from the seaboard, they produced no permanent effect. Experience has fully established the abnormal character of settlements in inner Africa which differentiates them from Coast Colonies such as Great Britain possesses. In the latter, administrative errors are repaired by the knowledge of the overwhelming force of H.M.'s fleet. In the Niger Territories, administrative influence has to rest on an entirely different basis.

The territories on each side of the Lower Niger and its numerous tributaries are divided among hundreds of tribes, occupying areas which vary from 10 to 10,000 square miles, and whose chiefs used to have no vent for their energies but inter-tribal wars, slave raids, and plunder. Prior to 1879, the rivalries of the petty British firms, the intrigues with the natives of their civilized coloured agents, and the want of unity of action even among their white agents, prevented rapid improvement. No real security for life or property could exist until some sound political system was established. At last, in 1879, all the interests in the Niger amalgamated into the "*United* African Company." From that fusion dates the effective administration of the Lower Niger. Previously, the traders' considerable political influence had only been spasmodically exercised in scattered districts; but, on the fusion, it was resolved to extend this administrative influence over the principal riverain tribes, and to exercise it effectively and continuously. It was also resolved to meddle as little as practicable with the internal administration of each tribe. This rule is still followed, unless the tribe itself appeals to the Company, or unless some outrage to humanity, or some strong political motive, urgently calls for interference. Thus, during the rebellion in Nupe, in 1881, the United African Company put forth all its strength (and successfully) to uphold Prince Omoru—not that it altogether approved of his methods of ruling, but because his overthrow would then have resulted in that province splitting up into petty and worse-governed tribes.

The United African Company proving highly successful, both commercially and politically, the next great step was made towards the close of 1881, when it was proposed to create a great public company, having a subscribed capital of a million sterling, with the following objects :—

First—Political and commercial development over the whole basin of the Lower and Middle Niger, and the acquisition of sovereign rights from the 237 tribes which occupy the country within reach of the main river and its numerous tributaries.

Second—To obtain, in accordance with a special provision of the Memorandum of Association, an international status for The Company by the grant of a Royal Charter from Her Majesty; or, in default of that, from some other power.   This step had become absolutely necessary in view of the principle of international law, that concessions of sovereign rights to a Company, unless endorsed by a civilized Government, have no validity as against the acts of another civilized Government.

Although the New Company would have been compelled, in the event of Great Britain refusing a Charter, to obtain it from a foreign State, the Directors naturally desired that this large section of Central Africa should fall within the British Empire, and it was therefore decided that the name to be adopted should be " The *National* African Company."   The Presidency was accepted by Lord Aberdare, then President of the Royal Geographical Society.   The public issue took place in 1882; and the New Company commenced by purchasing the commercial assets and the political goodwill of the United African Company. Almost immediately, however, two French houses which had entered the Niger regions—a matter of little importance to The Company, so long as they contented themselves with purely commercial work—commenced strenuous efforts to obtain political influence over the chiefs of some of the riverain tribes.

It soon became known that the larger of these French Companies had been specially formed under the auspices of M. Gambetta.   Schemes were propounded for uniting Tunis, Algiers, Senegal, the Central Sudan, and the Lower Niger into a Franco-African Empire; which, as Germany had not then annexed the Cameroons, would have extended continuously to the French Gaboon and to the basin of the Congo—of which France claimed the reversion from the Congo Free State.   However difficult of realisation in their entirety were these vast schemes, embracing one-half of the African Continent, they threatened to have, incidentally, one practical result, viz., the annihilation of The Company The French operations in the Lower Niger Basin developed so rapidly—the smaller house increasing its working capital in 1882 to 1½ million francs, and at the end of 1883 to 4 millions—that their factories actually exceeded in number, though, of course, not in importance, those of The Company, and in articles even in the British Press, the Lower Niger, which owed everything to British energy and British money, was spoken of as if it was a French river. The Company had previously attempted to carry through an amalgamation (under the British flag) with the smaller French house, with the view of subsequently buying out the other, whose capital was £600,000, some of which happily was employed on the Senegal coast.   But it had been found that national sentiment created a difficulty that had not stood in the way of the purely British amalgamation of 1879.   The Company therefore had only three courses open to it :—

1st.—The usual course in cases of risk, difficulty, and conflict of opinion: viz., to do nothing, or to adopt half measures. French influence over the native Princes and Chiefs would then have grown gradually at the expense of that of The Company, until a French protectorate could have been safely established, when those valuable regions would have doubtless fallen under the Upper Senegal colony, with its exclusive *régime*. British manufactures and British ships would have been excluded by the French bounty system or by differential duties, as in Tonquin or Madagascar. And as the French province of Senegal has developed latterly until it has ruined the trade of British Gambia, so France in the Niger could in a few months have cut off the entire trade of the "Oil Rivers."

2nd.—To amalgamate with its rivals under the French flag, and to throw those regions into the arms of France, which could naturally have been done on terms highly advantageous to the Company, on account of its strong hold on the natives, who know nothing of the European Powers, but who owe such wealth as they possess to The "Company," or,

3rd.—To eliminate foreign influence from the Niger at any cost.

The last course being adopted, it was pursued with the usual vigour and rapidity of The Company. While carefully avoiding the outlay of capital on permanent factories, temporary premises were occupied among a great number of tribes which had never before shared in the advantages of direct European trade, and their produce was thus cut off from the French houses. This involved a large increase of The Company's white staff, and a far larger increase of the civilized coloured servants, who are enlisted from Sierra Leone, Accra, and other British possessions on the West Coast. During the international struggle the barter rate of goods rose to nearly double its former rate. The Company's fleet was augmented, and subsidies were dealt out to native chiefs with no sparing hand. The amount of the loss to the larger French Company is not known, but the smaller Company, on its liquidation, admitted the loss of one-third of its Share Capital, and, if it had continued for another year, would have undoubtedly lost the remainder. Just before the meeting of the West African Conference at Berlin both French houses had disappeared from the Niger Basin, and Sir Edward Malet, at the opening of the Conference, was enabled to declare, on behalf of Great Britain, that no commerce whatever except hers existed in the Niger Basin.

But just before these, at one time, apparently insuperable difficulties were overcome, a new danger arose which no sagacity could have foreseen. Germany, for the first time in her history, resolved to start as a colonizing Power, and unexpectedly commenced her operations by concluding protectorate treaties with the native chiefs of the Cameroons, which is close to the Niger regions. The first White Book ever published at Berlin set forth with evident pride the despatch sent to Prince Bismarck by Dr. Nachtigal, describing how in the race for the Cameroons he had beaten the British vessels by running the German gunboat "Möwe" at full speed, *in spite of the bearings being heated.* The feeling against England was at that time excessively bitter in Germany. Her newspapers vehemently pressed her Government to make treaties in the Niger, and able German agents made persistent efforts to gain over the more important of the native powers. The Company, once more the sole European trader established in the Territories, was able to throw itself with undivided vigour into this new struggle, to conclude and uphold treaties with the remainder of the 237 tribes, and to exclude from district after district the intriguing agents.

Meanwhile, it was absolutely necessary to maintain the same heavy rate of expenditure.

Any attempt to reduce the barter rates from their extraordinary height during the period of international struggle to the normal rates before that struggle commenced, would have furnished a weapon to the foreign agents in inciting the less civilized of the native tribes to repudiate their treaties with The Company and enter into other engagements.   It is positively asserted that one agent went so far as to threaten a tribe that its towns would be burnt down if it did not break its treaty with The Company.

Any reduction of the enormously inflated staff of The Company to its normal strength would have thrown a number of dissatisfied and hostile agents, possessing great influence over the natives, into the hands of the foreigners.

Nor could steamers be laid up and their crews discharged, as all had to be incessantly employed in visits to the various riverain tribes, and in maintaining the threatened treaties.   The effect of the international struggle under this head may be best shown by the fact that The Company's steamers and launches, in 1885, numbered 27, as against 7 in 1882, just before that struggle commenced, whilst the tonnage of the trade had not quite doubled in the interval.

The Company's difficulties were greatly increased by the news from a sure source in France that a secret agent was being sent out by the Government to inquire into the causes of the sudden disappearance of French interests from the Niger.   The French press and the colonial party at that time were displaying great bitterness at the success of The Company in terminating French interests in those regions, and there can be little doubt that, had the agent in question been free to visit the various districts of the Niger, any patriotic action on his part would have received warm support from public opinion in France.

If the British Government had, at the time of The Company's formal petition in February, 1885, legalized the situation by granting the Charter, which was finally issued in July, 1886, the excessive political expenditure of the last seventeen months would have been spared; but The Company (having, rightly or wrongly, decided not to sell its treaties and influence to any foreign country) was compelled to try to prevent the Niger Territories falling, without its concurrence, into the hands of a foreign Government. Under the "free navigation" clauses of the Act of Berlin even partial foreign annexations would have rendered the necessary fiscal arrangements impossible, except at an insupportable cost for preventive service, owing to the physical peculiarities of the Lower Niger region, with its network of waterways.   Now that the whole navigable course (from the sea) of the Niger (and of its tributaries) is in the hands of The Company, this impossibility no longer exists.

In July, 1886, the Royal Charter was at last issued, with the cognizance and acquiescence (tacit or expressed) of Germany and France, and The Company immediately cabled its Agent-General to carry out the necessary reductions of stations, staff, &c.   Thus closed the international struggle for the Niger; but it may be easily imagined that the evil effects of this struggle did not cease at once.   The greater part of the country had relapsed into almost the same condition of anarchy and insecurity that existed prior to the formation of the United African Company in 1879.   Since the issue of the Charter the exertions of The Royal Niger Company have resulted in an extraordinary restoration of order, and although—until those vast regions are completely civilized—occasional local disturbances must be expected, there is reason to hope that an era of general peace and of great prosperity and development has commenced for the Niger Territories.

P

APPENDIX B.

ORTHOGRAPHY OF NATIVE NAMES OF PLACES.

With regard to the frequently perplexing matter of geographical nomenclature, the following notes on the system of orthography for native names of places, adopted by the Council of the Royal Geographical Society, will be useful for reference :—

1. No change will be made in the orthography of foreign names in countries which use Roman letters ; thus Spanish, Portuguese, Dutch, &c., names will be spelt as by the respective nations.

2. Neither will any change be made in the spelling of such names in languages which are not written in Roman characters as have become by long usage familiar to English readers ; thus Calcutta, Cutch, Celebes, Mecca, &c., will be retained in their present form.

3. The true sound of the word as locally pronounced will be taken as the basis of the spelling.

4. An approximation, however, to the sound is alone aimed at. A system which would attempt to represent the more delicate inflections of sound and accent would be so complicated as only to defeat itself. Those who desire a more accurate pronunciation of the written name must learn it on the spot by a study of local accent and peculiarities.

5. The broad features of the system are that vowels are pronounced as in Italian, and consonants as in English.

6. One accent only is used, the acute, to denote the syllable on which stress is laid. This is very important, as the sounds of many names are entirely altered by the misplacement of this "stress."

7. Every letter is pronounced. When two vowels come together each one is sounded, though the result, when spoken quickly, is sometimes scarcely to be distinguished from a single sound, as in *ai*, *au*, *ei*.

8. Indian names are accepted as spelt in Hunter's Gazetteer.

The amplification of the rules is given below :—

Letters.	Pronunciation and Remarks.	Examples.
a	*ah*, *a* as in *father*	Java. Banána.
e	*eh*, *e* as in *benefit*	Tel-el-Kebir, Olóleh, Yezo, Medina, Levúka, Peru.
i	English *e* ; *i* as in *ravine* ; the sound of *ee* in *beet*.	
	Thus, not *Feejee* ; but	Fiji, Hindi.
o	*o* as in *mote*	Tokio.
u	long *u* as in *flute* ; the sound of *oo* in boot.	
	Thus, not *Zooloo*, but	Zulu, Sumatra.
	All vowels are shortened in sound by doubling the following consonant.	Yarra, Tanna, Mecca, Jidda, Bonny.
	Doubling of a vowel is only necessary where there is a distinct repetition of the single sound.	Nuulúa, Oosima.

Letters.	Pronunciation and Remarks.	Examples.
ai	English *i* as in *ice*.	Shanghai.
au	*ow* as in *how*. Thus, not *Foochow*, but	Fuchau.
ao	is slightly different from above · · · ·	Macao.
ei	is the sound of the two Italian vowels, but is frequently slurred over, when it is scarcely to be distinguished from *ey* in the English *they*.	Beirût Beilûl.
b	English *b*.	
c	is always soft, but is so nearly the sound of *s* that it should be seldom used. If *Celebes* were not already recognised it would be written *Selebes*.	Celebes.
ch	is always soft, as in *church*. · · · ·	Chingchin.
d	English *d*.	
f	English *f*; *ph* should not be used for the sound of *f*.	
	Thus, not *Haiphong*, but	Haifong, Nafe.
g	is always hard. (Soft *g* is given by *j*.) · ·	Galápagos.
h	is always pronounced when inserted.	
j	English *j*; *dj* should never be put for this sound. · · · · · · ·	Japan, Jinchuen.
k	English *k*. It should always be put for the hard *c*.	
	Thus, not *Corea*, but	Korea.
kh	The Oriental guttural · · · · ·	Khan.
gh	is another guttural, as in the Turkish · ·	Dagh, Ghazi.
l		
m	} As in English.	
n		
ng	has two separate sounds, the one hard as in the English word *finger*, the other as in *singer*. As these two sounds are rarely employed in the same locality, no attempt is made to distinguish between them.	
p	As in English.	
q	should never be employed; *qu* is given as *kw*.	Kwantung.
r		
s		
t	} As in English.	
v	· · · · · · · · · ·	Sawákin.
w		
x		
y	is always a consonant, as in *yard*, and therefore, should never be used as a terminal, *i* or *e* being substituted.	Kikúyu.
	Thus, not *Mikindány*, but	Mikindáni.
	not *Kwaly*, but	Kwale.
z	English *z*.	Zulu.
	Accents should not generally be used, but where there is a very decided emphatic syllable or stress, which affects the sound of the word, it should be marked by an *acute* accent.	Tongatábu, Galápagos, Paláwan, Saráwak.

A SELECTED LIST

OF

GEORGE PHILIP & SON'S

PUBLICATIONS.

PHILIPS' IMPERIAL ATLAS OF THE WORLD. A Series of Eighty Maps, illustrating every aspect of geographical science. Each Map, with the exception of the Physical and Orographical Maps, accompanied by a Complete Index, in which the latitude and longitude of every place on it is given. Imperial Folio, half-bound russia or morocco, gilt edges, price £8; or full-bound russia or morocco, price Ten Guineas.

⁂ This Great Work, the production of which has occupied several years, is a Complete Atlas for all Purposes, and embodies an amount of information unequalled by any other Atlas, English or Foreign. Some idea of its magnitude and value may be gathered from the fact that the Latitude and Longitude of over 200,000 names of places are given in the Indexes.

PHILIPS' GENERAL ATLAS OF THE WORLD. A Series of Fifty Maps, based upon the latest surveys and the works of eminent travellers and explorers. Each Map is accompanied by a Complete Index, in which the latitude and longitude of every place on it is given. Imperial folio, half-bound russia or morocco, gilt edges, price £4.

⁂ The Plates in this Atlas have been drawn with the greatest care and precision from the most recent and reliable materials, and are distinguished by extreme clearness and accuracy, and worthily represent the present advanced state of Geographical knowledge and Cartographical skill.

GEORGE PHILIP & SON, LONDON AND LIVERPOOL.

PHILIPS' IMPERIAL ATLAS OF AUSTRALASIA. A Series of
Ten Imperial Maps, embodying the latest information. Each Colony is
shown separately in full detail, and there is a Complete Index to each
Map. Imp. folio, half-bound russia or morocco, gilt edges. price £1, 5s.

PHILIPS' ATLAS OF THE COUNTIES OF ENGLAND. A Series
of Maps reduced from the Ordnance Survey, with a Complete Index.
New and revised edition. Crown folio, half-bound morocco, gilt edges,
price Two Guineas.

PHILIPS' CLASSICAL, HISTORICAL, AND SCRIPTURAL ATLAS,
illustrating the Ancient Classics, Historians, and Poets. A Series of
Fifty-one Imperial Maps, forming a complete *vade mecum* to the Classical
Student, and an interesting book of reference to the lover of general litera-
ture. Imperial folio, half-bound morocco, gilt edges, price Two Guineas.

PHILIPS' HISTORICAL ATLAS, containing a Chronological Series of
Maps of Europe and other lands at successive periods, from the 5th to the
latter half of the 19th century, with Historical Memoirs to the Maps.
Crown folio, handsomely bound in cloth, gilt edges, price One Guinea.

PHILIPS' FAMILY ATLAS of Physical, General, and Classical Geography, with an Introductory Essay on Physical Geography, and a
copious Index. Imperial 4to, handsomely bound in cloth, gilt edges, 15s.

PHILIPS' LIBRARY ATLAS of Ancient and Modern Geography.
With an Index of upwards of 22,000 names. Imperial quarto, handsomely bound in cloth, gilt edges, 15s.

PHILIPS' CABINET ATLAS of Modern Geography. With a Complete
Index. Imperial quarto, handsomely bound in cloth, gilt edges, 10s. 6d.

PHILIPS' SELECT ATLAS of Modern Geography. With a Complete
Index. Imperial quarto, handsomely bound in cloth, gilt edges, 7s. 6d.

BALL'S ATLAS OF ASTRONOMY. A Series of 72 Plates, engraved
from Original Drawings. With Notes and Index. By Sir Robert
Stawell Ball, LL.D., F.R.S., F.R.A.S., Royal Astronomer of Ireland,
and Lowndean Professor of Astronomy in the University of Cambridge.
Crown 4to, handsomely bound in cloth, price 15s.

PHILIPS' HANDY-VOLUME ATLAS OF THE WORLD, an entirely new and enlarged Edition, containing new and specially engraved Plates, with Statistical Notes and Complete Index. By E. G. Ravenstein. F.R.G.S. Handsomely bound in cloth, size—6 in. by 4 in., rounded corners, price 5s.; or in French morocco, gilt edges, in box, price 7s. 6d.

PHILIPS' HANDY-VOLUME ATLAS OF THE BRITISH EMPIRE. A Series of 120 Maps and Plans. With Complete Index, and Geographical and Historical Notes to each Map, by J. Francon Williams, F.R.G.S. Handsomely bound in cloth, size—5¾ in. by 3½ in., rounded corners, price 3s. 6d. ; or in French morocco, gilt edges, in box, price 5s.

PHILIPS' HANDY ATLAS OF THE COUNTIES OF ENGLAND. A Series of Maps reduced from the Ordnance Survey, and coloured to show the present Parliamentary Divisions. With Complete Index. New Edition. Crown 8vo, cloth, 5s. ; in French morocco, 7s. 6d. *Tourist Edition*, limp cloth, rounded corners, 5s.

PHILIPS' HANDY ATLAS OF THE COUNTIES OF WALES. With Special Maps of the Snowdon and Cader Idris districts, and Index. Crown 8vo, cloth, 2s. 6d. ; French morocco, 4s. 6d. *Tourist Edition*, limp cloth, rounded corners, 2s. 6d.

PHILIPS' HANDY ATLAS OF THE COUNTIES OF SCOTLAND. With Index. Crown 8vo, cloth, 3s. 6d. ; French morocco, 6s *Tourist Edition*, limp cloth, rounded corners, 3s. 6d.

PHILIPS' HANDY ATLAS OF THE COUNTIES OF IRELAND. With Index. Crown 8vo, cloth, 3s. 6d. ; French morocco, 6s. *Tourist Edition*, limp cloth, rounded corners, 3s. 6d.

PHILIPS' HANDY-VOLUME ATLAS OF LONDON, containing a large-scale (3 inches to a mile) Street Plan, in 55 Sections, of London and its Suburbs, a number of Special Maps and Plans, and a Compendious Directory to the Public Buildings, Parks, &c., with a Complete Index. Imperial 16mo, cloth, rounded corners, 5s. ; French morocco, gilt edges, 7s. 6d.

GEORGE PHILIP & SON. LONDON AND LIVERPOOL.

PHILIPS' SYSTEMATIC ATLAS for Higher Schools and General Use. A Series of Physical and Political Maps of all the Countries of the World, with Diagrams and Illustrations of Astronomical and Physical Geography, specially drawn by E. G. Ravenstein, F.R.G.S. With Index. Imp. 8vo, cloth. Price, 15s. School Edition, price 10s. 6d.

PHILIPS' COMPREHENSIVE SCHOOL ATLAS of Ancient and Modern Geography. New edition, revised and enlarged, containing 60 Maps. With Index. Imperial 8vo, strongly half-bound, 10s. 6d.

PHILIPS' STUDENT'S ATLAS of Modern Geography, containing 48 Maps. With Index. Imperial 8vo, strongly bound in cloth, 7s. 6d.

PHILIPS' SELECT SCHOOL ATLAS. New and enlarged edition, containing 36 Maps, with Index. Imperial 8vo, strongly bound in cloth, 5s.

PHILIPS' INTRODUCTORY SCHOOL ATLAS. New and enlarged edition, containing 34 Maps. With Index. Imperial 8vo, cloth, 3s. 6d.

PHILIPS' YOUNG STUDENT'S ATLAS. New and enlarged edition, containing 36 Maps. With Index. Imperial 4to, cloth, 3s. 6d.

PHILIPS' ATLAS FOR BEGINNERS. New and enlarged edition, containing 60 Maps. With Complete Index. Crown 4to, cloth, 2s. 6d.

PHILIPS' FIRST SCHOOL ATLAS, containing 36 Maps, with Examination Questions on each Map. Crown 4to, limp cloth, 1s.

PHILIPS' "STANDARD" ATLAS, containing 24 coloured Maps and Diagrams of Geographical Terms. Small crown 4to, stiff cover, 6d.

PHILIPS' PREPARATORY ATLAS, containing 16 Maps, full coloured. Crown 4to, stiff cover, 6d.

PHILIPS' SHILLING ATLAS of Modern Geography, containing 24 Large-scale Imperial 4to Maps, printed in colours.

PHILIPS' EXCELSIOR ATLAS, containing 140 Maps and Plans. Stiff cover, illustrated, price 1s.

PHILIPS' FAVOURITE SIXPENNY ATLAS, containing 50 Maps, Plans, Diagrams, &c. Illustrated cover.

GEORGE PHILIP & SON, LONDON AND LIVERPOOL.

PHILIPS' "GRAPHIC" SCHOOL ATLAS. A Series of 110 Maps and Diagrams, specially designed for School Use. Edited by J. Francon Williams, F.R.G.S. Crown 8vo, stiff cover, 1s. ; cloth, lettered, 1s. 6d.

PHILIPS' ATLAS FOR JUNIOR CLASSES. A Series of Physical and Political Maps of all the Countries of the World. Medium 8vo, limp cloth, 1s. 6d.

PHILIPS' "UNIQUE" SIXPENNY ATLAS, containing 70 small but clear Physical and Political Maps. Stiff covers, illustrated.

PHILIPS' "UNIQUE" SHILLING ATLAS, containing 70 enlarged Maps, Physical and Political, and 10 Astronomical Diagrams. Demy 4to, stiff cover, illustrated.

PHILIPS' POPULAR ATLAS OF THE BRITISH EMPIRE, containing a Complete Series of Maps of the United Kingdom and of the British Colonies and Dependencies. With Statistical Tables. Crown 4to, illustrated cover, price 1s.

PHILIPS' SIXPENNY ATLAS OF THE BRITISH COLONIES, containing Maps of all the British Possessions and Dependencies. Crown 4to, illustrated cover.

PHILIPS' SCHOOL ATLAS OF PHYSICAL GEOGRAPHY, containing 20 Plates, with Explanatory Letterpress. Imperial 8vo, strongly bound in cloth. New and cheaper edition, 5s.

PHILIPS' PHYSICAL ATLAS FOR BEGINNERS, containing 12 Maps. New and cheaper edition, crown 4to, stiff cover, 1s. ; cloth, lettered, 1s. 6d.

PHILIPS' SCHOOL ATLAS OF CLASSICAL GEOGRAPHY. A Series of 18 Maps, with Index. Medium 4to, cloth, 3s. 6d.

PHILIPS' HANDY CLASSICAL ATLAS. A Series of 18 Maps. Medium 8vo, cloth, 2s. 6d.

PHILIPS' SCHOOL ATLAS OF SCRIPTURE GEOGRAPHY. A Series of 16 Maps. Crown 4to, illustrated cover, 1s. ; cloth, with Index, 1s. 6d.

PHILIPS' SMALLER SCRIPTURE ATLAS, containing 16 Maps. In illustrated cover, 6d. ; cloth, 1s.

GEORGE PHILIP & SON, LONDON AND LIVERPOOL.

PHILIPS' IMPERIAL MAPS. A Series of 80 Maps, drawn with the greatest care and precision from the most recent and reliable materials. The following is a complete list of the Maps in this unique series :—

World—Mercator's Projection.
The Atlantic Ocean.
Europe.
British Isles.
England and Wales.
England and Wales—North Sheet.
England and Wales—South Sheet.
*England and Wales—Complete Map.
Scotland.
Scotland—North Sheet.
Scotland—South Sheet.
*Scotland—Complete Map.
Ireland.
Ireland—North Sheet.
Ireland—South Sheet.
*Ireland—Complete Map.
The Mediterranean.
France.
Holland and Belgium.
Switzerland.
North-West Germany.
South-West Germany.
Prussia, and Smaller States of Northern Germany.
Austro-Hungarian Empire.
Denmark, Iceland, &c.
Sweden and Norway.
Russia in Europe.
Turkey in Europe.
Greece and the Archipelago.
Italy.
Italy—North Sheet.
Italy—South Sheet.
*Italy—Complete Map.
Spain and Portugal.
Asia.
Turkey in Asia.
Syria and the Sinai Peninsula.
Arabia, the Red Sea, and the Nile Valley.

Persia, Afghanistan, and Beluchistan.
India.
*India, with Burma and the Straits Settlements.
India—North Sheet.
India—South Sheet.
East Indian Archipelago.
Chinese Empire and Japan.
Asiatic Russia, and Russian Central Asia.
Africa.
North-West Africa.
North-East Africa.
Central and South Africa.
South Africa.
North America.
Canada—Sheet I., Nova Scotia, &c.
Canada—Sheet II., Ontario, &c.
Canada—Sheet III., Manitoba, &c.
Canada—Sheet IV., British Columbia, &c.
United States.
United States—North-Eastern Division.
United States—South-Eastern Division.
United States—Western Division.
Mexico and Central America.
West Indies, with Jamaica and British Guiana.
South America.
South America—North Sheet.
South America—South Sheet.
*South America—Complete Map.
Australia.
Victoria.
New South Wales.
Queensland.
South Australia.
Western Australia.
Tasmania—New Guinea.
New Zealand.
Oceania.

∴ The above Maps may be had (1) in Sheet, with Index, price 2s. 6d. ; and (2) mounted on cloth, and folded in case, with Index, price 3s. 6d. Those marked with an asterisk are two-sheet maps, price, in sheet, with Index, 5s. ; in case, with Index, 7s.

GEORGE PHILIP & SON, LONDON AND LIVERPOOL.

PHILIPS' SCHOOLROOM MAPS. A Series of 20 Wall Maps, 5 feet 8 inches by 4 feet 6 inches, mounted on rollers and varnished, price 14s. each. A detailed Prospectus of the series may be had on application.

    ∵ OUTLINE MAPS, uniform with the above in size and price. Now ready :—Europe—England and Wales.

    ∵ SUPPLEMENTARY MAPS :—The World, on Gall's Projection, price 21s. An Industrial Map of England and Wales, with part of Scotland, price 25s. The Dominion of Canada, from the Atlantic to the Pacific, price 21s.

PHILIPS' SMALLER SERIES OF SCHOOLROOM MAPS. A Series of 15 Maps, uniform in size, 3 feet by 2 feet 6 inches. Mounted on rollers and varnished, price 5s. each. List on application.

PHILIPS' SCHOOLROOM MAPS OF THE COUNTIES OF ENG-LAND. A New Series of Maps expressly prepared for use in Schools and Colleges. The following are now ready :—Cheshire, Derbyshire, Durham, Kent, Northumberland, Nottinghamshire, Warwickshire, and Worcestershire, each 7s. 6d. ; Lancashire and Yorkshire, each 10s. 6d. ; Lincolnshire, Staffordshire, and Surrey, each 12s. ; Lancashire and Cheshire, 16s. ; and Middlesex, 21s.

PHILIPS' SERIES OF HALFPENNY MAPS. A Series of 36 Crown 4to (10 in. by 8 in.) Maps of the Continents and Chief Countries. List on application. These Maps may be had in four different styles, viz. :—

    (1) *Full Coloured*, for reference, or as copies for drawing.
    (2) *Physical Outline Maps*, uncoloured, and without the names.
    (3) *Outline Maps*, with coast-line only.
    (4) *Blank Projections*, with lines of Latitude and Longitude only.

PHILIPS' LARGE PENNY MAPS. A Series of 40 Imperial 4to (13 in. by 11 in.) Maps of the Continents and Chief Countries. May be had in four different styles, similar to the Halfpenny Series. List on application.

PHILIPS' PENNY SCRIPTURE MAPS. A Series of 16 Crown 4to Maps, illustrating the geography of the Bible Lands. List on application.

PHILIPS' PENNY MAPS OF ANCIENT GEOGRAPHY. A Series of 18 Imperial 4to Maps, illustrating Classical Geography. List on application.

PHILIPS' PENNY COUNTY MAPS. A Series of Maps of the Counties of England, Wales, Scotland, and Ireland.

GEORGE PHILIP & SON, LONDON AND LIVERPOOL.

THE WORLD'S GREAT EXPLORERS and Explorations. A Series of Volumes dealing with the life and work of those heroic adventurers through whose exertions the face of the Earth has been made known to humanity. Edited by J. Scott Keltie, Assistant Secretary to the Royal Geographical Society, H. J. Mackinder, M.A., Reader in Geography at the University of Oxford, and E. G. Ravenstein, F.R.G.S.

The volumes are uniform in size (crown 8vo), and are handsomely bound in cloth, price 4s. 6d. each; cloth, gilt cover (with special design), gilt edges, 5s. ; half-bound polished morocco, marbled edges, 7s. 6d. The following are now ready :—

JOHN DAVIS, Arctic Explorer and Early India Navigator. By Clements R. Markham, C.B. With 24 Illustrations and 4 coloured Maps.

PALESTINE. By Major C. R. Conder, R.E., LL.D., Leader of the Palestine Exploring Expeditions. With 26 Illustrations and 7 Maps.

MUNGO PARK AND THE NIGER. By Joseph Thomson, F.R.G.S. With 24 Illustrations and 7 coloured Maps.

THE LIFE OF FERDINAND MAGELLAN, First Circumnavigator of the Globe. By Dr. F. H. H. Guillemard, Author of the "Cruise of the *Marchesa.*" With 20 Illustrations and 18 Maps.

SIR JOHN FRANKLIN AND THE NORTH-WEST PASSAGE. By Admiral Albert Markham, R.N. With 19 Illustrations and 4 Maps.

LIVINGSTONE AND THE EXPLORATION OF CENTRAL AFRICA. By H. H. Johnston, C.B., F.R.G.S., H.M. Commissioner and Consul-General in S.-E. Africa. With 22 Illustrations and 8 coloured Maps.

COLUMBUS. By Clements R. Markham, C.B. With 26 Illustrations and 8 coloured Maps.

·.· Other volumes are in preparation.

THE DEVELOPMENT OF AFRICA as a field for European enterprise. By A. Silva White, late Secretary to the Royal Scottish Geographical Society. With 14 Maps, specially designed by E. G. Ravenstein, F.R.G.S. New and cheap edition, demy 8vo, cloth, 7s. 6d.

ACROSS EAST AFRICAN GLACIERS, being an Account of the First Ascent of Mount Kilimanjaro. By Dr. Hans Meyer. Translated by E. H. Calder. One Volume. Super royal 8vo, with Maps and Illustrations, handsomely bound in cloth, price 32s.

GEORGE PHILIP & SON, LONDON AND LIVERPOOL.

A GIRL IN THE KARPATHIANS. By Ménie Muriel Norman (*née* Dowie). With upwards of 30 original Illustrations and a coloured Map. Cheap edition, crown 8vo, cloth, price 3s. 6d.

THREE YEARS IN WESTERN CHINA. A Narrative of Three Journeys in Ssŭ-Ch'uan, Kuei-Chow, and Yün-nan. By Alexander Hosie, M.A., F.R.G.S., H.B.M. Consular Service, China. With an Introduction by Archibald Little, F.R.G.S. Eight full-page Illustrations and a large coloured Map, showing the Author's routes. Demy 8vo, cloth, price 14s.

A NATURALIST AMONG THE HEAD HUNTERS, being an Account of Three Visits to the Solomon Islands in the years 1886, 1887, and 1888. By Charles Morris Woodford, F.R.G.S., F.Z.S. With 16 full-page Illustrations and 3 coloured Maps. Second edition, crown 8vo, cloth, price 8s. 6d.

TRAVELS IN THE ATLAS AND SOUTHERN MOROCCO. A Narrative of Exploration. By Joseph Thomson, F.R.G.S. With Maps and Illustrations. Cheap edition, crown 8vo, cloth, 7s. 6d.

THE UNKNOWN HORN OF AFRICA. An Exploration from Berbera to the Leopold River. By the late F. L. James, M.A. With 27 Illustrations and a Map. New and cheap edition, crown 8vo, cloth, 7s. 6d.

THE FIRST ASCENT OF THE KASAI, or Records of Service under the Lone Star. By C. S. L. Bateman. With 57 Illustrations and 2 original Maps. Cheap edition, medium 8vo, cloth, 10s. 6d. nett.

UP THE NIGER. Narrative of Major Claude MacDonald's Mission to the Niger and Benue Rivers, West Africa. By Captain A. F. Mockler-Ferryman, F.R.G.S., F.Z.S., Oxfordshire Light Infantry. To which is added a Chapter on Native Musical Instruments, by Captain C. R. Day. With Map, 15 Illustrations, and Appendix. Demy 8vo, cloth, bevelled edges, 16s.

HOME LIFE ON AN OSTRICH FARM. An Account of Life in the Interior of South Africa. By Mrs. Annie Martin. Cheap edition, crown 8vo, cloth, 3s. 6d.

DELAGOA BAY : Its Natives and Natural History. By Mrs. R. Monteiro. Illustrated with Plate of African Butterflies, hand-coloured. Crown 8vo, 12s. ; or with Plate of Butterflies uncoloured, 9s.

PARAGUAY : The Land and the People, Natural Wealth and Commercial Capabilities. By Dr. E. de Bourgade la Dardye. English edition, edited by E. G. Ravenstein, F.R.G.S. With a large-scale Map and 13 Illustrations. Crown 8vo, cloth, 7s. 6d.

GEORGE PHILIP & SON, LONDON AND LIVERPOOL.

THE ADVANCED CLASS-BOOK OF MODERN GEOGRAPHY—
Physical, Political, and Commercial. A Complete Manual of Geography
for Students in Training Colleges, Senior Pupils in Middle and Higher
Class Schools, Pupil Teachers, &c. By William Hughes, F.R.G.S.,
and J. Francon Williams, F.R.G.S. With Notes and Index. 850
pages. Crown 8vo, cloth, price 6s.

THE CLASS-BOOK OF MODERN GEOGRAPHY, with Examination
Questions, Notes and Index. By William Hughes, F.R.G.S. New
Edition, revised and enlarged by J. Francon Williams, F.R.G.S. 460 pp.
Crown 8vo, cloth, 3s. 6d.

THE ELEMENTARY CLASS-BOOK OF MODERN GEOGRAPHY.
With Examination Questions. New Edition, thoroughly revised and
considerably enlarged. 250 pp. Foolscap 8vo, cloth, 1s. 6d.

THE GEOGRAPHY OF THE BRITISH ISLES—Physical, Political,
and Commercial. By William Hughes, F.R.G.S., and J. Francon Williams, F.R.G.S. With 3 Coloured Maps. 106 pp. Crown 8vo, cloth,
1s. 6d.

THE GEOGRAPHY OF THE BRITISH COLONIES AND FOREIGN
POSSESSIONS. By William Hughes, F.R.G.S., and J. Francon
Williams, F.R.G.S. With Coloured Maps. Crown 8vo, cloth, 2s. 6d.

THE GEOGRAPHY OF EUROPE—Physical, Political, and Commercial.
By William Hughes, F.R.G.S., and J. Francon Williams, F.R.G.S.
With Coloured Map. 240 pp. Crown 8vo, cloth, 2s.

THE GEOGRAPHY OF ASIA—Physical, Political, and Commercial.
By William Hughes, F.R.G.S., and J. Francon Williams, F.R.G.S.
With 2 Coloured Maps. 128 pp. Crown 8vo, cloth, 1s. 6d.

THE GEOGRAPHY OF AFRICA—Physical, Political, and Commercial
By William Hughes, F.R.G.S., and J. Francon Williams, F.R.G.S.
With Map, coloured to show the present political partition of the Contin-
ent. 98 pp. Crown 8vo, cloth, 1s.

THE GEOGRAPHY OF AMERICA—Physical, Political, and Commercial. By William Hughes, F.R.G.S., and J. Francon Williams,
F.R.G.S. With 3 Coloured Maps. 120 pp. Crown 8vo, cloth, 1s. 6d.

THE GEOGRAPHY OF AUSTRALASIA AND POLYNESIA—
Physical, Political, and Commercial. By J. Francon Williams, F.R.G.S.
With 3 Coloured Maps. 122 pp. Crown 8vo, cloth, 1s. 6d.

GEORGE PHILIP & SON, LONDON AND LIVERPOOL.

GEOGRAPHICAL CLASS-BOOKS. 11

THE CLASS-BOOK OF PHYSICAL GEOGRAPHY, with Examination Questions, Notes, and Index. By William Hughes, F.R.G.S. New Edition, revised and enlarged by J. Francon Williams, F.R.G.S. 312 pp. Crown 8vo, cloth, 3s. 6d.

THE ELEMENTARY CLASS-BOOK OF PHYSICAL GEOGRAPHY, abridged from the larger Class-Book. New Edition, revised and enlarged. 134 pp. Foolscap 8vo, cloth, 1s. 6d.

THE GEOGRAPHY OF COAST LINES. By William Lawson, F.R.G.S. New and Revised Edition. 88 pp. Foolscap 8vo, cloth, 1s.

THE GEOGRAPHY OF RIVER SYSTEMS. By William Lawson, F.R.G.S. New and Revised Edition. 102 pp. Foolscap 8vo, cloth, 1s.

THE GEOGRAPHY OF THE OCEANS—Physical, Historical, and Descriptive, with Maps and Diagrams. By J. Francon Williams, F.R.G.S. Foolscap 8vo, cloth, 2s. 6d.

PHILIPS' ELEMENTARY ATLAS AND GEOGRAPHY. An Atlas and a Class-Book in one. Edited by J. Francon Williams, F.R.G.S Crown 4to, cloth, 3s. 6d.

THE GEOGRAPHY OF ENGLAND AND WALES—Descriptive, Physical, Industrial, and Historical. Designed for the use of Teachers and Students, and admirably adapted for reference in the preparation of lessons by Pupil Teachers, &c. By Thomas Haughton, late Head Master, Blue Coat School, Liverpool. Crown 8vo, cloth. *In the Press.*

PHILIPS' GEOGRAPHICAL READERS. A Series of 6 Reading Books in Geography. Illustrated by 800 Woodcuts and Maps. Edited by J. Francon Williams, F.R.G.S.

1. First Steps—Part I. . . 9d.
2. ,, Part II. . 10d.
3. England—Physical and Political . . . . . 1s.
4. British Isles, British North America, and Australia 1s. 6d
5. Europe—Physical and Political . . . 1s. 9d.
6. The World . . . 2s. 0d.

APPLIED GEOGRAPHY. By J. Scott Keltie, Assistant Secretary to the Royal Geographical Society. 170 pp. Crown 8vo, cloth, with Maps and Diagrams, 3s. 6d.

THE GOLDEN GATES OF TRADE, a Text-Book of Commercial Geography. With Examination Questions. By John Yeats, LL.D. 354 pp. Crown 8vo, cloth, 4s. 6d.

MAP STUDIES OF THE MERCANTILE WORLD, a Text-Book for the use of Students of Commerce. By John Yeats, LL.D. 336 pp. Crown 8vo, cloth, 4s. 6d.

GEORGE PHILIP & SON, LONDON AND LIVERPOOL.

PHILIPS' SCRIPTURE MANUALS.   A Series of Hand-Books for the use of Students preparing for Examination.   Uniform in size and style. Foolscap 8vo, cloth.

BY THE REV. CANON LINTON, M.A.

The Book of Genesis, with Map, 1s. 6d.
The Book of Numbers, with Map, 1s. 6d.
The Book of Jeremiah (Historical Chapters), with Map, 1s.
Haggai and Zechariah, with Map, 1s.
The Book of Nehemiah, with Plan of Jerusalem, 1s.
The First Epistle to the Corinthians, with Map, 1s.
The Second Epistle to the Corinthians, with Map, 1s.

BY THE LATE JAMES DAVIS.

Notes on Genesis, 9d.
Notes on Exodus, 9d.

Notes on Joshua, 9d.
Notes on Judges, 9d.
Notes on I. Samuel, 9d.
Notes on II. Samuel, 1s.
Notes on I. Kings, 1s.
Notes on II. Kings, 1s.
Notes on Ezra, 9d.
Notes on St. Matthew's Gospel, 1s. 6d.
Notes on St. Mark's Gospel, 9d.
Notes on St. Luke's Gospel, 1s.
Notes on St. John's Gospel, 1s. 6d.
Notes on the Acts of the Apostles, 1s.
Manual of the Church Catechism, 9d.
Manual of the Book of Common Prayer, 1s. 6d.

DAVIS'S HISTORICAL MANUALS.   A Series of Hand-Books for Students preparing for Examination.   Uniform in size and style.   Foolscap 8vo, cloth.

1.—From the death of Edward the Confessor to the death of King John (1066-1216), 2s.
2.—From the accession of Henry III. to the death of Richard III. (1216-1485), 2s.
3.—The Tudor Period (1485-1603), 2s.   } *These also embrace the* LITERATURE *of*
4.—The Stuart Period (1603-1689), 1s. 6d. } *the Tudor and Stuart Periods respectively.*
5.—From the reign of Charles I. to the end of the Commonwealth (1640-1660), 1s. 6d.
6.—From the restoration of Charles II. to the Revolution (1660-1688), 2s.
7.—From the accession of James I. to the Battle of the Boyne (1603-1690), 2s.
8.—From the Revolution of 1688 to the death of Queen Anne, 1714, 1s. 6d.
9.—From the accession of William III. to the accession of George III. (1689-1760), 2s. 6d.
10.—From the accession of George III. to the Battle of Waterloo (1760-1815), 2s. 6d.
11.—Manual of English Literature from 1760-1815, 2s.

PHILIPS' PICTURESQUE HISTORY OF ENGLAND, being the story of the English people—their growth, the development of their national institutions, the establishment of their world-wide Empire, the achievements of their great monarchs, warriors, statesmen, and writers, and the social and economic conditions of the nation at various epochs. With 300 Illustrations and Maps.   Edited by J. Francon Williams, F.R.G.S.   740 pages.   Crown 8vo, handsomely bound in cloth, 4s. 6d.; gilt edges, 5s.

GEORGE PHILIP & SON, LONDON AND LIVERPOOL.

THE STUDENT'S SUMMARY OF ENGLISH HISTORY. With Notes on Constitutional, Political, Civil, and Church History, selected from the works of eminent historians, and a Complete Index. By Thomas Haughton, late Head Master, Blue Coat School, Liverpool. Second Edition. 490 pp. Foolscap 8vo, cloth, 5s.

THE STUDENT'S SUMMARY OF ENGLISH HISTORY. By the same Author. *Without Notes.* 190 pp. Foolscap 8vo, cloth, 1s. 6d.

HAUGHTON'S SHILLING SUMMARY OF ENGLISH HISTORY. Containing all the principal events, with a list of the Sovereigns and Genealogical Tables. Foolscap 8vo, cloth, 1s.

A HISTORY OF ENGLAND AND WALES from the Roman to the Norman Conquest. With Notes and Tables, a full Index, and 3 Historical Maps. By T. Morgan Owen, M.A., H.M.I.S. Third edition, revised and enlarged. 238 pp. Cr. 8vo. cloth, 2s.; superior edition, 3s. 6d.

PHILIPS' HISTORICAL READERS. A Series of 4 Reading Books in English History, illustrated by 347 Woodcuts and 37 Maps and Tables. Edited by J. Francon Williams, F.R.G.S.

No. 1. Stories in English History, 1s. | No. 3. Middle England, 1s. 6d.
No. 2. Early England, 1s. | No. 4. Modern England, 1s. 6d.

THE ELEMENTS OF ELOCUTION, with special reference to the Literary Basis of Delivery, illustrated by Selections in Poetry and Prose for Reading and Recitation. By Charles E. Clegg. Crown 8vo, cloth, 2s. 6d.

ELOCUTIONARY SPECIMENS IN PROSE AND VERSE for Recitation and Reading. Selected and adapted from leading Writers, with general hints on Delivery. By Charles E. Clegg. Crown 8vo, cloth, 2s.

PHILIPS' ILLUSTRATED POETRY BOOK. A selection of Poems for Reading and Recitation. With Notes. 96 pp. Crown 8vo, cloth, 8d.

AN INTRODUCTION TO THE STUDY OF SHAKESPEARE AND MILTON. Containing classified Selections, with full Notes, Sketches of the Lives and Genius of the Poets, and Critical Analyses of their Language and Style. With Illustrations. Crown 8vo, cloth, 1s. 6d.; or in two parts, (1) Shakespeare, 6d.; (2) Milton, 6d.

GEORGE PHILIP & SON, LONDON AND LIVERPOOL.

THE ART OF TEACHING AND STUDYING LANGUAGES. By François Gouin. Translated by Howard Swan and Victor Bétis. Crown 8vo, cloth, 7s. 6d.

RALFS' RAPID ROAD TO SPANISH. A Manual for Self-Instruction. By J. W. Ralfs, Translator to the Board of Trade. Crown 8vo, cloth, 5s. PART II. of this work is now ready, price 5s., and a KEY to both Parts, price 5s.

GAILLARD'S COMPLETE FRENCH COURSE. Containing a Grammar of the French Language, and a Series of interesting Sketches, affording the means of conversing and composing on almost any subject. Second edition. Crown 8vo, cloth, 4s.

GAILLARD'S FRENCH FOR THE TIMES. A Series of Outline Narratives of Travel, Letters, &c. Crown 8vo, cloth, 3s. 6d.

ADVANCED ARITHMETIC for Schools and Colleges. By T. W. Piper, St. Katherine's Training College, London. Crown 8vo, cloth, 3s. 6d. KEY to the same, giving the full working of each example. Crown 8vo, cloth, 5s. nett.

ELEMENTARY ARITHMETIC for Schools and Colleges. By the same Author. Crown 8vo, cloth, 1s. 6d.

COMPLETE COURSE OF ARITHMETICAL EXAMPLES AND EXERCISES. By the same Author. With Answers. Cr. 8vo, cloth, 3s.

MENTAL ARITHMETIC for Schools and Colleges. By the same Author. New and enlarged edition. Foolscap 8vo, cloth, 2s. 6d.

INTRODUCTORY MENTAL ARITHMETIC. By the same Author. Foolscap 8vo, stiff cover, 6d.

THE ELEMENTS OF EUCLID for Schools and Colleges. Containing the First Six and parts of the Eleventh and Twelfth Books. With a Series of Geometrical Problems. By James Martin. Crown 8vo, cloth, 3s. 6d. Also separately—Book I., cheap edition, limp cloth, 6d.; superior edition, cloth, 1s. Books I. and II. in one volume, cloth, 1s. 6d.

A GRADUATED COURSE OF PROBLEMS in Practical, Plane, and Solid Geometry. By James Martin. Crown 8vo, cloth, 3s. 6d.

FIRST GRADE PRACTICAL GEOMETRY. By David Bain, F.R.G.S. Crown 8vo, stiff cover, 3d.

GEORGE PHILIP & SON, LONDON AND LIVERPOOL.

PHILIPS' SERIES OF DRAWING BOOKS. Designed and drawn by a Practical Teacher. Oblong 4to, printed on toned drawing paper. First Grade Drawing Books, a Series of 7 Books, price 2d. each.

EASY LANDSCAPES. First Series in 6 Books, oblong 4to, each 2d. Second Series in 6 Books, Oblong 4to, each 2d.

PHILIPS' SECOND GRADE TEST PAPERS. By John Carroll, Art Master, Hammersmith Training College. I.—FREEHAND, in Packets, each containing 48 Papers, price 2s. each packet. II.—PRACTICAL GEOMETRY, in Packets, each containing 48 Papers, price 2s. each.

THE "PRACTICAL" DRAWING CARDS. Prepared from and based upon the Drawings in the "Illustrated Drawing Syllabus" and "Dyce's Drawing Book." By Harry C. Wilcocks. Sets A. and B. for Kindergarten and Infant Schools. Sets 1 to 10 for Standards I. to VII. Each Set contains 24 Cards. Price per Set, 1s. *Specimen Cards sent post free.*

THE "PRACTICAL" DRAWING SHEETS, for Class Teaching and Examination Work. By Harry C. Wilcocks. Set A. for Kindergarten and Infant Schools. Set 1 to 7 for Standards I. to VI. Each Set contains 12 Designs, boldly printed in two colours, on tough white cards, size 20 by 15 inches, in strong wrapper, with elastic band. Price per Set, 3s. 6d. *Specimen Sheets sent post free on application.*

THE "PRACTICAL" HANDBOOK OF DRAWING. A Manual for the use of Teachers in Elementary Schools. Adapted to the latest requirements of the Science and Art Department. By Harry C. Wilcocks. With 250 Illustrations. Crown 8vo, cloth, price 3s. 6d.

SCIENCE LADDERS, a New Series of simple Reading Books in Elementary Science for the Young. By N. D'Anvers. In small crown 8vo, cloth, price 1s. each. Also in 6d. Parts, stiff covers. *List on application.*

ACOUSTICS, LIGHT, AND HEAT. An Introduction to the study of Physical Science. By T. W. Piper. Crown 8vo, cloth, 2s. 6d.

CLASS-BOOK OF INORGANIC CHEMISTRY. By D. Morris, B.A. New and enlarged edition. Crown 8vo, cloth, 2s. 6d.

CLASS-BOOK OF ELEMENTARY MECHANICS. By W. Hewitt, B.Sc., Crown 8vo, cloth, 3s. May be also had in Parts :—PART I., Matter, 1s. 6d. ; PART II., Force, 2s.

HANDBOOK OF SLOJD for English Teachers. Translated from the standard Swedish work of Herr Otto Salomon, the Director of the Nääs Training School, by Mary R. Walker and William Nelson. With over 130 Illustrations and Plates. Demy 8vo, cloth, 6s.

GEORGE PHILIP & SON, LONDON AND LIVERPOOL.

THE STUDENT'S TEXT-BOOK OF THE SCIENCE OF MUSIC. With full Tables, Glossaries, Index, &c. By John Taylor, Organist to Her Majesty the Queen at Kensington Palace. Crown 8vo, cloth, 6s.

MUSICAL THEORY COURSE. For Pupil Teachers, Training College Students, &c. By the same author. Second edition. Cr. 8vo, cloth, 1s. 6d.

A MANUAL OF VOCAL MUSIC. For use in Public Elementary Schools. Forming a Complete Guide to Singing at Sight from Notes. Crown 8vo, cloth, 1s. 6d.

THE YOUNG FOLKS' SONG BOOK. Containing 31 Original Songs, with suitable Actions. By H. Berkeley Score. STAFF NOTATION, crown 4to, cloth, price 2s. 6d.; TONIC SOL-FA EDITION, demy 8vo, limp cloth, price 2s.

THE TEACHER'S MANUAL.—"How to Teach Sight-Singing" on Taylor's Stave Sight-Singing Method. With graduated Exercises and music-type Illustrations from the best composers. Crown 4to, cloth, 5s.

PHYSICAL TRAINING. A Series of Handbooks for Teachers. By A. Alexander, F.R.G.S., Director of the Liverpool Gymnasium, and Founder of the National Physical Recreation Society. The following are now ready:—

MUSICAL DRILL, PART I.—FOR INFANTS. With 120 Illustrations and specially adapted Music. Third edition. Crown 8vo, cloth, 2s. 6d.

MUSICAL DRILL, PART II.—For Schools and Calisthenic Classes. With 132 Illustrations and specially adapted Music. Demy 8vo, cloth, 3s. 6d.

HEALTHFUL EXERCISES FOR GIRLS. Including 16 different Gymnastic Exercises, specially adapted for Ladies. With 200 Illustrations and Instructions for the Musical Accompaniments. Cr. 8vo, cloth, 2s. 6d.

MODERN GYMNASTIC EXERCISES, PART I. Elementary Exercises, suitable for Boys and Young Men. With Musical Accompaniments. Crown 8vo, cloth, 2s. 6d.

MODERN GYMNASTIC EXERCISES, PART II. Advanced Exercises, illustrated by 270 specially drawn Figures. With a complete Series of Musical Accompaniments for the Mass Exercises. Demy 8vo, full-bound cloth, 3.6d.

GEORGE PHILIP & SON, PUBLISHERS.
LONDON: 32 FLEET STREET, E.C.
LIVERPOOL: 45 TO 51, SOUTH CASTLE STREET.

www.ingramcontent.com/pod-product-compliance
Lightning Source LLC
Chambersburg PA
CBHW021106270326